Entropy, Water and Resources

Horst Niemes · Mario Schirmer

Entropy, Water and Resources

An Essay in Natural Sciences-Consistent Economics

Physica-Verlag

Dr. rer. pol. Horst Niemes
Helmholtz-Centre for Environmental
Research (UFZ) Leipzig
Permoserstraße 15
04301 Leipzig, Germany
e-mail: email@horst-niemes.de

Professor Dr. habil. Mario Schirmer
Eawag: Swiss Federal Institute of Aquatic
Science and Technology
Überlandstrasse 133
8600 Dübendorf, Switzerland
e-mail: mario.schirmer@eawag.ch

ISBN 978-3-7908-2415-5 e-ISBN 978-3-7908-2416-2
DOI 10.1007/978-3-7908-2416-2
Springer Heidelberg Dordrecht London New York

Library of Congress Control Number: 2010924287

Cover design: WMXDesign GmbH, Heidelberg

Printed on acid-free paper

Physica-Verlag is a brand of Springer-Verlag Berlin Heidelberg
Springer-Verlag is part of Springer Science+Business Media (www.springer.com)

Preface

This book lies at the intersection of natural sciences, economics, and water engineering and is in line with the long tradition of environmental economics at the University of Heidelberg. In the 1970s, the Neo-Austrian Capital Theory was developed using the fundamental laws of thermodynamics as a common language between the natural and social sciences. Niemes (1981) integrated the dynamic and irreversibility characteristics of the natural environment into the Neo-Austrian capital theory. Faber et al. (1983, 1987, 1995) then extended this interdisciplinary approach further to create a comprehensive, dynamic, environmental resource model.

Over the last 3 decades, the theoretical foundations of environmental economics have been modified and there have been an impressive variety of applications. This book aims to reduce the gaps between economic theory, natural sciences, and engineering practice. One of the reasons these gaps exist is because economic assumptions are used to construct dynamic environmental and resource models, which are not consistent with the fundamental laws of the natural sciences. Another reason for the gap might be the distance between academic theory and real world situations.

Based on an extended thermodynamic approach, the authors explain which economic assumptions are acceptable for constructing a dynamic model that is consistent with the natural sciences. In particular, the special role of water in the production and reproduction activities will be considered as an integral component. Water is generated in a separate water treatment process and is used to transport the unavoidable by-products of production and reproduction activities to a wastewater sector. In this respect, not only environmental protection aspects, but also the interrelation between the water requirements and the use of non-renewable resources for producing desired consumption goods will be highlighted. Special attention is given to show that the economies of developed countries, which still rely on the use of non-renewable natural resources, will be confronted with closely connected crises. The inter-temporal marginal costs of using the non-renewable resources also cause an increase in the costs of water and energy.

We will demonstrate how natural sciences consistent economic models are beneficial to the long-term study and realization of water supply and wastewater infrastructure projects and hydro-geological investigations. We use two case studies to develop an enhanced water infrastructure model. As part of the model, we introduce capital stocks for water use, water infrastructure, water treatment, water distribution, as well as wastewater collection and wastewater treatment. Although it takes serious effort to derive the optimal conditions for the extended dynamic model, the derived results confirm that close cooperation between theoretical work and practical experience can deliver a surplus of inside information that would otherwise not be achievable.

Special thanks are given to the UFZ-Helmholtz-Centre for Environmental Research – and Eawag-Swiss Institute of Aquatic Science and Technology – for financial support of our endeavor over a 3-year period, without which it could not have been realized. We also thank Mrs. Leslie Ferre (Arizona, USA) for her carefully proof-reading of this book, and Malte Faber and Frank Jöst (University of Heidelberg) and Gunter Stephan (University of Bern) for fruitful discussions and comments.

Last but not least, we emphasize their patience, assistance, and motivation and dedicate this book to Ingrid Niemes and Kristin Schirmer.

Leipzig
Zurich
April 2010

Horst Niemes
Mario Schirmer

Contents

1 **Introduction** 1

Part I The Water Use Model

2 **Conceptual Foundations: Thermodynamics and Capital Theory** . 7
2.1 Thermodynamics and Its Equivalency to Information Theory . 7
2.1.1 Entropy, Temperature and Heat 8
2.1.2 Entropy, Probability, and Information 8
2.1.3 Relations Between Work and Exergy 9
2.1.4 Exergy Far from the Thermodynamic Equilibrium . . 11
2.1.5 Relations Between Exergy and Information 12
2.1.6 Thermodynamics of Economic Transformation Processes 13
2.2 The Concept of Capital Theory 16
2.2.1 Neo-Austrian Capital Theory as Example 16
2.2.2 Capital Theory and Its Natural Sciences Consistency . 20

3 **General Design of Dynamic Models for Water Uses** 23
3.1 Model Structure and Economic Activities 23
3.2 Characteristics of Production Activities 26
3.2.1 Criteria for the Extraction and Use of Raw Materials . 26
3.2.2 Characteristics of Producing and Using Energy and Water 31
3.2.3 Characteristics of Wastewater Treatment Activities . . 35
3.3 Technological Progress and Human Labour Inputs 36

4 **Specifications for Constructing the Water Use Model** 39
4.1 Structure and Characteristics of the Water Use Model 39
4.2 Process Coefficients for the Water Use Model 46
4.2.1 Process Coefficients for the Production Sector 46
4.2.2 Process Coefficients for the Reproduction Sector . . . 53
4.2.3 The Wastewater Treatment Coefficients 55

5 **Constraints of the Water Use Model** 61
5.1 The Constraints for the Consumption Good Amounts 62

5.2 The Constraints for Extracting Raw Materials 62
5.3 Constraints for Water and Wastewater Amounts 63
5.4 Constraints for Free Energy 64
5.5 Constraints for Human Labour Inputs 68
5.6 Constraints for Sustaining and Developing the Capital Stock . 71
5.7 Aggregation of Processes to Sectors 72
5.7.1 Energy and Human Labour Inputs for the Production Sector 72
5.7.2 Energy and Human Labour Inputs for the Water Sectors 73

6 Optimality Conditions of the Water Use Model 75
6.1 The Optimization Concept 75
6.2 Optimality Conditions for the Demand Side 77
6.3 Optimality Conditions for the Production Side 78
6.3.1 Non-profit Conditions for the Production Sector . . . 78
6.3.2 Non-profit Conditions for the Water and Wastewater Sectors 87
6.4 Conclusions 100

Part II The Water Infrastructure Model

7 Case Studies Guiding the Integration of Water Infrastructure . . 105
7.1 The MTBE Contamination of the Leuna Aquifer 106
7.1.1 Characteristics of the MTBE Contamination Problem 106
7.1.2 Technical Solutions to Reduce MTBE Contamination 109
7.1.3 The Target Group of the Rehabilitation Measures . . . 111
7.1.4 Estimation of the MTBE Contamination Amounts . . 111
7.1.5 Estimation of Costs for Solving the MTBE Problem . 113
7.2 Water Infrastructure to Serve Adana in Turkey 117
7.2.1 Urbanization and Water Infrastructure of Mega-Cities 117
7.2.2 Private and Local Public Welfare Properties of Water . 118
7.2.3 Implementation Concept for Adana's Water Infrastructure 120
7.2.4 Dynamic Prime Costs of Adana's Water Infrastructure 128
7.3 Conclusions for Constructing the Water Infrastructure Model . 129

8 Specifications for Constructing the Water Infrastructure Model . 131
8.1 Structure of the Water Infrastructure Model 132
8.2 Process Coefficients of the Water Infrastructure Sectors 136
8.2.1 Coefficients of the CW and M&E Production Processes 136
8.2.2 Coefficients for the Water Infrastructure Processes . . 140
8.3 Reduction of Variables and Dynamics of the Capital Stocks . . 147

9 Constraints of the Water Infrastructure Model 151
9.1 Constraints for the Consumption Good Amounts 151

9.2 Constraints for Extracting Raw Materials 152
9.3 Constraints for the Water and Wastewater Amounts 152
9.4 Constraints for Free Energy 154
9.5 Constraints for the Human Labour Input Amounts 155
9.6 Constraints for the Capital Stocks 157
9.7 Constraints for Reduced Variables 158
9.7.1 Human Labour Constraint for Reduced Variables . . . 158
9.7.2 Energy Constraint for Reduced Variables 159
9.7.3 Water and Wastewater Constraints for Reduced Variables 161
9.8 Aggregation of Process Inputs to Sector Inputs 162
9.8.1 Aggregation of Processes to the Production Sector . . 162
9.8.2 Aggregation of Processes to the Water Sectors 163

10 Optimality Conditions of the Water Infrastructure Model 171
10.1 Optimality Conditions for the Demand Side 172
10.2 Optimality Conditions for the Production Side 173
10.2.1 Non-profit Conditions for the Production Sector . . . 173
10.2.2 Non-profit Conditions for the Water Sectors 174
10.3 Conclusions and Perspectives 175
A Marginal Costs for Water Treatment 178
A.1 Marginal Human Labour Costs for Water Treatment . 178
A.2 Marginal Energy Costs for Water Treatment 184
B Marginal Costs for Water Distribution 190
B.1 Marginal Human Labour Costs for Water Distribution 190
B.2 Marginal Human Energy Costs for Water Distribution 193
C Marginal Costs for Wastewater Collection 197
C.1 Marginal Human Labour Costs for Wastewater Collection 198
C.2 Marginal Energy Costs for Wastewater Collection . . 200
D Marginal Costs for Wastewater Treatment 204
D.1 Marginal Human Labour Costs for Wastewater Treatment 204
D.2 Marginal Energy Costs for Wastewater Treatment . . 207

References 213

Index 217

Chapter 1
Introduction

An Essay in Natural Sciences Consistent Economics

Abstract This book at the intersection of natural sciences, economics, and water engineering aims to reduce the gaps between economic theory, natural sciences, and engineering practice. Based on an extended thermodynamic approach, the authors explain which economic assumptions are acceptable for constructing a dynamic model that is consistent with the natural sciences. In particular, the special role of water in the production and reproduction activities will be considered as an integral component. Water is generated in a separate water treatment process and is used to transport the unavoidable by-products of production and reproduction activities to a wastewater sector. In this respect, not only environmental protection aspects, but also the interrelation between the water requirements and the use of non-renewable resources for producing desired consumption goods will be highlighted.

Environmental and Ecological Economics might be considered unique multi-disciplinary scientific fields, for which the two headings are used synonymously because both disciplines deal with resource and environmental problems from an economics perspective. Faber (2007:18) reflected on and characterized both his own work as well as the scientific work of the impressive number of scientists dealing with environmental and resource problems at the University of Heidelberg over a period of 3 decades, and drew a clear dividing line between ecological and environmental economics as follows:

> While the representatives of ecology have few, if any difficulties with Ecological Economics, the relationship between mainstream Economics and Ecological Economics is not quite so harmonious. Why is this? The answer is: the mainstream economist views nature as a subsystem of the economy, whereas the ecological economist takes quite the opposite view.

This differentiation should be examined in the context of the development of the scientific work of Faber and his colleagues at the University of Heidelberg, (for more details see Faber and Winkler 2006). After dynamics and irreversibility of the natural environment were integrated by Niemes (1981) in the Neo-Austrian capital theory, Faber et al. (1983) extended this interdisciplinary approach further to create a comprehensive dynamic environmental-resource model. The English versions by

H. Niemes, M. Schirmer, *Entropy, Water and Resources*,
DOI 10.1007/978-3-7908-2416-2_1, © Springer-Verlag Berlin Heidelberg 2010

Faber et al. (1987, 1995), especially motivated an impressive number of scientists to deal with specific environmental and resource problems, and to develop the dynamic model in different directions.

Although the economic kernel, the Neo-Austrian capital theory approach for characterizing the dynamics of these environmental resource models, has been modified and there is an impressive variety of applications, a more strict application of the thermodynamic laws for the description of ecologic and economic transformation processes has been shown by Baumgärtner (2000) and Baumgärtner et al. (2001, 2002, 2006). Special attention is given to prove (see e.g. Baumgärtner et al. 2006:63–65) that joint production, which is often assumed to be a special case in economics, must be considered the normal case for ecologic and economic processes to preserve consistency with the essential laws of the natural sciences. It was in this context that the famous physicist R.P. Feynman stated (Feynman 1999) "economics is no science at all in case of inconsistency of the economic models with natural scientific laws".

The general objective of our contribution is to develop this approach further by considering not only the energy and material but also the information characteristics of natural and economic transformation processes consistent with the laws of modern physics. Compared with classic physics, the modern disciplines in natural science, (quantum physics and information theory) introduce "subjective" aspects in the observations of transformation processes and systems. This modern view, focusing principally on the equivalence of the material, energy and information components, helps to close the gap between natural and social sciences, and to strengthen interdisciplinary approaches. Our specific objective in this book will be to develop a dynamic model consistent with natural sciences, which can be applied to natural resources, environmental and specific water problems, e.g. for the dynamic analysis of the water infrastructure of urban centers. We will illustrate, at least graphically, that this modern view requires essential revisions to the concept of the capital theory used by Faber and his colleagues at the University of Heidelberg for the construction of the dynamic resource and ecologic models. Addressing the problems of finding convincing measurement units for the information component and its subjective interpretation, however, would be too ambitious for this book (for more details about modern view of natural sciences and the problems of how to measure abstract information and its subjectivism see Görnitz and Görnitz 2006 and Baeyer 2005).

Our book is divided into two parts: in Part I, only water uses are integrated into a basic dynamic model; capital stocks for the water supply and wastewater sectors will be introduced in Part II. Since thermodynamics and information theory are used as a common language for describing natural and economic transformation processes, an introduction to these fields is given in Chap. 2 of Part I, where the interrelation between thermodynamics and the concept of the capital theory is emphasized.

Within the framework of this theoretical background, the general design of an economic model that is consistent with natural science and is intended for natural resources, especially for water uses and water protection measures, is the subject of Chap. 3. For the construction of our basic model for water uses, some additional assumptions and restrictions will be introduced in Chap. 4, which lead us to formulate the model constraints in Chap. 5. In Chap. 6, we derive the optimality conditions expressed in terms of non-profit conditions or marginal costs for both the production and water sectors.

Being guided by two case studies in Chap. 7 of Part II, our basic dynamic model will be extended in Chap. 8 to include essential water infrastructure components. To come to an applicable dynamic model, some additional assumptions, restrictions, and aggregations of processes to sectors are required. Based on the model constraints in Chap. 9, the optimal conditions being formulated in the form of non-profit conditions and marginal costs are derived in Chap. 10. Apart from possible modifications, extensions and a generalization of the dynamic water use and infrastructure models, our conclusions and perspectives will be summarized.

Part I
The Water Use Model

Chapter 2
Conceptual Foundations: Thermodynamics and Capital Theory

Abstract Economic transformation processes, specifically the extraction of non-renewable natural resources for production and reproduction activities, are irreversible. The entropy notion of classic thermodynamics and its equivalent in information theory can be applied to derive the relations between free energy, useful work (exergy), unusable work (anergy) and changes in the concentrations of desired raw materials and undesired residuals that are being discharged into the natural environment. Capital theory is a corner stone in economic theory and ecological economics for analysing the dynamics of environmental and resource problems. It will be shown that information theory can be used to extend thermodynamics and allow an interpretation of capital theory that is consistent with the natural sciences.

The description of transformation processes within the ecological system and its embedded sub-systems (e.g. economic) can be based upon a generalized entropy concept. Section 2.1 discusses entropy in the contexts of classic thermodynamics as well as information theory. Section 2.2 contains a short description of the capital theory and its thermodynamic implications.

2.1 Thermodynamics and Its Equivalency to Information Theory

In this section, special attention is given to show that only a portion of the external available free energy for shifting a system from its thermodynamic equilibrium to a new status can do useful work, while the other portion of free energy can not be used because of irreversible processes within the system. Furthermore, it will be derived that the entropy-free portion (the exergy) is comprised of a materialistic term and an informational term. For a biological system, with its internal information storage capabilities, it is not surprising that it is necessary to distinguish between the materialistic and informational characteristics of the system. We will show, however, that this extended interpretation of thermodynamics makes it necessary to distinguish between the materialistic and informational components for chemical systems, which are less, evolved than biological systems.

H. Niemes, M. Schirmer, *Entropy, Water and Resources*,
DOI 10.1007/978-3-7908-2416-2_2, © Springer-Verlag Berlin Heidelberg 2010

2.1.1 Entropy, Temperature and Heat

As explained in more detail by Niemes (1981:3–12), Clausius' (1850) introduction of entropy led to the clarification that heat is not a special matter but an energy form similar to the others. Marginal changes in heat dQ are defined in Eq. (2.1) as the product of the intensive variable (temperature T) and the extensive variable (marginal change of entropy dS).

$$dQ := TdS. \tag{2.1}$$

Entropy, S, is introduced as the integral of marginal changes in heat between status 1 and 2 for reversible processes

$$\int_{rev.1}^{2} \frac{dQ}{T} := S(2) - S(1). \tag{2.2}$$

For a circular process in a closed system, where the change from status 1 to status 2 is irreversible and the change back from status 2 to status 1 is reversible, we receive

$$\oint_{rev.} \frac{dQ}{T} = 0 \quad \wedge \quad \oint_{irrev.} \frac{dQ}{T} < 0 \Rightarrow \int_{irrev.1}^{2} \frac{dQ}{T} + \int_{rev.2}^{1} \frac{dQ}{T} < 0. \tag{2.3}$$

With the definition of entropy from Eq. (2.2) for the case of a completely closed system, we obtain the so-called Second Law of classic thermodynamics

$$S(2) - S(1) > \int_{irrev.1}^{2} \frac{dQ}{T} \quad \Rightarrow \quad S(2) > S(1). \tag{2.4}$$

These equations show that the entropy of a closed system with irreversible processes never decreases. This can be extended to any system by extending it sufficiently. It also has the implication that an irreversible process can only be returned in the opposite direction when free energy is available for the sub-system where irreversible processes are occurring.

2.1.2 Entropy, Probability, and Information

Whereas entropy in classic thermodynamics formulates the macro state of a system, Boltzmann's formula connects the entropy, S, and the thermodynamic probability of microstates, W, of the system with the so-called Boltzmann constant k

$$S = k \ln W. \tag{2.5}$$

This relationship means that the Second Law of Thermodynamics, which is stated in terms of an increase in microstates, can also be expressed in terms of information changes in the microstates of the system. In the case where the system contains N particles (with $h = 1\ldots$n properties), which do not depend on each other and can be combined in different ways, we get

$$S = -kN \sum_{h=1}^{n} p_h \ln p_h, \tag{2.6}$$

Where the frequencies $p_h = N_h/N$ is the probabilities of given particles having h sets of properties (see e.g. Jörgensen and Svirezhev 2004: Chap. 4 for more details).

Shannon's formula (2.7) for measuring information (or the information entropy) can be interpreted as a loss of information during a transfer of coded information, which has the characteristics of an irreversible process.

$$I = -N \sum_{h=1}^{n} p_h \log_2 p_h. \tag{2.7}$$

Consequently, free energy is needed to avoid potential irreversible loss of information.

The formal similarity between entropy and information has a very deep meaning. Entropy is a deficiency of information for the full description of the system. The proof of the equivalency of information and entropy, however, has been seen in the context of the discovery by Bekenstein (1974) and its theoretical foundation by Hawkins (1975), which confirms that black holes of the universe also have entropy increase, or loss of information. Equation (2.8) means that the entropy increase (equal to the related loss of information) depends mainly on the mass of the black hole, M_{BW}, and on the mass of the object, m_{Obj}, disappearing into the black hole (for more details see Görnitz and Görnitz 2006:385).

$$\Delta S = \Delta I = konst. \cdot M_{BW} \cdot m_{Obj}. \tag{2.8}$$

2.1.3 Relations Between Work and Exergy

Assume a system is in thermodynamic equilibrium with its environment. A certain amount of work is required by the environment to change the system from its initial state, 0, to another state, 1. For the reverse transition, from state 1 to the initial state 0, the system must do work on the environment. For the direct forced transition, which requires the environment to supply the minimum work, $min\,(\delta_{01}) = \delta A_{min}$ there is a corresponding transition where the system does maximum work, $max\,|\,\delta A_{10}| = |\,\delta A_{max}|$. The latter transition is started when the forcing action stops and the system begins moving spontaneously towards its thermodynamic equilibrium. It is evident that the magnitudes of δA_{min} and $|\,\delta A_{max}|$ are identical. Equation (2.9) (cf. Jörgensen and Svirehev 2004: Chap. 15) shows that, for a reversible process, the external work equals the sum of the changes in internal energy, heat, compression, and chemical potentials

$$\delta A \geq \Delta U - T_0 \Delta S + p_0 \Delta V - \sum_{h=1}^{n} \mu_h^0 \Delta N_h. \tag{2.9}$$

The value of minimum work is given by the formula

$$\delta A_{min} = |\delta A_{max}| = \Delta U - T_0 \Delta S + p_0 \Delta V - \sum_{h=1}^{n} \mu_h^0 \Delta N_h. \tag{2.10}$$

When using Gibbs' equation, $dU = TdS - pdV + \sum_{h=1}^{n} \mu_h^0 \Delta N_h$,

Equation (2.10) can be written in the form

$$dA_{min} = |dA_{max}| = (T - T_0)\ dS - (p - p_0)\ dV - \sum_{h=1}^{n} \mu_h^0\ dN_h. \tag{2.11}$$

Another definition can be applied to the maximum work by using the following Eq. (2.12) (cf. Jörgensen and Svirehev 2004:98–99)

$$\delta A_{max} = T_0 \Delta S_{tot} = T_0 \left(S_{tot} - S_{tot}^{eq} \right) = \Delta U - T_0 \Delta S + p_0 \Delta V - \sum_{h=1}^{n} \mu_h^0 \Delta N_h. \tag{2.12}$$

In Eq. (2.12), ΔU, ΔS, ΔV and ΔN_h express the differences from thermodynamic equilibrium between the values of energy, entropy, volume, and number of particles in the system. This equation expresses how much the entropy of the closed super system (e.g. the whole universe) differs from its maximum possible value if the system is not in equilibrium with its environment. Since the concept of maximum work plays a very important role, a special term, exergy, has been introduced. Exergy, Ex, is defined as the amount of entropy-free energy that a system can perform when it is brought into thermodynamic equilibrium with its environment, i.e. $|A_{max}| = Ex$. As it is formulated by Jörgensen and Svirezhev (2004:100)

> It seems more useful to apply exergy than entropy to describe the irreversibility of real processes as it has the same unit as energy and is an energy form, while the definition of entropy is more difficult to relate to concepts associated with our usual description of reality. Furthermore, exergy facilitates the differentiation between low-entropy energy and high-entropy energy, as exergy is entropy-free energy.

Therefore, it is helpful (cf. Cerbe and Hoffmann 1996:470) to represent free energy E as the sum of the entropy-free exergy, Ex, and the anergy, An, caused by the entropy of the irreversible processes of the system:

$$E = Ex + An. \tag{2.13}$$

2.1.4 Exergy Far from the Thermodynamic Equilibrium

Now that various thermodynamic equations have been introduced, let us approach the transition of a chemical system from its initial state, 0, thermodynamic equilibrium, to another state, 1, that is far from the thermodynamic equilibrium. When assuming that the infinitesimal change of exergy *d(Ex)* takes place in an infinitesimally short time, dt, so that the environment is not able to change, we obtain from (2.12) the simplified Eq. (2.14) (Jörgensen and Svirezhev 2004:106)

$$\frac{d(Ex)}{dt} = \sum_{h=1}^{n} (\mu_h - \mu_h^0)\frac{dN_h}{dt}. \tag{2.14}$$

When using the definition of the chemical potential

$$\mu_h = \mu_h(0) + RT \ln N_h, \; h = 1, \ldots, n,$$

Where N_h can be considered as the molar concentrations of corresponding chemical substances h, and R is the gas constant, we can rewrite Eq. (2.14) for the case where $T = T_0$ as

$$\frac{d(Ex)}{dt} = RT_0 \sum_{h=1}^{n} \ln \frac{N_h}{N_h^0}\frac{dN_h}{dt}. \tag{2.15}$$

By integrating both sides of Eq. (2.15) with respect to time and taking into account that $Ex(t_0) = 0$, we get

$$\begin{aligned} Ex(t) = RT_0 \int_{t_0}^{t} \sum_{h=1}^{n} \ln \tfrac{N_h(t)}{N_h^0}\tfrac{dN_h}{dt}dt = RT_0 \sum_{h=1}^{n} \int_{t_0}^{t} \ln \tfrac{N_h(t)}{N_h^0}\tfrac{dN_h}{dt}dt = \\ RT_0 \sum_{h=1}^{n} \int_{N_i^0}^{N_i} (\ln N_h - \ln N_h^0)\, dN_h = T_0 \sum_{h=1}^{n} \left[N_h \ln \tfrac{N_h}{N_h^0} - (N_h - N_h^0) \right]. \end{aligned} \tag{2.16}$$

We can see that $Ex(t) > 0$ for any $N_h > 0$, except $N_h = N_h^0$, $h = 1, \ldots, n$, if $Ex \equiv 0$.

When using N to represent the total number of particles in the system, Eqs. (2.14) and (2.15) can also be written in concentration form

$$\frac{d(Ex)}{dt} = NRT_0 \sum_{h=1}^{n} \ln \frac{c_h}{c_h^0}\frac{dc_h}{dt}, \text{ with } N{:}= \sum_{h=1}^{n} N_h \tag{2.17}$$

$$\begin{aligned} Ex(t) = NRT_0 \int_{t_0}^{t} \sum_{h=1}^{n} \ln \tfrac{c_h(t)}{c_h^0}\tfrac{dc_h}{dt}dt = NRT_0 \sum_{h=1}^{n} \int_{t_0}^{t} \ln \tfrac{c_h(t)}{c_h^0}\tfrac{dc_h}{dt}dt = \\ NRT_0 \sum_{h=1}^{n} \int_{c_h^0}^{c_h} (\ln c_h - \ln c_h^0)\, dc_h = NRT_0 \sum_{h=1}^{n} \left[c_h \ln \tfrac{c_h}{c_h^0} - (c_h - c_h^0) \right]. \end{aligned} \tag{2.18}$$

We obtain the relation that entropy-free work done on the system by its external environment for transitions from one state to another, called dissipative work, must be the negative value of exergy

$$Diss(c^0 \to c^1) = Diss(c^1 \to c^0) = -Ex\,(c^1, c^0). \tag{2.19}$$

Equation (2.19) implies that (cf. Jörgensen and Svirezhev 2004:109)

1. Work done on the system by its external environment in the forced transition $c^0 \to c^1$ is equal to $|A_{01}| = -\text{Diss}(c^0 \to c^1)$.
2. In the course of the transition, the system accumulates exergy, in an amount equal to the absolute value of the work done on the system $Ex(c^1, c^0) = |A_{01}| = -\text{Diss}(c^0 \to c^1)$.
3. When the system is closed, it returns spontaneously to the stable state, c^0, and dissipating exergy in the process. The transmission is complete when the system reaches c^0: specifically, at this moment the work $|A_{10}| = -Diss(c^1 \to c^0) = Ex(c^1, c^0)$ and all of the exergy has been exhausted.

2.1.5 Relations Between Exergy and Information

By introducing the variable $p_h = N_h/N = c_h$, we can rewrite Eq. (2.16) for the exergy as

$$Ex = N \sum_{h=1}^{n} p_h \ln \frac{p_h}{p_h^0} + \left[N \ln \frac{N}{N_0} - (N - N_0) \right]. \tag{2.20}$$

The vector of the intensive variables $\vec{p} = \{p_1, \ldots, p_n\}$ describes the composition of the system, while N is the extensive variable for the status of the system.

The value $K = \sum_{h=1}^{n} p_h \ln \left[p_h / p_h^0 \right]$ with the constant factor $1/\ln 2$ is the Kullback measure of the increment of information, which was introduced in Eq. (2.6). That is, we can present the expression for exergy as the sum of an informational and a materialistic component

$$Ex = Ex_{inf} + Ex_{mat}. \tag{2.21}$$

The two terms:

$$Ex_{\text{inf}} = NK\left(p, p^0\right) \geq 0 \text{ and } Ex_{mat} = N \ln\left(N/N_0\right) - (N - N_0) \geq 0$$

Represent the structural changes in the system and the change of the total mass of the system. Consequently, the application of thermodynamic laws extended to include information aspects results in useful work, or the entropy-free energy (exergy), being the sum of materialistic and informational components for chemical systems.

Since biological systems have internal information storage capabilities, whereas chemical systems do not, chemical systems are considered to be on a lower evolutionary level than biological systems. These concepts must be considered when describing economic transformation processes from the extended thermodynamic point of view.

Equation (2.21) also shows that the evolution of chemical systems on earth, which started more than four billion years ago, has both materialistic and informational components. Three billion years was required for biological systems to evolve with internal information processes for survival and sexual reproduction. Approximately 10 millions ago, the evolution process began which resulted in human intelligence, which has developed to the cultural level that supports permanent reproduction through modern education and information systems (for more details see e.g. Ulmenschneider 2006:Chap. 6 and Jörgensen and Svirezhev 2004:109). Equation (2.21) can be extended to Eq. (2.22) by adding terms for the evolution from the chemical to the biological system and then on to intelligent life and its social systems.

$$Ex = Ex_{inf}^{soc} + Ex_{inf}^{bio} + Ex_{inf}^{chem} + Ex_{mat}^{soc} + Ex_{mat}^{bio} + Ex_{mat}^{chem}. \qquad (2.22)$$

Ex_{inf}^{soc} Stands for the cumulative information of the individual and social education process i.e. (human intelligence). The other terms represent the information and material exergy components needed for the survival of the biological and chemical processes of this specific system. Such an extended interpretation of economic transformation processes and its interrelation with the natural environment are the contributions of Binswanger (1992), Beard and Luzada (2000) and Lozada (2004).

2.1.6 Thermodynamics of Economic Transformation Processes

The entropy and exergy concept developed previously will now be used to describe the thermodynamic structure of modern industrial production in terms of mass, entropy, and exergy. Modern industrial production is described as the joint production of desired goods with high exergy (entropy-free energy) and by-products with low exergy (high anergy) caused by the entropy of irreversible processes. As illustrated in Figs. 2.1 and 2.2, economic activities must always be seen as joint production in accordance with the laws of thermodynamics. Consequently, joint production is the normal case and not a special case, as is often assumed in economic theory. According to Eq. (2.13), in the case of fixed inputs of free energy and materials, an increase in exergy of the desired outputs requires a decrease in anergy of the system.

In the practise of economics, contradictions between natural laws and common economic assumptions are often not known, ignored, or "corrected" by using strange assumptions. One such assumption is free disposal costs for unavoidable by-products. Another common, yet contradictory, assumption is that the marginal production of an additional output of the desired good is a decreasing function

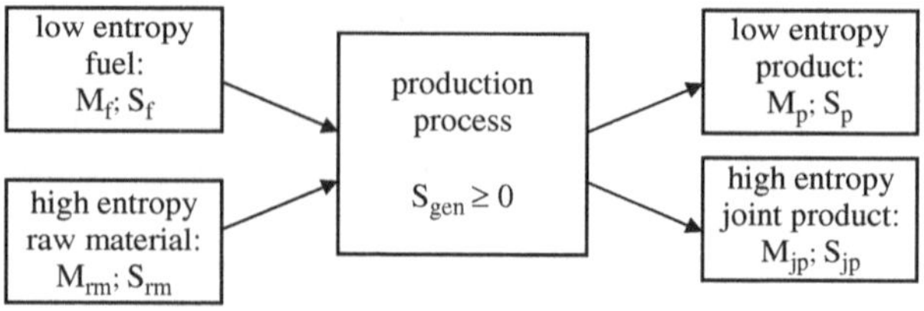

Fig. 2.1 Thermodynamic structure of industrial production in terms of mass and entropy (adapted from Baumgärtner et al. 2006:51)

of the required production inputs. This assumption requires that the anergy of the by-product must increase because the total amount of all energy forms must remain constant.

So far, however, the parameters for the external relations of the system with the surrounding environment have not been considered. Consequently, the description of economic transformation in Figs. 2.1 and 2.2 is incomplete. The transitions from state 0 to state 1 of a chemical or economic system must be formulated by an extended set of differential equations $dc_h/dt = f(c_1, \ldots, c_n; \alpha_1, \ldots, \alpha_n)$. The parameters α_h ; $h = 1, \ldots, n$ of the vector $\vec{\alpha} = (\alpha_1, \ldots, \alpha_n)$ represent the external intervention for changing the composition of the system. The concentration vector $\vec{c} = (c_1, \ldots, c_n)$ includes the flow variables. The vector $\vec{\alpha} = (\alpha_1, \ldots, \alpha_n)$ can be interpreted as stock variables that keep the system far from thermodynamic equilibrium within a time period, *T*. To be complete, Fig. 2.2 should include the relationship between the economic sub-system and its environment. This is done in Fig. 2.3.

In Fig. 2.3, the role and characteristics of capital and public goods for production and consumption activities are shown for two follow-up time periods, where we strictly distinguish between flow and stock variables for the transformation processes. The active role of capital and public goods in physical terms is limited to delivering physical inputs for the transformation processes. This is measured as deterioration, or exergy losses, of the stock variables and must be permanently compensated by equivalent flow inputs for the maintenance of the capital stock. The embedded information of the capital stock and public goods and its interaction with the embedded information of the human labour is the main driving force of the

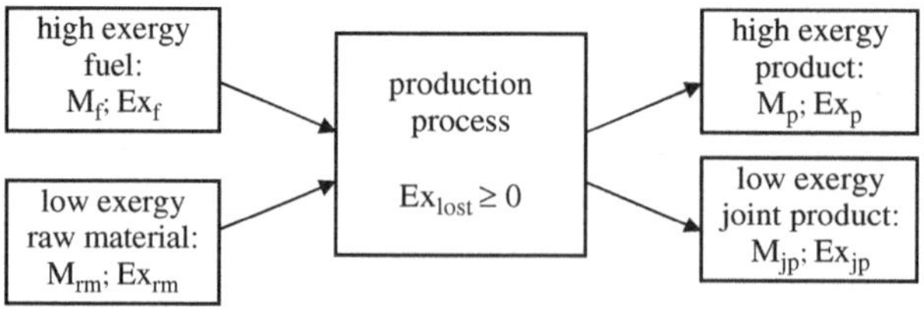

Fig. 2.2 Thermodynamic structure of industrial production in terms of mass and exergy (adapted from Baumgärtner et al. 2006:51)

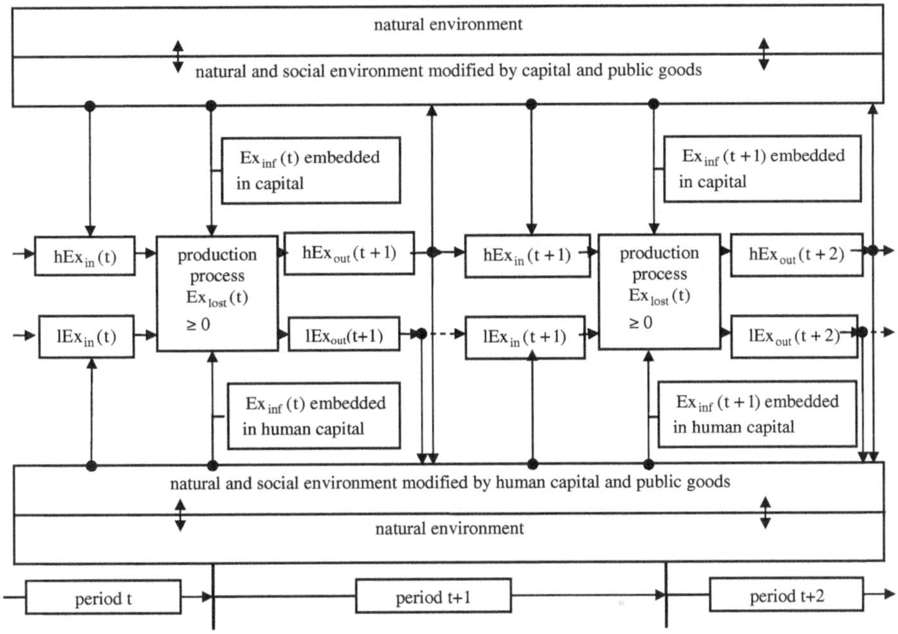

Fig. 2.3 Extended thermodynamic structure and time characteristics of industrial production in terms of (specific) exergy

transformation processes. Recalling that exergy and information are equivalent, the exergy contribution for economic activities has a physical and informational component. The industrial revolution of the last two centuries can be generalized as having reduced the hard physical work of labour by embedding this less comfortable coordinative work into the capital and public goods. This then allows more time to be spent on reproduction activities and the improvement of the levels of education and qualifications of the human labourers.

The preceding has explained the physical and informational components of the human labour. Although the total labour input for production may have remained constant or even reduced during the industrial evolution process, the physical portion of the human labour force is, on average, a declining function when it is compared with the increase in the coordinative and creative portion that is required for operating and managing capital stocks. Essential indicators for this evolution process are the efforts, time and resources spent on education and professional qualification, which are needed to be an efficient participant in both the production and reproduction processes. Therefore, we have to distinguish the human labour force into its physical (or flow related) and informational (or stock related) components. The rapid change in production and reproduction technologies requires a permanent change in the education and qualification stock, which has to be seen as a complementary "twin set" of the capital stock.

2.2 The Concept of Capital Theory

Although we cannot cover economic theory in detail, a short description of capital theory will be the subject of this section. Aside from other advantages, when compared with the neoclassical capital theory (for more details see Faber 1979), the Neo-Austrian capital theory, with some modifications, might be a more flexible approach for constructing disaggregated models. To the extent that they provide sufficient understanding of the rationale behind inter-temporal optimization problems, only graphical illustrations will be provided.

2.2.1 Neo-Austrian Capital Theory as Example

When deriving optimal conditions under model constraints, it is common in economics to use individual utility or social welfare functions. This practise reflects the usual properties of a preference order system for the comparison of consumption bundles (for more details see e.g. Breyer 2004:117–123). When we start (for more details see e.g. Stephan 1995:69) with a fictitious central planning economy and assume that a central planning authority maximizes an aggregate, strictly quasi-concave, strictly monotonic and differentiable welfare function, the conventional dynamic characteristics of the inter-temporal welfare are given by Eq. (2.23).

Assumption 2.1 (for the inter-temporal welfare function)

$$W(\vec{Q}(1),\ldots,\vec{Q}(T)) = \sum_{\tau=1}^{T}(1+\delta)^{1-\tau}W_\tau(\vec{Q}(\tau)), \tag{2.23}$$

Where the discount rate $\delta > -1$ measures the society's time preference and

$$\mathrm{W}_\tau(\vec{\mathrm{Q}}(\tau)) \quad \tau\in\{1,\ldots,\mathrm{T}\} \wedge \vec{\mathrm{Q}}(\tau) := (\mathrm{Q}_1\tau),\ldots,(\mathrm{Q}_\mathrm{q}(\tau))$$

Are welfare functions in interval τ for consumption vectors of q-types of goods used for consumption and reproduction purposes. These functions are assumed to be identical for a different interval, $\tau' \neq \tau$, and to be quasi-concave, strictly monotonic, and differentiable functions for the N-dimensional consumption vector $\vec{\mathrm{Q}}(\tau)$ for each of the periods $\tau \in \{1,\ldots T\}$. *W*, therefore, also has these properties.

When using Assumption 2.2, which is also common in economics,

Assumption 2.2 (for producing feasible consumption bundles) There are technology sets that are used to produce feasible consumption bundles for each of the τ-intervals, as well as between the intervals τ and $\tau + 1$. These sets form a convex and closed *NT*-dimensional vector space.

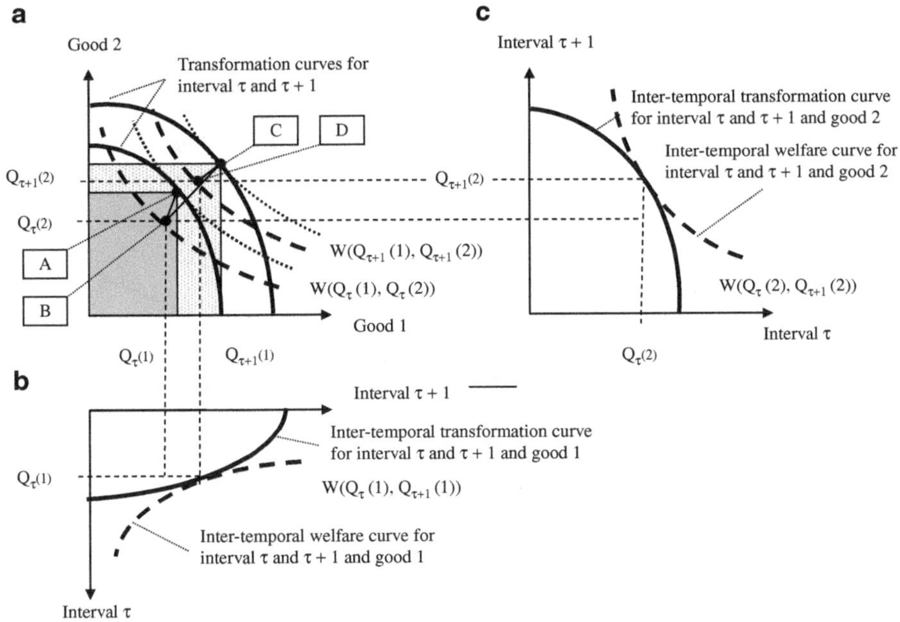

Fig. 2.4 Illustration of the static and inter-temporal optimization process in the case where investments for future welfare increase

Static transformation curves for each of the τ-intervals (as shown in Fig. 2.4a) and inter-temporal transformation curves between the different intervals following each other exist.

The transformation curves outline the border between feasible consumption programmes when optimizing static and inter-temporal welfare functions, which are shown in Fig. 2.4. At point A in Fig. 2.4a, the static transformation curve and the welfare function (marked as a dotted line) have a tangential contact point. This point shows the optimal welfare level for the case where there is no saving or investment in interval τ for the future. If some portion of the desired goods is saved for future investment, the welfare level (the broken line of the welfare function curve for interval τ) is reduced by choosing the consumption bundle, point B, in Fig. 2.4a. The technology set for the production of feasible consumption bundles can then be extended to the area outlined by the transformation curve of the next interval $\tau + 1$. Point C in Fig. 2.4a shows the maximal welfare level in interval $\tau + 1$. If saving and investment for the future will be continued over the next intervals, we would choose a lower welfare level for the consumption bundle at point D in interval $\tau + 1$. We would then continue the inter-temporal optimization process in the next interval within the finite (or infinite) planning horizon in the frame of the future inter-temporal transformation and welfare opportunities, which are shown in Figs. 2.4b, c for two goods and the intervals τ and $\tau + 1$.

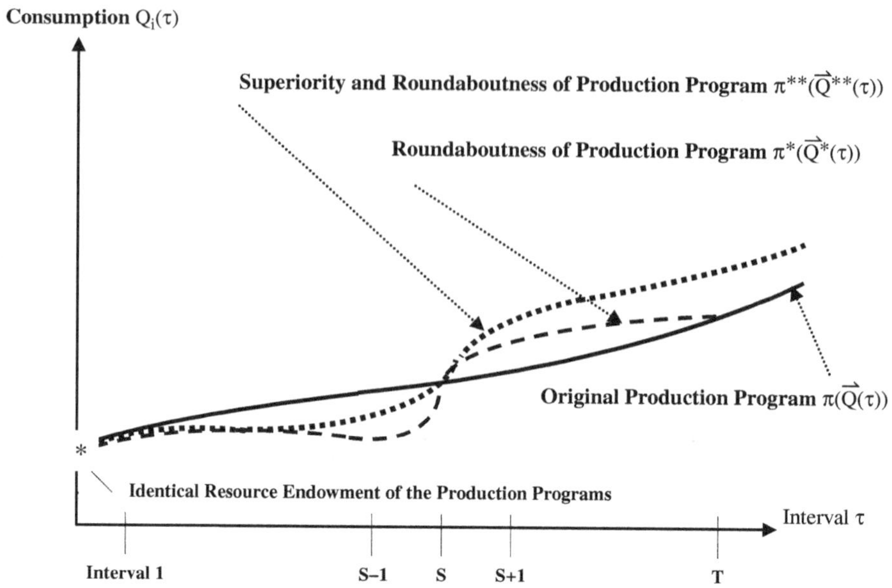

Fig. 2.5 Superiority and roundaboutness of production programs (adapted from Stephan 1995:56)

Inter-temporal welfare levels are permanently increasing and even the future welfare levels might be less attractive than the present one, which is expressed by the discount rate in Eq. (2.23). The saving and investment strategy finally leads to an inter-temporal welfare optimum, if some essential characteristics and necessary conditions for production and reproduction are met.

The characteristics of the production side that are essential to Neo-Austrian capital theory are the concepts of roundaboutness and superiority of feasible production programs. These are illustrated in Fig. 2.5 for one selected good $h \in \{1,,,,N\}$ of a NT dimensional consumption vector of feasible production programs.

Generally speaking, roundaboutness means that a comparable lower consumption is honoured by an increase of consumption opportunity later. If the reduction of the total consumption is then honoured by the increase of the total consumption amounts later, then we would say that roundaboutness and superiority of production programs exist.

When considering only one good, these two properties are clearly defined. In the case of N goods, however, these properties could be valid for all, only one, or several components of the N dimensional consumption vector by keeping the other components flexible or under a ceteris paribus assumption. Although these weaknesses in the Neo-Austrian theory are often discussed, and should therefore be known, it is surprising that even Stephan (Stephan 1995:53–61, 94–99) is not addressing these problems for the multi-good case being also been analysed e.g. by Jaksch (1975) and Belloc (1980).

One way of solving these problems could be to introduce the idea of global roundaboutness and superiority.

Definition 2.1 (for global roundaboutness)an antipsychotic drug with a neuroleptic effect. Let $\left\{\pi\left(\vec{Q}(\tau)\right), \tau = 1,\ldots,T\right\}$ and $\left\{\pi^*\left(\vec{Q}^*(\tau)\right), \tau = 1,\ldots,T\right\}$ be inter-temporal production programs, which are feasible from the same resource endowment and let $\left\{\vec{Q}(\tau), \tau = 1,\ldots,T\right\}$ and $\left\{\vec{Q}^*(\tau), \tau = 1,\ldots,T\right\}$ denote the corresponding consumption programs. The production program $\left\{\pi^*\left(\vec{Q}^*(\tau)\right), \tau = 1,\ldots,T\right\}$ labelled as globally roundabout, compared to, $\left\{\pi\left(\vec{q}(\tau)\right), \tau = 1,\ldots,T\right\}$ if an activity interval S $(S \leq T)$ exists such that:

$$\begin{aligned} \vec{Q}(\tau) &\geq \vec{Q}^*(\tau) && \text{for } \tau < S, \\ Q(S) &\geq \vec{Q}^*(S), \\ \vec{Q}(\tau) &\leq \vec{Q}^*(\tau) && \text{for } \tau > S \end{aligned} \tag{2.24}$$

Definition 2.2 (for global superiority) Let $\left\{\pi\left(\vec{Q}(\tau)\right), \tau = 1,\ldots,T\right\}$ and $\left\{\pi^*\left(\vec{Q}^*(\tau)\right), \tau = 1,\ldots,T\right\}$ be inter-temporal production programs, which are feasible from the same resource endowment and let $\left\{\vec{Q}(\tau), \tau = 1,\ldots,T\right\}$ and $\left\{\vec{Q}^*(\tau), \tau = 1,\ldots,T\right\}$ denote the corresponding consumption programs. The consumption program $\left\{\pi^*\left(\vec{Q}^*(\tau)\right), \tau = 1,\ldots,T\right\}$ is labelled globally superior, compared to, $\left\{\pi\left(\vec{q}(\tau)\right), \tau = 1,\ldots,T\right\}$ if an activity interval S $(S \leq T)$ exists such that

$$\sum_{\tau=1}^{T} \vec{Q}^*(\tau) > \sum_{\tau=1}^{T} \vec{Q}(\tau) \tag{2.25}$$

The concept of partial roundaboutness and partial superiority of feasible production programs could be used if only one or several goods have these properties under flexible or under ceteris paribus assumption for the other goods. Another way to overcome these problems is to use subjective norms. Winkler (2005), for example, characterizes production programs by welfare functions and uses welfare roundaboutness and superiority for the comparison of production programs.

2.2.2 Capital Theory and Its Natural Sciences Consistency

Though this concept of partial and global roundaboutness, superiority and inter-temporal welfare functions is going in the right direction for dealing with multi-good problems, it has the disadvantage that objective judgements of production programs are replaced by normative welfare judgements. As a compromise, thermodynamic exergy can be used to characterize production programs using the following definitions.

Definition 2.3 (Exergy roundaboutness) Let $\left\{\pi\left(\vec{Q}(\tau)\right), \tau = 1,\ldots,T\right\}$ and $\left\{\pi^*\left(\vec{Q}^*(\tau)\right), \tau = 1,\ldots,T\right\}$ be inter-temporal production programs, which are feasible from the same resource endowment and let $\left\{\vec{Q}(\tau), \tau = 1,\ldots,T\right\}$ and $\left\{\vec{Q}^*(\tau), \tau = 1,\ldots,T\right\}$ denote the corresponding consumption programs. The set of production programs $\left\{\pi^*\left(\vec{Q}^*(t)\right), t = 1,\ldots,T\right\}$ is labelled as exergy roundabout, compared to $\left\{\pi\left(\vec{q}(\tau)\right), \tau = 1,\ldots,T\right\}$, if an activity interval S $(S \leq T)$ exists such that, for exergy of consumption bundles, we have

$$
\begin{aligned}
&Ex\vec{Q}(\tau)) \geq Ex(\vec{Q}^*(\tau)) \qquad && \textit{for } \tau < S,\\
&Ex(\vec{Q}(S)) \geq Ex(\vec{Q}^*(S)), &&\\
&Ex(\vec{Q}(\tau))) \leq Ex(\vec{Q}^*(\tau))) && \textit{for } \tau > S.
\end{aligned}
\tag{2.26}
$$

Definition 2.4 (Exergy superiority) Let $\left\{\pi\left(\vec{Q}(\tau)\right), \tau = 1,\ldots,T\right\}$ and $\left\{\pi^*\left(\vec{Q}^*(\tau)\right), \tau = 1,\ldots,T\right\}$ be inter-temporal production programs, which are feasible from the same resource endowment. Let $\left\{\vec{Q}(\tau), \tau = 1,\ldots,T\right\}$ and $\left\{\vec{Q}^*(\tau), \tau = 1,\ldots,T\right\}$ denote the corresponding consumption programs. The set of production programs $\left\{\pi^*\left(\vec{Q}^*(\tau)\right), \tau = 1,\ldots,T\right\}$ is labelled exergy superior, compared to, $\left\{\pi\left(\vec{q}(\tau)\right), \tau = 1,\ldots,T\right\}$ if an activity interval S $(S \leq T)$ exists such that:

$$
\sum_{\tau=1}^{T} Ex(\vec{Q}^*(\tau))) > \sum_{\tau=1}^{T} Ex(\vec{Q}(\tau))). \tag{2.27}
$$

There is one final aspect worth mentioning, because it differs from conventional Neo-Austrian capital theory. The concept of exergy roundaboutness and superiority characterize technical innovations in production activities. The dynamics of the economic system are initiated by the increase of knowledge and information of the human individuals and society in the past and present about its natural and social environment. When considering that it takes more than 20 years to educate a child to the point where they can be integrated into the production processes and social life, roundaboutness and superiority of production activities must be considered to be closely connected to the existence of roundaboutness and superiority of reproduction activities. As already mentioned in the description of the economic transformation processes, social evolution has resulted in the transfer of part of the less attractive physical labour so it is done by technical facilities. As a result, the requirements for more intelligent labour inputs and the efforts of education are permanently increasing. Therefore, the welfare optimization procedure within our model must consider both the desired goods consumed in each of the intervals as well as the durable goods (or capital goods) for reproduction and education activities, closely connected with capital stock of the production activities. The integration of durable goods into the welfare function leads to a more complicated welfare function, because the portions of the durable goods that are not consumed in an interval τ influences the welfare level of the follow-up periods. In order to avoid this complication, we develop Assumption 2.3.

Assumption 2.3 (for durable goods forming the capital stock) Durable goods are not considered to be arguments in the inter-temporal welfare function, but are used for constructing a closely connected capital stock; there is one for production and one for reproduction activities.

Aside from other social services, one part of the capital stock might be required for education. The high exergy human labour output of the reproduction sector might need further specifications of the following simplified expression

$$Ex(L(\tau))) = Ex(L_{phys}(\tau)) + Ex(L_{inf}(\tau)). \tag{2.28}$$

Such a revision to the capital theory would be particularly important in the case where there is more than one consumable good.

Chapter 3
General Design of Dynamic Models for Water Uses

Abstract Water plays a special role in dynamic water use and water infrastructure models because it is involved in both production and reproduction activities. The extraction of required raw materials, which are combined to create desired consumption and capital goods, generates inter-temporal concentration changes of non-renewable resources in the natural environment. Increasing amounts of water are needed to transport the undesired residuals that are created by the economic activities to the wastewater treatment sector. It will be shown that the use of free energy and human labour inputs for water production and wastewater treatment are closely connected to the changes in concentration of the residuals that are generated by the different processes in the economic system. The general design of dynamic models includes the model structure as well as the characteristics of the different processes that make up the water production and wastewater treatment sectors. The dynamics of the model are determined by the development of the capital stocks, which carry innovation effects in terms of human labour and energy inputs.

In Chap. 2, we applied the extended thermodynamic theory to describe economic transformation processes. We also introduced capital theory and discussed its relation to thermodynamic theory. We will now present the general design of an economic model that is consistent with natural sciences. Furthermore, we will integrate water uses and water infrastructure into the model.

3.1 Model Structure and Economic Activities

Figure 3.1 shows the model structure that will be used in the general design of the model.

The model, in general, contains three main sectors

Sector 1 reproduction (or consumption) activities

Sector 2 production activities composed of the following four sub-sectors

Sub-sector 21 producing free energy for maintaining the exergy of the different sub-systems and transforming low exergy inputs to high exergy outputs

H. Niemes, M. Schirmer, *Entropy, Water and Resources*,
DOI 10.1007/978-3-7908-2416-2_3,

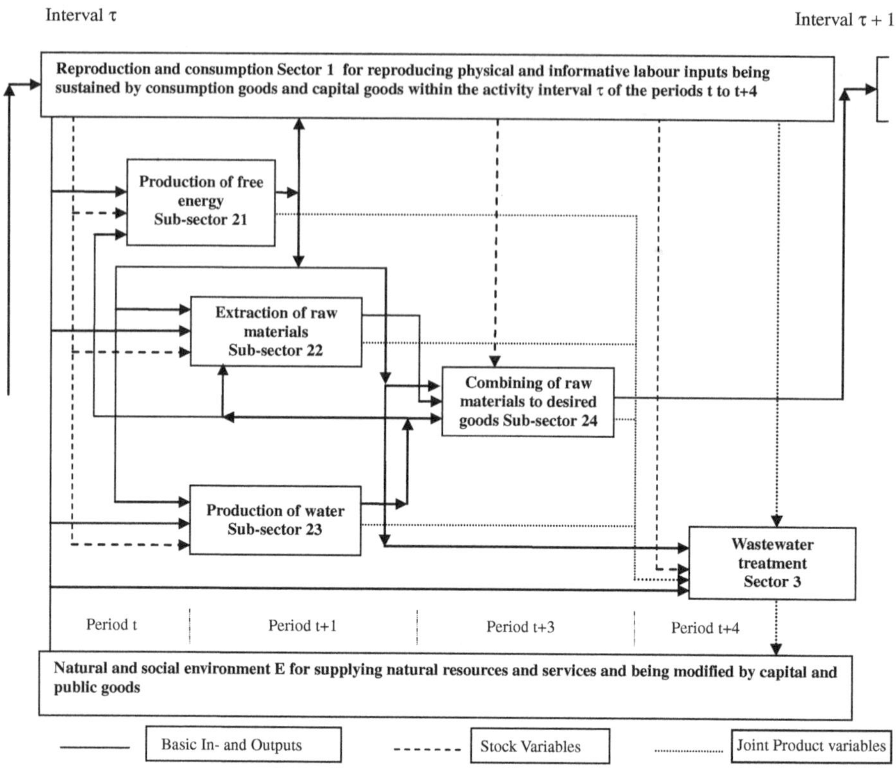

Fig. 3.1 Structure of the model to be designed

Sub-sector 22 extracting raw materials from the natural environment

Sub-sector 23 producing water used for the production and reproduction activities

Sub-sector 24 combining of raw materials into desired goods

Sector 3 environmental protection activities

Some assumptions are necessary for the time structure of the model, (for more details see e.g. Stephan 1995:29–31).

Assumption 3.1 (for time discretization): Time is discrete and, therefore, is depicted using a series of arbitrary periods of equal length $t = 1,...,T$.

Assumption 3.2 (for the duration of the transformation processes): The transformation processes uniformly require one time period. Therefore, if an input-vector, $x(t)$, is employed at time t, then the technically producible output will not be available until time $t+1$ and the output must be dated accordingly. The transformation processes being technically feasible and an element of the technology set G_t or the technical knowledge about the transformation process in the period t, always take the form

$$[X(t), Y(t+1)] \in G_t \tag{3.1}$$

Assumption 3.3 (for the time direction of the transformation processes): For all periods t and technology sets G_t it is true that

$$\text{If } [X(t), Y(t+1)] \in G_t, \textit{ then } [Y(t+1), X(t)] \notin G_t \tag{3.2}$$

In Fig. 3.1, we only differentiate between the unavoidable by-products from flow and stock variables graphically. Flow and stock variables are the capital and durable goods and permanently reproduced labour inputs. In the context of model construction, however, we must distinguish more precisely between these variables. The free energy produced in sub-sector 21 is required for the reproduction and consumption activities in sector 1 and for the production activities in sub-sectors 22, 23 and 24. In these sub-sectors, low exergy inputs are transferred and combined into high exergy outputs and low exergy unavoidable by-products. The joint by-products, marked as dotted lines in Fig. 3.1, are transported to sector 3, central wastewater treatment, by the clean water produced in sub-sector 23. Clean water is used in sub-sectors 21, 22, 24 as well as in sector 3 and polluted there by the joint by-products. The amount of water needed will be calculated by multiplying specific coefficients with production and reproduction activity levels in the corresponding sectors and sub-sectors. The dynamics of the model are determined by a subjective decision that is made in the reproduction and consumption sector 1. The decision involves choosing which portion of the produced goods are used for daily consumption and which portion will be saved as capital and durable goods for the mid-term and long-term development of the production and reproduction sector. The surplus portions are shown as broken lines in Fig. 3.1. The consumption amounts must be sufficient to survive the activity interval τ and its four consumption and production periods, within which the consumption and capital goods are produced for the next activity interval $\tau + 1$. For simplicity of presentation, we will focus on time intervals instead of using the precise time periods. We use the precise time periods only for timing within the activity intervals.

The transformation processes of raw materials to desired goods in our model requires at least three periods. In the first period, t, free energy is made available to extract raw materials and water from the environment in period t+1. The conversion of the extracted raw materials to desired goods and the separation and transportation of the unavoidable by-products by water to the environmental protection sector will be completed at the end of period t+2. If we argue that the consumption of the desired goods also reproduces human labour as an input for production activities, the whole reproduction process last 4 periods if the environmental protection activities are also taken into consideration.

3.2 Characteristics of Production Activities

3.2.1 Criteria for the Extraction and Use of Raw Materials

The extraction of raw materials in sub-sector 22, which will be converted to desired goods in sub-sector 24, will be described by the following assumptions

Assumption 3.4 (for the concentrations of the extracted raw materials and unavoidable by-products) The $m \geq 1$ types of desired raw materials form chemical systems $i \in \{1,..., m\}$ containing o-types of undesired, unavoidable by-products. In total, n-types ($n = m + o$) of raw materials with low exergy characteristics or low concentrations in accordance to Eq. (2.17) and (2.18) are produced. For simplification, each concentration input vector $\vec{c}^{\,i}(t)$ for the chemical systems $i \in \{1,..., m\}$ at time t might contain only one type $i \in \{1,..., m\}$ of raw material, but $j \in \{m + 1,..., n\}$ types of undesired, unavoidable by-products. The concentration vectors $\vec{c}^{\,i}(t)$ at time t have the following form and concentration values

$$\begin{aligned}
\vec{c}^{\,i}_{i}(t) &:= \left(0,...., c^{\,i}_{i}(t),..., 0\ ;\ c^{\,i}_{m+1}(t),..., c^{\,i}_{n}(t)\right) \wedge \sum_{h=1}^{n} c^{\,i}_{h}(t) = 1 \\
\wedge\ c^{\,i}_{h}(t) &= 0 \text{ if } h \neq i \in \{1,..., m\} \\
\wedge\ c^{\,i}_{h}(t) &> 0 \text{ if } h = i \\
\wedge\ c^{\,i}_{h}(t) &\geq 0 \text{ if } h = j \in \{m + 1,..., n\}.
\end{aligned} \tag{3.3}$$

For the general design of the dynamic water use and water infrastructure model, we make additional assumptions.

Assumption 3.5 (for the relation between the extracted raw materials and desired goods): For the production of a high exergy output vector $\vec{Y}\ (t + 2) = \left(Y_1(t + 2),..., Y_q(t + 2)\right)$ of q-types ($q \geq 1$) of desired goods at time t+2, m-types of raw material extracted from the chemical system $i \in \{1,..., m\}$) are required at time t+1. For the production of one unit of desired goods k, $k \in \{1,..., q\}$, a specific amount of each raw material is defined by a simple linear relation

$$N^{\,i}_{ik}(t + 1) := \kappa_{ik}\, Y_k(t + 2) \wedge N_i(t + 1) := \sum_{k=1}^{q} \kappa_{ik}\, Y_k(t + 2). \tag{3.4}$$

Assumption 3.6 (for the use of the desired goods as capital or consumption goods) The desired goods of the high exergy output vector produced in period t+2 in sub-sector 24 containing q processes $\vec{Y}_P(t{+}2) = \left(Y_{P1}(t + 2),..., Y_{Pq}(t + 2)\right)$ are used in the next period, t+3, for consumption and reproduction activities in sector 1, for the maintenance and development of the capital stock in sub-sectors 21, 22, 23 and 24, and for the environmental protection activities, *EP*, in sector 3. The total amounts of the desired goods produced are equal to the sum

$$\vec{Y}_{P=24}(t+2) = \sum_{r=1}^{4} \vec{X}_{P=2r}(t+3) + \vec{Q}_{RP=1}(t+3) + \vec{Z}_{EP}(t+3)\,. \qquad (3.5)$$

Assumption 3.7 (for the deterioration and consumption rates of the desired goods used as capital and/or consumption goods) When using the desired goods of the high exergy output vector $\vec{Y}^{t+2!}_{P=24} = \left(Y^{t+2}_{P1},...,Y^{t+2}_{Pq}\right)$ for the consumption and reproduction activities in sector 1, the maintenance and development of the capital stock in different production sub-sectors *Pr* ($r = 21, 22, 23, 24$), and the environmental protection sector in the next period $t + 3$, some portion of the desired goods might be completely consumed or required for maintenance of the capital stocks. The remaining portion, however, will be available for the follow-up periods. The portions of the desired goods that is not available for the next period will be determined by good-specific deterioration (or consumption) rates $0 < d_k \leq 1,\ k \in \{1,..., m\}$.

Assumption 3.8 (for the use of high exergy natural resources) The o -types of unavoidable raw materials with $j = m + 1,..., n$ and the portions of the m -types of desired goods with $k \in \{1,..., q\}$ that are used for consumption and production purposes in period t and not available for the next periods are mixed together and transported out of the system by high exergy natural resources such as water, soil, and air. These natural resources are also required for the production and reproduction activities. The pollution input vector for sector 3 is determined by the quantities of unavoidable by-products, desired goods used for consumption in sector 1, and desired goods not available for the next periods because of stock variable deterioration as well as high exergy natural resources required for production and consumption purposes.

Assumption 3.9 (for the integration of the general resource problem in the model) In cases where general resource problems will be considered in the dynamic model, we assume that the extraction of raw materials leads to a reduction in the concentrations of the types $i \in \{1,...\ m\}$ of required raw materials and therefore to an increase of the concentrations of the types $j \in \{m + 1,..., n\}$ of unavoidable by-products of these raw materials between the time periods t and $t + 1$

$$\left\{ c^{\,i}_{i}(t) \geq c^{\,i}_{i}(t+1) \right\} \wedge \left\{ c^{\,i}_{j}(t) \leq c^{\,i}_{j}(t+1) \right\}. \qquad (3.6)$$

It is important to highlight some of the implications of these assumptions. For example, the maximum total number of desired goods, q, which is theoretically possible by the combinations of the m raw materials, can be estimated by combinatory relations. The number of sub-sets with r elements of a set with m elements is equal to

$$C(m, r) = \binom{m}{r} = \frac{m!}{r!\ (m-r)!} = \frac{m(m-1)\dots(m-(r-1))}{r!}.$$

Consequently, the total number of combinations of all possible sub-sets is equal to

$$C(m, r = 0, \dots, m) = \sum_{r=0}^{m} \binom{m}{r}.$$

When using the binomial theorem for all real numbers, a and b, and all natural numbers, *m*, we obtain, from the general equation

$$(a+b)^m = a^m + \binom{m}{1} a^{m-1}b + \dots + \binom{m}{r} a^{m-r}b^r + b^m = \sum_{r=0}^{m} \binom{m}{r} a^{m-r}b^r.$$

For the special case of $a = b = 1$, the total number of combinations is equal to

$$C(m, r = 0, \dots, m) = \sum_{r=0}^{m} \binom{m}{r} = 2^m.$$

An empty sub-set ($r = 0$) has no economic meaning and so this combination should not be included. The number of economically reasonable combinations of raw materials $i \in \{1, \dots, m\}$ is equal to

$$C(m, i = 1, \dots, m) = \sum_{i=1}^{m} \binom{m}{i} = 2^m - 1. \tag{3.7}$$

The number of theoretically possible combinations is a rapidly increasing sequence, even for a low number of raw materials. A selected sequence of the numbers of raw materials, for example, has the following corresponding values:

$$\begin{pmatrix} 1 & 3 & 7 & 15 & 31 & 63 & 127 & 255 & 511 & 1\,024 & & 1\,048\,575 & & 1\,073\,740\,000 & \\ 1, & 2, & 3, & 4, & 5, & 6, & 7, & 8, & 9, & 10, & \dots, & 20 & \dots, & 30 & \dots \end{pmatrix}.$$

In reality, however, the number of possible combinations is limited by technological constraints such as the physical and chemical properties of the raw materials. In the context of an environmental protection strategy, an increase in the number of unavoidable by-products (as a portion of the desired goods used for production and reproduction activities) must be taken into consideration because the used desired goods will also be transported by water to the environmental protection sector.

Within the framework of the natural science foundations introduced in Chap. 2, another implication of our assumptions should be discussed. The exergy change of shifting raw materials from low to high concentration within a short time is given in Eq. (2.17). If only m-types of raw materials are required and o-types are undesired, unavoidable by-products, the target concentrations necessary for completely separating the raw materials from the chemical system $i \in \{1,\dots, m\}$ would have the values $c_i^i(t+1) = 1$.

The concentrations of the o-types of undesired raw materials, which will be mixed together, are also influenced by the separation of the desired raw materials. Since the desired raw materials are not included in this sub-system, we get the following relations for the concentrations of the undesired materials

$$c_j^i(t+1) = \frac{N_j^i(t)}{\left(\sum_{h=1}^{n} N_h^i(t)\right) - N_i^i(t)} = \frac{N_j^i(t)}{\left(\sum_{h=1}^{n} N_h^i(t)\right)\left(1 - \frac{N_i^i(t)}{\sum_{h=1}^{n} N_h^i(t)}\right)} = \frac{c_j^i(t)}{1 - c_i^i(t)}$$

$$\Leftrightarrow \frac{c_j^i(t+1)}{c_j^i(t)} = \frac{1}{1 - c_i^i(t)}. \tag{3.8}$$

$$1 - c_i^i(t) = \sum_{j=m+1}^{n} c_j^i(t). \tag{3.9}$$

We can see that the exergy lost for structural change of undesired raw materials is less relevant when the average concentration of desired raw materials is low.

With discrete time intervals, the differential Eq. (2.17) in Chap. 2 can be modified to the difference equation

$$\begin{aligned} \frac{\Delta(Ex)}{\Delta t} &= N^0 RT_0 \sum_{h=1}^{n} \ln \frac{c_h(t+1)}{c_h(t)} \frac{\Delta c_h}{\Delta t} \\ \Leftrightarrow \sum_{h=1}^{n} \frac{\Delta Ex_h}{\Delta c_h} &= RT_0 \sum_{h=1}^{n} N_h \ln \frac{c_h(t+1)}{c_h(t)}. \end{aligned} \tag{3.10}$$

Next, let us take one of the chemical systems $i \in \{1,\dots,m\}$, from which type i of the desired raw material with initial concentration $c_i^i(t)$ will be extracted and brought to target concentration $c_i^i(t+1)$ from a composition with n-m types of unavoidable by-products with the initial concentrations $c_j^i(t)$. Equations (3.9) and (3.10) and the following calculation steps

$$\begin{aligned} \left.\frac{\Delta Ex^i}{\Delta c^i}\right|_{c^i(t)}^{c^i(t+1) \wedge c_i^i(t+1)=1} &:= \left.\frac{\Delta Ex_i^i}{\Delta c_i^i}\right|_{c_i^i(t)}^{c_i^i(t+1)=1} + \sum_{j=m+1}^{n} \left.\frac{\Delta Ex_{ij}}{\Delta c_{ij}}\right|_{c_j^i(t)}^{c_j^i(t+1)} \\ &= RT_0 \left\{ N_i^i(t) \ln \frac{c_i^i(t+1)}{c_i^i(t)} + \sum_{j=m+1}^{n} N_j^i(t) \ln \frac{c_j^i(t+1)}{c_j^i(t)} \right\} \\ &= RT_0 \left\{ N_i^i(t) \ln \frac{1}{c_i^i(t)} + \sum_{j=m+1}^{n} N_{ij}(t) \ln \frac{1}{1 - c_i^i(t)} \right\} \\ &= RT_0 N_i^i \left\{ \ln \frac{1}{c_i^i(t)} + \frac{N^i(t) \sum_{j=m+1}^{n} N_j^i(t)}{N^i(t)\, N_i^i(t)} \ln \frac{1}{1 - c_i^i(t)} \right\}, \end{aligned}$$

Lead to Eq. (3.11), which describes the exergy change for a complete separation of type i from the required raw material

$$\frac{\Delta Ex^i}{\Delta c^L i}\Bigg|_{c^i(t)}^{c^i(t+1)\,\wedge\, c_i^i(t+1)=1} = -RT_0 N_i^i(t)\left\{\ln c_i^i(t) + \frac{1-c_i^i(t)}{c_i^i(t)}\ln\left(1-c_i^i(t)\right)\right\}. \tag{3.11}$$

The absolute value of the exergy change for separating the required raw material is larger when concentrations of the raw materials are low, which is the typical situation. The first term in Eq. (3.11) is the exergy change for forming a separate sub-system with only the raw material. Equations (3.8) and (3.9) show that the separation of desired raw materials is associated with a slight increase in the concentrations of unavoidable by-products. The separation of the raw material is combined with an increase in information (or exergy) about the complementary sub-system, namely that it does not contain the separated raw material, only unavoidable by-products.

Equation (3.11) was derived within the framework of an extended thermodynamic approach and is identical to the special case used by Faber et al. (1983:114–115) for integrating resource problems in their dynamic model. Equation (3.11) shows that the exergy change of the system is equal to the product of an intensive variable (number of moles to be extracted) and an extensive variable (change in concentration). Irreversible processes typically proceed in the direction of high concentration to the lower concentration level of thermodynamic equilibrium without any external intervention on the system. These irreversible processes, however, are returned to a new status far from thermodynamic equilibrium by the use of free energy, which contains an entropy-free energy (called exergy) and a surplus energy component (called anergy). The surplus energy component is ultimately discharged into the surrounding environment.

According to Faber et al. (1995:125–126), the limit of the second term in Eq. (3.11) is equal to -1. We therefore approximate the exergy change as

$$\begin{aligned}&\frac{\Delta Ex^i}{\Delta c^i}\Bigg|_{c^i(t)}^{c^i(t+1)\,\wedge\, c_i^i(t+1)=1}\\ &\approx -RT_0 N_i^i\left\{\ln c_i^i(t) - 1\right\} \le -RT_0 N_i^i\left\{\ln c_i^i(t)\right\}.\end{aligned} \tag{3.12}$$

The total exergy change of a complete separation of raw materials is equal to the sum

$$\frac{\Delta Ex^i}{\Delta c^i}\Bigg|_{c^i(t)}^{c^i(t+1)\,\wedge\, c_i^i(t+1)=1} = -RT_0\sum_{i=1}^{m} N_i^i\left\{\ln c_i^i(t) + \frac{1-c_i^i(t)}{c_i^i(t)}\ln\left(1-c_i^i(t)\right)\right\}. \tag{3.13}$$

If the initial concentrations are of the magnitude $\bar{c}^0 = c_i^0,\ i \in \{1,\ldots,m\}$, Eq. (3.13) can be simplified to

$$\frac{\Delta Ex^i}{\Delta c^i}\bigg|_{c^i(t)}^{c^i(t+1)\,\wedge\, c_i^i(t+1)=1} = -RT_0\left\{\ln\bar{c}(t) + \frac{1-\bar{c}(t)}{\bar{c}(t)}\ln\left(1-\bar{c}^0\right)\right\}\sum_{i=1}^{m} N_i^i. \tag{3.14}$$

Equation (3.14) leads us to the approximation for comparable low concentrations

$$\begin{aligned} &\frac{\Delta Ex^i}{\Delta c^i}\bigg|_{c^i(t)}^{c^i(t+1)\,\wedge\, c_i^i(t+1)=1} \\ &\approx -RT_0\{\ln\bar{c}(t) - 1\}\sum_{i=1}^{m} N_i^i \le -RT_0\{\ln\bar{c}(t)\}\sum_{i=1}^{m} N_i^i. \end{aligned} \tag{3.15}$$

3.2.2 Characteristics of Producing and Using Energy and Water

Thermodynamic laws require high exergy inputs such as free energy and water to shift a system from its initial status to a status far from thermodynamic equilibrium. The characteristics of the two high exergy inputs for the different sub-systems of the model are given by

Assumption 3.10 (for the production and use of free energy) The economic transformation processes of the production, reproduction and environmental protection sectors require the availability of free energy E. If the economic transformation processes are joined with irreversible processes, which is usually the case, a process specific portion $\varepsilon_\upsilon := Ex_\upsilon / E_\upsilon$, $0 < \varepsilon_\upsilon < 1$ of the free energy amount E is available in the form of entropy free energy or exergy Ex. The following relation between the free energy, exergy, and anergy in accordance with Eq. (2.13) must be considered for the different processes

$\upsilon \in \{RP = 1;\ P = 21,\ 22,\ 23_i$ with $i \in \{1,\ldots,m\}$, 24_k with $k \in \{1,\ldots,m\}$; $EP = 3\}$

$$\begin{aligned} &\varepsilon_\upsilon := \frac{Ex_\upsilon}{E_i} \ \wedge\ E_\upsilon = Ex_\upsilon + An_\upsilon \\ &\Leftrightarrow E_\upsilon = \varepsilon_\upsilon E_\upsilon + An_\upsilon \Leftrightarrow \frac{An_\upsilon}{E_\upsilon} = 1 - \varepsilon_\upsilon. \end{aligned} \tag{3.16}$$

Assumption 3.11 (for water use in production and reproduction activities) Water is the sole method for transporting the unavoidable by-products and the consumed portions of the desired goods to the wastewater treatment sector, where the treated water and other residuals are discharged into the surrounding ecological system. The amounts of water required for production and reproduction activities are considered to be linear functions of the appropriated material N^i and the consumed portion of the desired goods.

With the concentration of the desired raw material $c_i^i(t) := N_i^i(t)/N^i(t)$ and the specific water coefficients $\tilde{\psi}_i^P(t)$ defined as follows

$$\tilde{\psi}_i^P(t) := \frac{W_i^P(t)}{N^i(t)} \Leftrightarrow W_i^P(t) := \tilde{\psi}_i^P(t) N^i(t) = \tilde{\psi}_i^P(t) \frac{N_i^i(t+1)}{c_i^i(t)}, \tag{3.17}$$

For the water amounts, $W_i^P(t)$ for one and all extraction processes, we receive the expressions

$$W_i^P(t) := \frac{\tilde{\psi}_i^P(t)}{c_i^i(t)} \sum_{k=1}^{q} N_{ik}(t+1) \ . \tag{3.18}$$

$$W^P(t) := \sum_{i=1}^{m} \frac{\tilde{\psi}_i^P(t)}{c_i^i(t)} \sum_{k=1}^{q} N_{ik}(t+1) = \sum_{i=1}^{m} \frac{\tilde{\psi}_i^P(t)}{c_i^i(t)} \sum_{k=1}^{q} \kappa_{ik} Y_k(t+2) \ , \ \tilde{\psi}_i^P(t) \geq \ 0. \tag{3.19}$$

A similar relation can be defined for the consumption (or reproduction) activities. Although most of the water is needed for transporting residuals from the reproduction sector, it is acceptable to assume that the basic needs for water independent of the consumption bundle are included in the specific water consumption coefficients that are defined as follows

$$W_k^{RP}(t) = \psi_k^{RP}(t) Q_k(t+1) \ \wedge \ W^{RP}(t) := \sum_{k=1}^{q} \psi_k^{RP}(t) Q_k(t+1), \ \psi_k^{RP}(t) \geq \ 0. \tag{3.20}$$

These assumptions and specifications have some implications and restrictions. First, we implicitly assume that the water needed for the extraction process is also used for the combination process, which means that no additional water is required for combining the raw materials and the desired goods. Secondly, we assume the high exergy portion of the free energy is produced in the separate energy production sub-sector, 21. This sub-sector could contain nuclear, coal, fuel, wind, water, or solar power plants for the production of electricity or other energy forms. In the case where energy with a low exergy portion is produced by combining the raw materials and the desired goods, we implicitly assume that this energy, mainly in form of heat, will not be used and therefore belongs to the anergy output vector for the transformation processes. Aside from the unusable energy (or anergy), other unavoidable by-products will always be generated in this sub-sector, 21, and mixed together with the unavoidable by-products generated in the other production sub-sectors as well as in the reproduction sector. For purposes of simplification, these residuals from the energy sub-sector are considered less relevant and will therefore be ignored.

Assumption 3.11 characterizes the special role that water plays. It transports the unavoidable by-products to the environmental protection sector, it discharges the residuals into the wastewater treatment sector, and then after treatment into the surrounding environment. The water quality change, or pollution, caused by the undesired unavoidable by-products can be derived by also considering the linear relation (3.4) of Assumption 3.5. The total amount of required raw materials,

$i \in \{1,..., m\}$, for producing an amount $Y_k(t+1)$ of all desired goods is equal to $N_i^i(t) = \sum_{k=1}^{q} \kappa_{ik} Y_k(t+1)$. Since these raw materials are extracted from chemical systems, which also contain unavoidable by-products, the total amounts, N^i, of all types of the material of these chemical systems can be determined by the equation: $N^i = N_i^i / c_i^i$ between the amounts of the required raw material and its initial concentration c_i^i. As a result, the amounts of the by-products $j \in \{m+1,..., n\}$ of one or all of the chemical systems $i \in \{1,..., m\}$ are determined by the relations

$$N_j^i = c_j^i N^i = \frac{c_j^i N_i^i}{c_i^i} = \frac{c_j^i}{c_i^i} \sum_{k=1}^{q} \kappa_{ik} Y_k \wedge N_j = \sum_{i=1}^{m} \frac{c_j^i}{c_i^i} \sum_{k=1}^{q} \kappa_{ik} Y_k. \qquad (3.21)$$

If the total amount of water for production activities is determined by Eq. (3.21), the expression for the water concentrations of the by-products would be equal to

$$c_j^P := \frac{N_j}{\sum\limits_{j=m+1}^{n} N_j + W^P} = \frac{\sum\limits_{i=1}^{m} \frac{c_j^i}{c_i^i} \sum\limits_{k=1}^{q} \kappa_{ik} Y_k}{\sum\limits_{j=m+1}^{n} \sum\limits_{i=1}^{m} \frac{c_j^i}{c_i^i} \sum\limits_{k=1}^{q} \kappa_{ik} Y_k + \sum\limits_{i=1}^{m} \frac{\tilde{\psi}_i^P}{c_i^i} \sum\limits_{k=1}^{q} \kappa_{ik} Y_k} \approx \frac{\sum\limits_{i=1}^{m} \frac{c_j^i}{c_i^i} \sum\limits_{k=1}^{q} \kappa_{ik} Y_k}{\sum\limits_{i=1}^{m} \frac{\tilde{\psi}_i^P}{c_i^i} \sum\limits_{k=1}^{q} \kappa_{ik} Y_k}. \qquad (3.22)$$

Even though this expression looks quite complicated, we can immediately see that the water concentrations are mainly determined by the initial concentrations and water specific coefficients. The water concentrations vary in how they depend on the amounts and structure of the produced desired goods. For the interpretation of Eq. (3.22), it is very helpful to assume that the initial concentrations are in the ranges $\bar{c}_m = c_i^i, \wedge \ \bar{c}_{n-m} = c_j^i$.

For this special case, from Eq. (3.16), we get

$$1 - \bar{c}_m = \sum\nolimits_{j=m+!}^{n} \bar{c}_{n-m} = (n-m)\bar{c}_{n-m} \Leftrightarrow \bar{c}_{n-m} = \frac{1-\bar{c}_m}{n-m}.$$

That leads us to a simplified expression for Eq. (3.22)

$$c_j^P < \frac{(1-\bar{c}_m)}{\bar{c}_m} \frac{m}{(n-m)} \frac{\sum\limits_{k=1}^{q} \kappa_{ik} Y_k}{\sum\limits_{i=1}^{m} \frac{\tilde{\psi}_i^P}{c_i^i} \sum\limits_{k=1}^{q} \kappa_{ik} Y_k}, \qquad (3.23)$$

By using the following calculation steps

$$c_j^P := \frac{\sum_{i=1}^{m} \frac{c_j^i}{c_i^i} \sum_{k=1}^{q} \kappa_{ik} Y_k}{\sum_{j=m+1}^{n} \sum_{i=1}^{m} \frac{c_j^i}{c_i^i} \sum_{k=1}^{q} \kappa_{ik} Y_k + \sum_{i=1}^{m} \frac{\tilde{\psi}_i^P}{c_i^i} \sum_{k=1}^{q} \kappa_{ik} Y_k}$$

$$= \frac{\frac{(1-\bar{c}_m)m}{\bar{c}_m(n-m)} \sum_{k=1}^{q} \kappa_{ik} Y_k}{\frac{(1-\bar{c}_m)m}{\bar{c}_m} \sum_{k=1}^{q} \kappa_{ik} Y_k + \sum_{i=1}^{m} \frac{\tilde{\psi}_i^P}{c_i^i} \sum_{k=1}^{q} \kappa_{ik} Y_k}.$$

$$c_j^P = \frac{\frac{(1-\bar{c}_m)m}{\bar{c}_m(n-m)} \sum_{k=1}^{q} \kappa_{ik} Y_k}{\frac{(1-\bar{c}_m)m}{\bar{c}_m} \sum_{k=1}^{q} \kappa_{ik} Y_k + \sum_{i=1}^{m} \frac{\tilde{\psi}_i^P}{c_i^i} \sum_{k=1}^{q} \kappa_{ik} Y_k}.$$

$$c_j^P = \frac{1}{(n-m) + \frac{\sum_{i=1}^{m} \frac{\tilde{\psi}_i^P}{\bar{c}_m} \sum_{k=1}^{q} \kappa_{ik} Y_k}{\frac{(1-\bar{c}_m)m}{\bar{c}_m(n-m)} \sum_{k=1}^{q} \kappa_{ik} Y_k}} < \frac{1}{\frac{\frac{1}{\bar{c}_m} \sum_{i=1}^{m} \tilde{\psi}_i^P \sum_{k=1}^{q} \kappa_{ik} Y_k}{\frac{(1-\bar{c}_m)m}{\bar{c}_m(n-m)} \sum_{k=1}^{q} \kappa_{ik} Y_k}}.$$

If the water coefficient for extracting the raw materials is of the same magnitude $\bar{\Psi}^P = \tilde{\Psi}_i^P, i = 1,\ldots,m$, Eq. (3.23) can be simplified to

$$c_j^P < (1-\bar{c}_m) \frac{m}{(n-m)} \frac{\sum_{k=1}^{q} \kappa_{ik} Y_k}{m\bar{\psi}^P \sum_{k=1}^{q} \kappa_{ik} Y_k} = \frac{(1-\bar{c}_m)}{(n-m)\bar{\psi}^P}. \qquad (3.24)$$

From Eq. (3.24), we see that the concentrations of the by-products in the water mixture increase if the initial concentrations of the required raw materials are low. This increase will be lower if the number of types of raw materials is low when compared to the number of types of unavoidable by-products. The concentration of the unavoidable by-products is lower if the quotient of the amounts of the raw material and the required water amounts is low. This also means that there is a high dilution effect with water used for production. The concentrations of contaminants in the water are normally in the range of micro- and milligrams per litre of water. The dilution factor is therefore relatively large and can be considered as the dominant effect in this context (for more details see e.g. Hoekstra and Chapagain 2007:42). As a result, water and the other natural resources play an important role in industrial production and consumption.

3.2.3 Characteristics of Wastewater Treatment Activities

The water input vector for wastewater treatment activities in sector 3 has not only *n-m* components of unavoidable by-products, but also m components of the portions of the desired goods that are consumed or used in the period t and not available for the next periods.

Assumption 3.12 (for the wastewater treatment strategy) The wastewater generated by production and consumption activities will be mixed and will form a wastewater input vector. The vector will contain *n-m* components for unavoidable by-products and q components for residuals of the desired goods that are lost in the production sector or consumed in the reproduction sector. For each of the wastewater components $q + (n - m)$, an emission target in the form of an acceptable concentration will be determined exogenously. For each of the wastewater components, an emission target in the form of an acceptable concentration $c_h^*, h \in \{1,...,n\}$ will also be determined exogenously. If the concentrations of some or all of the residuals in the wastewater exceed the emission standard $c_h^{P+RP}(t) > c_h^*(t)$, then these residuals have to be reduced to the level $c_h^{P+RP}(t) \leq c_h^*(t)$.

Consequently, Eq. (3.22) must be extended by the domestic wastewater amounts to

$$c_j^P \approx \frac{N_j}{W^{RP} + W^P} = \frac{\sum_{i=1}^{m} \frac{c_j^i}{c_i^i} \sum_{k=1}^{q} \kappa_{ik} Y_k}{\sum_{k=1}^{q} \psi_k^{RP} Q_k + \sum_{i=1}^{m} \frac{\tilde{\psi}_i^P}{c_i^i} \sum_{k=1}^{q} \kappa_{ik} Y_k}, \; j \in \{m+1,..., n\}. \tag{3.25}$$

Let the vector $\vec{D}(t)$ represent the residuals that are generated by the deterioration of the cumulated capital stock vector $\vec{K}(t)$ at period t. In the case where $D(t)$ is a relevant part of the mixture with water, then $D(t)$ must be considered along with the amounts of the consumed desired goods

$$\vec{D}_k(t) := d_k \vec{K}(t) = d_k \left(\vec{K}_k^{EP}(t) + \sum_{r=1}^{4} \vec{K}_k^{P=2r}(t) \right), \; k \in \{1,..., q\}.$$

If a unique, good-specific deterioration rate is assumed, there is a similar relationship for approximating the concentrations of the residuals of the desired goods used for production and reproduction purposes

$$c_k^{RP+P+EP}(t) \approx \frac{D_k(t) + Q_k(t)}{W^{RP}(t) + W^P(t)} = \frac{d_k \left(K_k^{EP}(t) + \sum_{r=1}^{4} K_k^{P=2r}(t) \right) + Q_k(t)}{\sum_{k=1}^{q} \psi_k^{RP} Q_k(t) + \sum_{i=1}^{m} \frac{\tilde{\psi}_i^P}{c_i^i} \sum_{k=1}^{q} \kappa_{ik} Y_k(t)}. \tag{3.26}$$

The integration of these residuals would result in extended expressions. For simplification, we invoke Assumption 3.13.

Assumption 3.13 (for the discharge of the deteriorated capital goods) The residuals generated by the deterioration of the capital goods are considered a solid waste disposal problem, and therefore not a pollution component of the water used for the production and reproduction activities.

In this context, it is important to address the problem of how to determine the environmental targets for the polluted water. For this purpose, we must incorporate the relationship between the emission of the treated wastewater and its influences on the quality of environmental goods that are components of the welfare function in our model. This approach (for more details Faber et al. 1995:44–46) would require two components: first, integration of dynamic diffusion and self-purification for the environmental system; and second, definition of a damage function between the quality of the environmental goods and the remaining residuals of the wastewater independent of the wastewater treatment activities. We do not want to introduce additional fictitious equations, which are difficult to find, so we have chosen the so called emission standard approach as our wastewater treatment strategy.

The polluted water contains not only the unavoidable by-products, but also the consumed desired goods. The maximum number of exogenously-given emission standards for the unavoidable by-products in the form of concentration limits and the consumed desired goods would be equal to

$$C(n,\ h = 1, \dots, n) = \sum\nolimits_{h=1}^{n} \binom{n}{h} = 2^n - 1\ ,$$

in the case where no synergetic processes between the water pollution components exist. Such synergetic processes can lead to the generation of more dangerous substances than the risks caused by residuals of the production and consumption activities (for details see Schirmer et al. 2007).

3.3 Technological Progress and Human Labour Inputs

In Chap. 2, we discussed the modification of the concepts of roundaboutness and superiority. In this context, the dynamics of the model depend on the normative decision about which portions of the desired goods will be consumed and which portions will be saved for further development of production, reproduction and environmental protection activities. The dynamics of the model are also determined by the way that human labour and capital goods are connected with each other. The primary motivation of human individuals and the social society for taking care of their future development is not only to produce consumption goods for survival, but also to acquire knowledge and experience from the natural and social environment in order to make life more safe and comfortable. Consequently, the output of high exergy consumption goods is used to reproduce human labour inputs for the production and reproduction activities. Different from the conventional approach of the Neo-Austrian capital theory, we consider the human labour inputs not as exogenously given but as being endogenously reproduced.

The social society, traditional family and alternative social formation, is to be served by desired goods for optimizing its welfare status within the chosen planning horizon. When we consider a social reproduction unit, the consumption of high exergy goods generates a low exergy output vector of unavoidable by-product and a high exergy output vector of human labour. Consistent with the chosen inter-temporal welfare function, the generation of the human labour outputs of the individual social reproduction units are also aggregated to the human labour output vector of the whole society. The number of social reproduction units could increase, decrease, or remain nearly constant, which is typical for modern, developed societies. Instead of high biological reproduction rates, the dominating individual interest in developed societies might be to have a more safe, comfortable and sustainable life. When we assume.

Assumption 3.14 (for the total exergy of the human labour inputs) The total exergy of the human labour input available for production and reproduction activities follows an exponential function of the type $L(t) = L(0)\, e^{\beta t}$. However, the exergy is constant, which means that $\beta = 0$.

We must keep in mind that our dynamic model could be used for a society with a growing number of reproduction units, which may struggle to attain a satisfying consumption level for survival and development. What a more safe, comfortable, and sustainable life in developed societies means can be formalized by using the Eqs. (2.21) and (2.22) with some modifications that reflect the objective of the cultural evolution process. Specifically, cultural evolution means having more time for education, cultural development, and a higher quality of life by reducing the exergy requirements of human labour inputs for production activities. This can primarily be achieved by reducing the exergy of the physical inputs for producing one unit of the desired good (or output), but also by reducing the less attractive exergy of informative inputs, and should therefore also be considered as embedded in the capital goods.

Equation 2.21 can be applied to define the labour coefficients, l, instead of labour values, L, by distinguishing between the portions of the physical and informative exergy of human labour inputs. Human labour inputs still have to be done by humans, and those portions are embedded in the capital goods, which are then invested and added to already existing capital stock in the next period. These social aspects of reproducing human labour inputs for production will not be detailed subject in this contribution. For simplification reasons, we therefore assume.

Assumption 3.15 (for the exergy of the physical human labour inputs) The success of the evolution process in the context of economic activities leads to a reduction in the human labour input for producing one unit of the desired good, for which capital goods are required.

Permanent knowledge increase may happen by chance. When considering the complexity of the industrial production, however, we should assume that this evolution process is mainly a result of a systematic individual and social education process, which requires private and public investment into the education system. Consequently, the desired goods that are produced cannot be used solely for consumption and production purposes. A certain portion must be reserved for developing the education system that is part of the reproduction sector. We therefore assume.

Assumption 3.16 (for connecting exergy roundaboutness and exergy superiority of production and reproduction activities) Exergy roundaboutness and superiority in the production sector are closely connected with the existence of exergy roundaboutness and superiority in the reproduction sector.

One way of integrating this connection into our model would be to develop a capital stock for the reproduction sector, in addition to the capital stock of the production sector. The capital stocks are used for generating the physical and informative exergy of human labour inputs for production and reproduction activities. For simplification purposes, however, we develop a joint capital stock with a double function, one for developing the human labour inputs and a second one for modifying the capital stock for production. In contrast to the conventional approach chosen by the Neo-Austrian capital theory, we will introduce the technological changes as a permanent evolution process. The extension of the capital stock will be connected with an increase of knowledge, which leads to a reduction of the physical human labour component combined with an increased qualification level for production and reproduction activities.

Chapter 4
Specifications for Constructing the Water Use Model

Abstract Some restrictions are introduced based on the theoretical foundations and the general design. These restrictions deal with the timing and model structure, the number of desired goods and un-avoidable by products, and the process coefficients. Apart from the energy requirements for extracting raw materials from the natural environment and for producing water and treating wastewater, the relation of the capital stocks to the human labour inputs determines the dynamics of the water use model. Since the capital stock is only built up for the production sector, no capital stocks for the water and wastewater sector exist in the basic model. Instead of using dilution and damage functions for the relation between emissions and the environmental targets, the water and wastewater treatment requirements are determined by exogenously given water quality standards.

Chapter 2 covered the thermodynamic and capital theory foundations. Chapter 3 discussed the general design of the dynamic model for water uses. In Chap. 4, we will provide additional specifications concerning the model structure and characteristics. We will also introduce process coefficients that represent the production and reproduction activities.

4.1 Structure and Characteristics of the Water Use Model

There are four production steps, or periods, that form the activity interval τ that is sustained by the consumption goods and reproduces human labour inputs. The four steps are: production of free energy and water; extraction of raw materials; combination of the raw materials to create the capital and consumption goods that are used to build up the capital stock and for consumption, respectively; and finally, the treatment of the polluted water. The handling of time indices is simpler if we ignore the precise timing of the different processes within the activity intervals. We therefore have the following assumption.

Assumption 4.1 (for the time structure) The whole planning span will be divided into time intervals $\tau = 1,\ldots,T$, where each interval contains discrete time periods

H. Niemes, M. Schirmer, *Entropy, Water and Resources*,
DOI 10.1007/978-3-7908-2416-2_4,

that can be considered very short for distinguishing between them. The consumption and capital goods produced in the activity interval τ sustain the requirements for the next time intervals. For example, the consumption good for the activity interval $\tau + 1$ and the capital good for the future activity intervals depend on the deterioration rate of the capital stock up to the end of the planning horizon.

Assumption 4.2 (for the number of desired goods) Two desired goods, 1 and 2, will be produced in the activity interval τ in the amounts $Y_1(\tau)$ and $Y_2(\tau)$. Good 1, in the amount $Q(\tau) = Q_1(\tau)$, will be used exclusively for consumption and reproduction purposes in the next activity interval $\tau + 1$. Good 2, in the amount $K(\tau) = Q_2(\tau)$, will be invested for maintaining and building up the capital stock that is used for operating for a selected number of production processes. In the activity interval, τ, good 1 will therefore be called the consumption good and good 2 will be called the capital good.

Assumption 4.3 (for the number of processes that require capital goods) Capital goods will only be required for one production process, combining the extracted raw materials with the desired consumption good.

Instead of distinguishing between the consumption and capital good, we could also have assumed to have only one good, which can be used for consumption and production purposes. As an example, let us look at harvested potatoes. We could choose to consume all of the potatoes or we could reserve some of them to produce more potatoes in the next season. Saving some potatoes carries a risk because some of them can be lost for various reasons, such as bad weather. Our choice depends on technical losses, the time preference, and the likelihood of taking the risk for this investment in the future.

For the number and concentrations of required raw materials, and free energy and unavoidable by-products we have

Assumption 4.4 (for the number and amounts of required raw materials) Within the activity interval τ only four different raw materials are required: two for producing the consumption $Y^Q(\tau)$ and two for the capital good $Y^K(\tau)$. The four different extraction processes are marked with the indices $\overset{Q}{_{i=1,2}} \wedge \overset{K}{_{\mu=1,2}}$.

The number of materialistic unavoidable by-products could be arbitrarily chosen. The lowest number could be only one unique by-product or a mixture of a finite number of by-products, j, that are generated by the extraction processes of the different raw materials. For the four extraction processes we have

Assumption 4.5 (for the number and concentrations of the by-products) A unit mixture of unavoidable by-products, j, will be generated. The concentrations $\overset{Q}{\bar{c}}{}_j^{i=1,2}(\tau) \wedge \overset{K}{\bar{c}}{}_j^{\mu=1,2}(\tau)$ of these mixtures are determined by the concentrations of the raw material $c^Q_{i=1,2}(\tau) \wedge c^K_{\mu=1,2}(\tau)$ required to produce the consumption Q or capital good K through the relations

$$\overset{Q}{\bar{c}_j^{i=1,2}}(\tau) = 1 - c^Q_{i=1,2}(\tau) \wedge \overset{K}{\bar{c}_j^{\mu=1,2}}(\tau) = 1 - c^K_{\mu=1,2}(\tau). \tag{4.1}$$

The production of the free energy could also be accompanied by the same mixture of a certain number of by-products *j*. The average concentration of the mixture of unavoidable by-products $\bar{c}_j^E(\tau)$ would be determined by the energy concentrations $c^E(\tau)$ of the material used for producing free energy through the relation $\bar{c}_j^E(\tau) = 1 - c^E(\tau)$. For simplification reasons, we have

Assumption 4.6 (for the residuals from producing energy and the deterioration of the capital stock) The residuals from producing free energy and the residuals generated by the deterioration of the capital stock are not considered as water pollution components.

Taking into account the residuals from the consumption good and the mixture of the unavoidable by-products from extracting the raw materials, we obtain only two types of by-products if the unusable or unused energy portions (called anergy) of the economic activities are not taken into consideration. Apart from free energy and raw materials, water plays a special role in our model. The water supply and wastewater sector could be constructed in such a way that all essential components of the urban water infrastructure would be addressed. These essential components are water extraction, treatment, transport, distribution, wastewater collection, wastewater treatment, sludge treatment, and sludge disposal. We know from Niemes et al. (2007), that the dynamics of the urban water infrastructure is a central future application field and we will therefore use it to guide our development of the dynamic water use and water infrastructure model. In the first step, however, we will concentrate on integrating the water uses into the dynamic model. In the second step, Part II, this basic model will be extended to include urban water infrastructure components. For simplification reasons, we will temporarily use

Assumption 4.7 (for water production) Since the water quality, measured as concentration $c^W(\tau)$, $\tau = 1, \ldots, T$, of the required water amounts is higher (or better) than a given quality standard $c^W(\tau) \geq \bar{c}^W$, $\tau = 1, \ldots, T$, no treatment is required and no unavoidable by-products have to be considered. Consequently, only the water pollution by the residuals of production and reproduction activities is relevant for the wastewater treatment sector.

Assumption 4.8 (for the usage of water in production and reproduction activities) The produced water will be used to transport the unavoidable by-products from the different processes of the production and the reproduction sectors to the wastewater treatment sector. The wastewater generation rates are defined as the quotient between the wastewater amounts and the water amounts. These rates are used for the different economic transformation processes and are set to equal 1.

The large amounts of water that are required in the energy production process for cooling the water will not be considered. Separate water-cooling facilities are often constructed and occasionally contain separate wastewater treatment facilities. Consequently, the water coefficient of the energy production process addresses only the water required to transport the unavoidable by-products to the wastewater treatment sector.

Assumptions 4.1–4.8 are introduced for simplification purposes. The following assumptions define the dynamics of the model.

Assumption 4.9 (for the con centrations of the materials taken from the ecological system and used for extracting the required raw materials) The concentrations of the raw materials in the natural environment are assumed to be a declining function within the planning horizon, which means that the renewable rate of the ecological system is too low (or even zero) to compensate for the use of the natural resources for the economic activities.

In order to avoid unnecessary mathematical formalism due to discontinuities in how the concentration of the raw materials in the natural environment might actually develop, we will apply the slightly modified declining function chosen by Faber et al. (1995:136–139). The raw materials $i = 1, 2$ in the natural environment are used for producing the consumption good Q and the capital good K in period $\tau = 1, ..., T$. The concentrations of the raw materials decrease with a constant coefficient $\varphi^{Q}_{i=1,2} \geq 0 \wedge \varphi^{K}_{\mu=1,2} \geq 0$ in relation to the total amount of the natural resource already extracted in the previous activity intervals

$$\begin{aligned} c(\bar{N}^{Q}_{i}(\tau)) &= c(\bar{N}^{Q}_{i}(0))\, e^{-\varphi^{Q}_{i}\bar{N}^{Q}_{i}(\tau)} >> \underline{c}^{Q}_{i} > 0 \wedge \bar{N}^{Q}_{i}(\tau) = \sum_{t=1}^{\tau-1} N^{Q}_{i}(t) \\ c(\bar{N}^{K}_{\mu}(\tau)) &= c(\bar{N}^{K}_{\mu}(0))\, e^{-\varphi^{K}_{\mu}\bar{N}^{K}_{\mu}(\tau)} >> \underline{c}^{K}_{\mu} > 0 \wedge \bar{N}^{K}_{\mu}(\tau) = \sum_{t=1}^{\tau-1} N^{K}_{\mu}(t)\,. \end{aligned} \tag{4.2}$$

The concentrations should always be much larger than the upper concentration boundary $\underline{c}^{Q}_{i} \wedge \underline{c}^{K}_{\mu}$. At concentrations higher than the boundary, recycling processes become attractive when compared to the concentrations of the natural raw material stocks being degraded step-by-step. The raw materials are given back to the natural environment over time and the law of mass conservation is fulfilled, but always with a certain time lag. The law of mass (and energy) conservation is therefore a useful instrument for static, but not dynamic models. The application of the second law of the thermodynamics and information theory ensures that the dynamic models are consistent with the natural sciences.

Similar to the production and use of raw materials, the appropriation and separation of the material can be considered as a two-step process similar to the manner in which the other raw materials are produced. For the energy production process, however, a third step must be added, namely the generation of useful energy from raw material. We assume

Assumption 4.10 (for producing free energy for the economic activities) Material $N^{E}(\tau)$ is extracted from the natural environment for producing free energy in the energy sector E during the production interval τ. The average concentration $c(\bar{N}^E(\tau))$of material $N^E(\tau)$ should be a declining function similar to the one in Eq. (4.2). One unit of raw material for energy production, however, should have a time-invariant, embedded, specific energy content $\sigma^E(N_e^E)$, which leads us to the simple linear relation between energy and raw materials shown in Eq. (4.3).

$$E_e^E(\tau) \;=\; \sigma^E\,(N_e^E(\tau)) \;=\; \frac{\sigma^E\,(N_e^E(\tau))}{c(\bar{N}^E(\tau))} \;\; with \;\; c(\bar{N}^E(\tau)) = \frac{N_e^E(\tau)}{N^E(\tau)}. \tag{4.3}$$

Non-renewable energy resources from the natural environment are used in the different processes of the economic system during production interval. The concentration of these non-renewable resources will also be described using a decreasing function, with a constant coefficient $\phi_e \geq 0$, in relation to the total amount of the materials already extracted in the periods before

$$c(\bar{N}^E(\tau)) \;=\; c(\bar{N}^E(0))\,e^{-\varphi_e\,\bar{N}^E(\tau)} \;>>\; \underline{c}^E \;>\; 0 \;\wedge\; \bar{N}^E(\tau) = \sum_{t=1}^{\tau-1} N^E(t). \tag{4.4}$$

The concentration of the energy raw material should always be much larger than the lower concentration boundary $\underline{c}^E$. That boundary could be determined, for example, by using solely solar energy and other regenerative energy forms as a base line. For simplification, we furthermore assume

Assumption 4.11 (for the quality of the produced free energy) The produced free energy should have the highest qualitative level attainable. It should also be available in form of electricity, and should be the dominating energy form for transferring the low exergy inputs to the high exergy outputs of the economic transformation processes.

Human labour is mainly used to assist in extracting the raw material from the ecological system and to manage the entire process. The hard physical work and the separation of the raw materials use free energy, or more precisely its exergy portion, because the unusable portion (the anergy) will be discharged back to the ecologic system. Models where only human labour is considered the primary physical and informative input (see Faber et al. 1995) have the disadvantage that the free energy for economic transformation processes must also be delivered from the human labour inputs. Such a simplification goes too far and does not reflect reality. Furthermore, it can make the model inconsistent with natural laws. Economic transformation processes often require a huge amount of free energy apart from human labour inputs. Management and information aspects of the transformation processes become more important when a certain stage of development has been achieved. Therefore, the availability of free energy must be taken into account apart from the human labour input.

As discussed in Chap. 2, the human labour input can be separated into a physical and information component. For simplification, however, certain social and education aspects of the development of human labour inputs will not be considered in detail. We will assume

Assumption 4.12 (for the human labour inputs of the consumption good process) The human labour input coefficient for producing the consumption good, which requires the capital good, is a declining function. The function is closely connected to the newly acquired investment in the capital stock, which permanently modifies the interface between the economical and ecological system.

The human labour input coefficient of the other processes in the economic system could also change, but are assumed to remain constant. For simplification reasons, however, the human labour input coefficient has an initial value that is determined by the initial capital stock $\bar{K}(0)$ and is a declining function of the type

$$l(\bar{K}^Q(0), \bar{Y}^Q_K(\tau)) = l(\bar{K}(0))\, e^{-\varphi_l \bar{Y}^Q_K(\tau)} > 0 \text{ with } \bar{Y}^Q_K(\tau) := \sum_{t=1}^{\tau-1} Y^Q_K(t). \qquad (4.5)$$

The total amount of human labour available for the production activities could also change over time. If the total labour amount for the production activities decreases, the human labour amount available for the reproduction process would increase. These social components of the structural changes in human labour amounts are a very interesting subject, which could also be integrated into our model. However, we will not consider them and, therefore, have

Assumption 4.13 (for the total human labour inputs for the production and reproduction activities) The total human labour input for production activities, $L^P(\tau)$, remains constant and is equal to

$$L^P(\tau) = L(0)\ ,\ \tau \in \{1,...,T\}\,. \qquad (4.6)$$

The Assumptions 4.12 and 4.13 together mean that the same amount of human labour input achieves higher transformation process outputs because the labour input coefficients are declining functions. Additionally, the increase in consumption by the society has no influence on the total amount or structure of the human labour input that is available for the production and reproduction activities. Less developed societies, where population growth, consumption, and reproduction of human labour are still correlated can be modelled by simply releasing these assumptions.

A strict interpretation of Assumption 4.12 implies that the capital good produced in interval τ has a completely different property than the existing capital stock that is used in the previous intervals. As a result, in each activity interval, new capital goods with different properties are added to the already existing capital stock. We could furthermore argue that, in reality, the development of the new capital good requires an innovation time interval of several periods. These strict and more precise interpretations of Assumption 4.12 would immediately lead to a very complex structure of the capital stock. In order to avoid going too deep into capital theory, the general design of the capital stock will be simplified by

Assumption 4.14 (the restriction for building up the capital stock) The capital stock will be maintained and evolved to such a degree that the properties of the existing capital stock and the new invested capital goods used for building up the capital stocks are only marginally different.

For two discrete time intervals, capital dynamics are typically modelled in the form

$$\bar{K}^Q(\tau+1) \;=\; \bar{K}^Q(\tau) \;+\; Y_K^Q(\tau) \;-\; d\bar{K}^Q(\tau), \tag{4.7}$$

Where $\bar{K}^Q(\tau+1)$ denotes the capital stock in interval $\tau+1$, $Y_K^Q(\tau)$ is the investment into the capital stock, and $d\bar{K}^Q(\tau)$ is the capital depreciation, which can be assumed to be proportional to the capital stock $\bar{K}^Q(\tau)$. A second form for interpreting capital stock dynamics is given by Eq. (4.8), where the term $\bar{K}^Q(\tau)(1 \;-\; d)$ denotes the residual amount of the cumulated capital stock in past intervals

$$\bar{K}^Q(\tau+1) \;=\; \bar{K}^Q(\tau)\;(1-d) \;+\; Y_K^Q(\tau). \tag{4.8}$$

If all of the capital stock is applied for the production of the consumption good, the amount of consumption good available in the next interval will determine the amounts of capital stock input. In this case, the relations $\bar{K}(\tau) \;=\; k^Q Y^Q(\tau+1)$ where k^Q is the input coefficient of the capital stock for producing one unit of the consumption good is used to determine the capital stock input. We get the following relationship between the newly acquired investment and the change in desired consumption good amounts

$$\begin{aligned} Y_K^Q(\tau) &= \bar{K}^Q(\tau+1) \;-\; \bar{K}^Q(\tau)(1\;-\;d) \\ &= k\left\{\left(Y^Q(\tau+2) - Y^Q(\tau+1)(1-d)\right)\right\} \;=\; kY^Q(\tau+1)\left\{\frac{\Delta Y^Q|_{\tau+1}^{\tau+2}}{Y^Q(\tau+1)} \;+\; d\right\} \\ &= \bar{K}^Q(\tau)\left\{\frac{\Delta Y^Q|_{\tau+1}^{\tau+2}}{Y^Q(\tau+1)} \;+\; d\right\}. \end{aligned} \tag{4.9}$$

If there is no change in consumption amounts between the two intervals, which means that $\Delta Y^Q|_{\tau+1}^{\tau+2} \;=\; 0$, the investment only has to cover the depreciation of the capital stock that is used for the production of the consumption good amount $Y^Q(\tau+1)$. If the change in the consumption amounts is positive, then more than the depreciation of the capital stock must be invested. In the case where the capital stock is used completely, but no new investment into the capital stock will be done $Y_K^Q(\tau) \;=\; 0$, the relative change of consumption amounts is equal to $\Delta Y^Q|_{\tau}^{\tau+1}/Y^Q(\tau) \;=\; -d$ and limited by the negative value of the depreciation rate. This case can be interpreted as a limited disinvestment situation.

A third, more detailed form of modelling the capital stock dynamics is to base the development of the capital stock on the activities of the past. In this case, the capital stock, $\bar{K}^{Q}(\bar{\tau})$, is equal to the residual amount of a selected interval $\bar{\tau}$, the initial capital stock invested in time interval $R^{Q}_{\bar{\tau}}(0) = \bar{K}^{Q}_{\bar{\tau}}(0)\,(1-d)^{\bar{\tau}}$ plus the sum of the residuals of the new investments into the capital stock in the past, which is equal to $\sum_{t=1}^{\bar{\tau}} R_{\tau}{}^{Q}(t) := \sum_{t=1}^{\bar{\tau}-1} Y^{Q}_{t}(t)\,(1-d)^{\bar{\tau}-t}$. For the capital dynamics together with the acquired investment $Y^{Q}_{K}(\bar{\tau})$ of this selected interval, we obtain

$$\begin{aligned} \bar{K}^{Q}(\bar{\tau}+1) &= R^{Q}_{\bar{\tau}}(0) + \sum_{t=1}^{\bar{\tau}-1} R^{Q}_{\tau}(t) + Y^{Q}_{K}(\bar{\tau}) \\ &= \bar{K}^{Q}_{\tau}(0)\,(1-d)^{\bar{\tau}} + \sum_{t=1}^{\bar{\tau}-1} Y^{Q}_{K}(t)\,(1-d)^{\bar{\tau}-t} + Y^{Q}_{K}(\bar{\tau}). \end{aligned} \tag{4.10}$$

This form shows that the development of the capital stock is determined by the initial capital stock and the investments of the past. It also means that the residual amounts of the capital stocks available for future activity intervals $\tau^{*} \in \{\bar{\tau}+1, ..., T\}$ are determined by the residual amount of the capital stock of a selected interval plus the newly acquired investment into the capital stock.

In the dynamic water use model, only the production process for the consumption good requires the capital good as an input factor. When our basic model is actually applied (e.g. for urban water infrastructure or a specific water pollution problem), additional process-specific capital goods will have to be introduced to reflect the characteristics of the application. The introduction of process-specific additional capital goods is the more successful way for broadening the application of dynamic models. The alternative would be to use a unique multi-usable capital good. However, this would lead to extraordinarily complicated expressions without a convincing increase in information.

4.2 Process Coefficients for the Water Use Model

4.2.1 Process Coefficients for the Production Sector

Based on previous assumptions, we now introduce the input and output coefficients for the different economic transformation processes. The different processes and corresponding input and output coefficients, will be specified using the following indices

$$g \in \left\{ {}^{E}, {}^{W}, {}^{Q}_{1}; {}^{Q}_{2}, {}^{K}_{1}; {}^{K}_{2}, {}^{Q}, {}^{K}, {}^{RP}, {}^{WWT} \right\}$$

where the energy and water processes, the four extraction processes of the raw materials, the production processes for combing the raw materials to the consumption and the capital good, the reproduction process and finally the wastewater treatment process.

4.2.1.1 The Coefficients for the Extraction Processes for the Raw Material

The extraction processes of the raw materials have a more complicated when compared to the other economic processes. Figure 4.1 illustrates the most suitable extraction process for further explaining the characteristics of the major input and output coefficients. It is very helpful to divide the extraction process into two steps: the appropriation of the material from the natural environment; and the separation of the desired raw material. If we temporarily assume that the concentration of the desired raw material already has a target concentration of $c^* = 1$, then only the appropriation would be required. The high exergy inputs for the appropriation of one unit of raw materials, $i = 1, 2 \wedge \mu = 1, 2$, for producing the consumption good, Q, and the capital good, K, with the target concentration $c^* = 1$ are the free energy $e(c^*)$, human labour $l(c^*)$, water $\psi(c^*)$, and capital $k(c^*)$ inputs. Capital inputs for the extraction processes will not be considered in our basic model.

Let the concentration of the desired raw material $N_i^Q(\tau) \wedge N_\mu^K(\tau)$ in the natural environment be described as $c(N_i^Q(\tau)) \wedge c(N_\mu^K(\tau))$. If this concentration is lower than the assumed target concentrations, $c^* = 1$, then, in the first step of the extraction process, more material in the amount of the reciprocal value of the concentrations $1/c(N_i^Q(\tau)) \wedge 1/c(N_\mu^K(\tau))$, must be appropriated from the natural environment for separating one unit of the desired raw material in the second step. Consequently, for the appropriation of the materials used to separate one unit of desired raw materials, the following coefficients for human labour, water, and energy inputs will be used

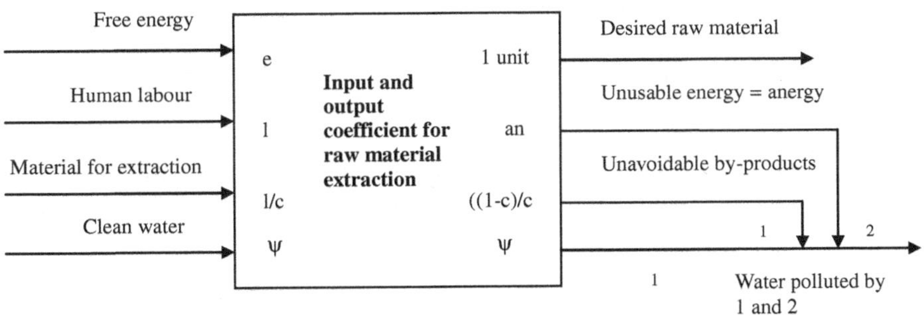

Fig. 4.1 Input and output coefficients for extracting one unit of required raw materials

$$
\begin{aligned}
&l^{app}(c(\bar{N}_i^Q(\tau)) := \frac{l^{app}(c_i^*)}{c(\bar{N}_i^Q(\tau))};\ \psi^{app}(c(\bar{N}_i^Q(\tau)) := \frac{\psi^{app}(c_i^*)}{c(\bar{N}_i^Q(\tau))};\\
&e^{app}(c(\bar{N}_i^Q(\tau)) := \frac{e^{app}(c_i^*)}{c(\bar{N}_i^Q(\tau))}\\
&l^{app}(c(\bar{N}_\mu^K(\tau)) := \frac{l^{app}(c_i^*)}{c(\bar{N}_\mu^K(\tau))};\ \psi^{app}(c(\bar{N}_\mu^K(\tau)) := \frac{\psi^{app}(c_i^*)}{c(\bar{N}_\mu^K(\tau))};\\
&e^{app}(c(\bar{N}_\mu^K(\tau)) := \frac{e^{app}(c_i^*)}{c(\bar{N}_\mu^K(\tau))}.
\end{aligned}
\tag{4.11}
$$

The separation step also requires labour and water inputs, but mainly based on the availability of free energy input in accordance to the relations between the exergy and concentration. Therefore, the coefficients for the labour and water capital input of the whole extraction process are set equal to

$$
\begin{aligned}
&l^{ext}(c(\bar{N}_i^Q(\tau))) = \frac{l_i^{app}}{c(\bar{N}_i^Q(\tau))};\ \psi^{ext}(c(\bar{N}_i^Q(\tau))) = \frac{\psi_i^{app}}{c(\bar{N}_i^Q(\tau))}\\
&l^{ext}(c(\bar{N}_\mu^K(\tau))) = \frac{l_\mu^{app}}{c(\bar{N}_\mu^K(\tau))};\ \psi^{ext}(c(\bar{N}_\mu^K(\tau))) = \frac{\psi_\mu^{app}}{c(\bar{N}_\mu^K(\tau))}.
\end{aligned}
\tag{4.12}
$$

Because the concentrations are changing over time, we obtain for Eq. (4.12) in detail

$$
\begin{aligned}
&l^{ext}(c(\bar{N}_i^Q(\tau)) = \frac{l_i^{app}}{c(\bar{N}_i^Q(0))}\, e^{\varphi_i^Q \bar{N}_i^Q(\tau)};\ \psi^{ext} c(\bar{N}_i^Q(\tau)) = \frac{\psi_i^{app}}{c(\bar{N}_i^Q(0))}\, e^{\varphi_i^Q \bar{N}_i^Q(\tau)}\\
&l^{ext}(c(\bar{N}_\mu^K(\tau)) = \frac{l_\mu^{app}}{c(\bar{N}_\mu^K(0))}\, e^{\varphi_\mu^K \bar{N}_\mu^K(\tau)};\ \psi^{ext} c(\bar{N}_\mu^K(\tau)) = \frac{\psi_\mu^{app}}{c(\bar{N}_\mu^K(0))}\, e^{\varphi_\mu^K \bar{N}_\mu^K(\tau)}.
\end{aligned}
\tag{4.13}
$$

The energy input coefficient for the extraction process must be extended by the additional energy requirements of the separation step

$$
\begin{aligned}
&e^{ext}(c(\bar{N}_i^Q(\tau)) := \frac{e_i^{Q\,app}}{c(\bar{N}_i^{Q,K}(\tau))} + e^{sep}\, c(\bar{N}_i^Q(\tau))\\
&e^{ext}(c(\bar{N}_\mu^K(\tau)) := \frac{e_\mu^{K\,app}}{c(\bar{N}_\mu^K(\tau))} + e^{sep}\, c(\bar{N}_\mu^K(\tau)).
\end{aligned}
\tag{4.14}
$$

Certain portions of the free energy, the so-called anergy, will not be available for the extraction process. From the following relations and calculation steps in between

$$
\begin{aligned}
&an^{ext}(c(\bar{N}_i^Q(\tau))) = an^{app}(c(\bar{N}_i^Q(\tau))) + an^{sep}(c(\bar{N}_i^Q(\tau)))\\
&\text{with } an^{app}(c(\bar{N}_i^Q(\tau))) := an^{app} / c(\bar{N}_i^Q(\tau))\\
&an^{ext}(c(\bar{N}_\mu^K(\tau))) = an^{app}(c(\bar{N}_\mu^K(\tau))) + an^{sep}(c(\bar{N}_\mu^K(\tau)))\\
&\text{with } an^{app}(c(\bar{N}_\mu^K(\tau))) := an^{app} / c(\bar{N}_\mu^K(\tau))
\end{aligned}
$$

$$
\begin{aligned}
e^{ext}(c(\bar{N}_i^Q(\tau)) :=& \frac{ex_i^{app}+an_i^{app}}{c(\bar{N}_i^Q(\tau))} + ex^{sep}(c(\bar{N}_i^Q(\tau))) + an^{sep}(c(\bar{N}_i^Q(\tau))) \\
=& \frac{ex_i^{app}}{c(\bar{N}_i^Q(\tau))} + ex^{sep}(c(\bar{N}_i^Q(\tau)))) + \frac{an^{app}}{c(\bar{N}_i^{Q,K}(\tau))} + an^{sep}(c(\bar{N}_i^Q(\tau))) \\
e^{ext}(c(\bar{N}_\mu^K(\tau)) :=& \frac{ex_i^{app}+an_i^{app}}{c(\bar{N}_\mu^K(\tau))} + ex^{sep}(c(\bar{N}_\mu^K(\tau))) + an^{sep}(c(\bar{N}_\mu^K(\tau))) \\
=& \frac{ex_i^{app}}{c(\bar{N}_\mu^K(\tau))} + ex^{sep}(c(\bar{N}_\mu^K(\tau))) + \frac{an^{app}}{c(\bar{N}_\mu^K(\tau))} + an^{sep}(c(\bar{N}_\mu^K(\tau))),
\end{aligned}
$$

We obtain the expressions for the energy coefficients

$$
\begin{aligned}
e^{ext}(c(\bar{N}_i^Q(\tau))) &= \frac{ex_i^{app}}{c(\bar{N}_i^Q(\tau))} + ex^{sep}(c(\bar{N}_i^Q(\tau))) + an^{ext}(c(\bar{N}_i^Q(\tau))) \\
e^{ext}(c(\bar{N}_\mu^K(\tau))) &= \frac{ex_i^{app}}{c(\bar{N}_\mu^K(\tau))} + ex^{sep}(c(\bar{N}_\mu^K(\tau))) + an^{ext}(c(\bar{N}_\mu^K(\tau))).
\end{aligned}
\tag{4.15}
$$

When taking into account the relation $ex^{ext} = e^{ext} - an^{ext}$ between the useable energy (exergy), free energy, and anergy, we alternatively obtain for the exergy coefficients

$$
\begin{aligned}
ex^{ext}(c(\bar{N}_i^Q(\tau))) &= \frac{ex_i^{app}}{c(\bar{N}_i^Q(\tau))} + ex^{sep}(c(\bar{N}_i^Q(\tau))) \\
ex^{ext}(c(\bar{N}_\mu^K(\tau))) &= \frac{ex_\mu^{app}}{c(\bar{N}_\mu^K(\tau))} + ex^{sep}(c(\bar{N}_\mu^K(\tau))).
\end{aligned}
\tag{4.16}
$$

In the case of low concentrations of raw materials in the natural environment, the approximation of the exergy and the introduction of a mole-specific exergy for the separation of one unit of desired raw material are defined as follows

$$
\begin{aligned}
&\left.\frac{\Delta Ex_{i,\mu}^{Q,K}}{\Delta c(\bar{N}_i^{Q,K}(\tau))}\right|_{c_i^{Q,K}(\tau)}^{1} \approx -RT_0 N_{i,\mu}^{Q,K} \ln c(\bar{N}_{i,\mu}^{Q,K}(\tau)) \\
\wedge\ & ex^{sep}(c(\bar{N}_{i,\mu}^{Q,K}(\tau))) := \frac{1}{N_{i,\mu}^{Q,K}} \left.\frac{\Delta Ex_{i,\mu}^{Q,K}}{\Delta c(\bar{N}_{i,\mu}^{Q,K}(\tau))}\right|_{c_i^{Q,K}(\tau)}^{1} \approx -RT_0 \ln c(\bar{N}_{i,\mu}^{Q,K}(\tau)),
\end{aligned}
$$

This leads us to the relations

$$
\begin{aligned}
ex^{sep}(c(\bar{N}_i^Q(\tau))) &= -\varepsilon \ln c(\bar{N}_i^Q(\tau)) \text{ with } \varepsilon := RT_0 \\
ex^{sep}(c(\bar{N}_\mu^K(\tau))) &= -\varepsilon \ln c(\bar{N}_\mu^K(\tau)).
\end{aligned}
\tag{4.18}
$$

When using these relations, we get the detailed expression for the exergy input coefficients of the extraction processes

$$
\begin{aligned}
e^{ext}(c(\bar{N}_i^Q(\tau))) &= \frac{ex_i^{app}}{c(\bar{N}_i^Q(\tau))} - \varepsilon \ln c(\bar{N}_i^Q(\tau)) + an^{ext}(c(\bar{N}_i^Q(\tau)) \\
e^{ext}(c(\bar{N}_\mu^K(\tau))) &= \frac{ex_\mu^{app}}{c(\bar{N}_\mu^K(\tau))} - \varepsilon \ln c(\bar{N}_\mu^K(\tau)) + an^{ext}(c(\bar{N}_\mu^K(\tau)).
\end{aligned}
\tag{4.19}
$$

$$ex^{ext}(c(\bar{N}_i^Q(\tau))) = \frac{ex_i^{app}}{c(\bar{N}_i^Q(\tau))} - \varepsilon \ln c(\bar{N}_i^Q(\tau))$$
$$ex^{ext}(c(\bar{N}_\mu^K(\tau))) = \frac{ex_\mu^{app}}{c(\bar{N}_\mu^K(\tau))} - \varepsilon \ln c(\bar{N}_\mu^K(\tau)) \,. \tag{4.20}$$

Since the concentrations are changing over time, we finally obtain

$$ex^{ext}(c(\bar{N}_i^Q(\tau))) = \frac{ex_i^{app}}{c(\bar{N}_i^Q(0))} e^{\varphi_i^Q \bar{N}_i^Q(\tau)} - \varepsilon \ln c(\bar{N}_i^Q(0)) + \varepsilon\, \varphi_i^Q \bar{N}_i^Q(\tau)$$
$$ex^{ext}(c(\bar{N}_\mu^K(\tau))) = \frac{ex_\mu^{app}}{c(\bar{N}_\mu^K(0))} e^{\varphi_\mu^K \bar{N}_\mu^K(\tau)} - \varepsilon \ln c(\bar{N}_\mu^K(0)) + \varepsilon\, \varphi_\mu^K \bar{N}_\mu^K(\tau). \tag{4.21}$$

These equations confirm that the extraction of the raw materials in the past influences the actual and future exergy requirements. The extraction of non-renewable resources causes inter-temporal negative externalities. If the amounts of resources being extracted are compensated by natural processes in the environment, which means that $\varphi_{i,\mu}^{Q,K} = 0$, then the exergy requirements remain constant. These formulae seem to be identical to the one developed by Faber et al. (1995:140). However, if we are to use these formulae to interpret the relationship between the energy and concentration of the raw material, we must correct them by using the exergy, not the energy, notation. If the energy efficiency coefficient $\eta(c(\bar{N}_{i,\mu}^{Q,K}(\tau)))$, $0 < \eta(c(\bar{N}_{i,\mu}^{Q,K}(\tau))) \leq 1$, is used, which defines how much of the free energy can do useful work as follows

$$\eta(c(\bar{N}_{i,\mu}^{Q,K}(\tau))) := \frac{ex(c(\bar{N}_{i,\mu}^{Q,K}(\tau)))}{e(c(\bar{N}_{i,\mu}^{Q,K}(\tau)))} \Leftrightarrow e(c_{i,\mu}^{Q,K}(\tau)) := \frac{ex(c(\bar{N}_{i,\mu}^{Q,K}(\tau)))}{\eta(c(\bar{N}_{i,\mu}^{Q,K}(\tau)))}, \tag{4.22}$$

Then we finally receive the detailed version of Eq. (4.20) that is used in our dynamic model

$$e_i(c(\bar{N}_i^Q(\tau))) = \frac{\dfrac{ex_i^{app}}{c_i(\bar{N}_i^Q(0))} e^{\varphi_i^Q \bar{N}_i^Q(\tau)} - \varepsilon \ln c_i(\bar{N}_i^Q(0)) + \varepsilon\, \varphi_i^Q \bar{N}_i^Q(\tau))}{\eta_i(c_i(\bar{N}_i^Q(\tau))}$$
$$e_\mu(c(\bar{N}_\mu^K(\tau))) = \frac{\dfrac{ex_\mu^{app}}{c_\mu(\bar{N}_\mu^K(0))} e^{\varphi_\mu^K \bar{N}_\mu^K(\tau)} - \varepsilon \ln c(\bar{N}_\mu^K(0)) + \varepsilon\, \varphi_\mu^K \bar{N}_\mu^K(\tau))}{\eta_\mu(c_\mu(\bar{N}_\mu^K(\tau)))}. \tag{4.23}$$

This equation shows the necessity to distinguish between the energy and exergy notion. An energy saving strategy, for example, would focus on increasing the efficiency coefficient of energy $\eta(c(\bar{N}_{i,\mu}^{Q,K}(\tau)))$.

The material appropriated from the natural environment contains a low concentration $c\,(\bar{N}^{Q,K}_{i,\mu}(\tau))$ of the required raw materials and the unavoidable by-products at a concentration of $(1 - c\,(\bar{N}^{Q,K}_{i,\mu}(\tau)))$. The low exergy level of the raw material will be shifted to the high exergy level represented by the target concentration $c^* \approx 1$. The raw materials, desired goods, and remaining capital stock are all high exergy outputs. The other outputs, unusable anergy, unavoidable by-products, deterioration of the capital, and the previously clean water that was polluted by the different residuals will have a lower exergy status.

4.2.1.2 The Coefficients for Producing Energy

We will not analyse energy saving strategies in our basic model, so we therefore assume temporarily

Assumption 4.15 (for the energy efficiency of the economic processes) The energy efficiency of all economic transformation processes remains unchanged over time.

The energy process can be divided into three steps: the appropriation of the material amount $N^E(\tau)$ from the natural environment; the separation of the raw material amount $N^E_e(\tau)$; and the transformation of the embedded energy into useful free energy in the amount $E^E_e(\tau) = \sigma^E_e N^E_e(\tau)$. The approximation of the exergy input coefficients can be used to estimate explicitly the exergy and energy input coefficients for the first two steps of the energy process with a concentration $c(N^E_e(\tau)) = N^E_e(\tau)/N^E(\tau)$ and the mole-specific energy content $\sigma(N^E_e) = \sigma^E_e$ for the raw materials. The concentration and energy content of the raw materials used to generate free energy are both assumed to be constant over time. This assumption is necessary because our dynamic water use and infrastructure models do not take into account the problems associated with using non-renewable resources for energy production.

The implicit way to determine the energy requirement of the energy production process is as follows. The appropriated and separated raw material amount $N^E_e(\tau)$ has a time-invariant specific energy content $\sigma^E_e(\tau) = \sigma^E_e$ that will be brought to a concentration value of 1. We then we obtain for the free energy amount, $E^{gE}_e(\tau)$, the linear relation $E^{gE}_e(\tau) = \sigma^E_e N^E_e(\tau)$, which we call gross free energy. The term gross free energy is used because one portion will be used for the energy production process and the other portion, which we call net free energy $E^{nE}_e(\tau)$, will be available to cover the energy demand of the other economic transformation processes. As a result, the gross free energy amount is equal to the sum $E^{gE}(\tau) = E^{EP}(\tau) + E^{nE}(\tau)$ for the net free energy amount $E^{nE}_e(\tau)$ and the energy requirement of the energy process $E^{EP}_e(\tau)$.

If we ignore what is happening inside the energy production process, then the internal exergy demand and exergy losses (anergy) can both be considered from outside as anergy because these components will not be available for doing useful work in the other economic transformation processes. The energy efficiency coefficient can be defined as the ratio of the net and gross free energies $\tilde{\eta}^E := E^{nE}/E^{gE}$. With that definition, we obtain, from Eq. (4.19) for the energy demand of the energy process the expression

$$E^{EP}(\tau) = E^{gE}(\tau)(1 - \tilde{\eta}^E) = \sigma^E c^E(N^E(\tau))\, N^E(\tau)(1 - \tilde{\eta}^E). \qquad (4.24)$$

Equation (4.24) shows that the free energy amounts must be generated by the raw material with the specific energy content and concentration of the material as it was appropriated from the natural environment.

Since the activity level of the energy process is determined by the total material amounts, $N^E(\tau)$, taken from the natural environment, we also use the following expressions for the so-called gross free energy amount and for the total input amount of the material

$$E^{gE}(\tau) = \frac{E^{nE}(\tau)}{\tilde{\eta}^E} \wedge N^E(\tau) = \frac{E^{nE}}{\sigma^E c^E(N^E(\tau))\, \tilde{\eta}^E}. \qquad (4.25)$$

The required net free energy input amounts of the other economic transformation processes also determine the activity level and the input amounts of the energy process.

Based on the different assumptions and specifications made to form the basic model for water use, the following input and output coefficients can now be summarized for the economic transformation processes of the production sector as follows. The time period, t, only symbolically marks those coefficients that change over time.

4.2.1.3 Summary of the Process Coefficients for the Production Sector

Coefficients for the extraction processes for the raw material $i = 1,2 \wedge \mu = 1,2$

$e(c(N_{i,\mu}^{Q,K}(t))$ units of energy for extracting the raw material i, μ
$\oplus\ l(c(N_{i,\mu}^{Q,K}(t))$ units of human labour
$\oplus\ 1/c(N_{i,\mu}^{Q,K}(t))$ units of the material containing the raw material i, μ
$\oplus\ \psi(N_{i,\mu}^{Q,K}(t))$ units of clean water

$\mapsto$ 1 unit of the raw material i, μ
$\oplus\ \psi(N_{i,\mu}^{Q,K}(t))$ units of polluted water
$\oplus\ (1 - c(N_{i,\mu}^{Q,K}(t))/c(N_{i,\mu}^{Q,K}(t))$ units of the by – products

Coefficients for the energy production process E

$e^E(c^E(t))$ units of energy
$\oplus\ l^E((c^E(t)))$ units of human labour
$\oplus\ 1/(c^E)$ units of material with a specific energy content

$\mapsto$ 1 unit of free energy
$\oplus$ $(1 - c^E(t))/c^E(t)$ units of the by – products

Coefficients for the water production process W

e^w units of energy
$\oplus$ $l^w(\bar{K}(t))$ units of human labour
$\oplus$ $\psi^w = 1$ units of water extracted from the enviornment

$\mapsto$ $\psi^w = 1$ unit of clean water

Coefficients for combining the raw materials $i = 1,2$ to consumption good Q

e_i^Q units of energy
$\oplus$ $l_i^Q(\bar{K}^Q(t))$ units of human labour
$\oplus$ κ_i^Q units of material containing the raw material $i = 1,2$
$\oplus$ k^Q units of the capital stock

$\mapsto$ 1 unit of desired goods Q
$\oplus$ $(1-d)k^Q$ units of remaing capital stock
$\oplus$ dk^Q units of deteriorated capital stock

Coefficients for combining the raw material $\mu = 1,2$ to the capital good K:

e_μ^K units of energy
$\oplus$ l_μ^K units of human labour
$\oplus$ κ_μ^K units of material containing the raw material $\mu = 1,2$

$\mapsto$ 1 unit of the capital good Y^K

The input coefficient of the capital good for combining the raw materials $i = 1,2$ to consumption good Q addresses the relation between the activity levels (or the produced consumption good amounts) and the capital stock variable. The other input coefficients describe relations between the activity level and the flow variables, namely the human labour and energy inputs.

4.2.2 Process Coefficients for the Reproduction Sector

The consumption good Q contributes to sustaining the whole reproduction process, in particular the reproduction of the human labour inputs. The same is also true for the capital good with a time lag of one period because the building up of the capital stock contributes to an increase in the amount of the consumption goods in the future.

Similar to production processes, the reproduction process can also be described by the input and output coefficient as follows

Coefficients for the reproduction process

e^{RP} units of energy for reproducing one unit of human labour
$\oplus\ l^{RP}$ units of human labour required for reproduction
$\oplus\ q^{RP}$ units of the consumption good
$\oplus$ 1 units of human labour to be restored
$\oplus\ \psi^{RP}$ units of clean water

$\mapsto$ 1 unit of human labour restored
$\oplus\ \psi^{RP}$ units of the generated wastewater
$\oplus\ q^{RP}$ unit of the used consumption good transported to the water treatment sector

We furthermore add the following assumptions.

Assumption 4.16 (for the human labour and energy inputs of the reproduction process) The fact that the reproduction activities also need human labour and energy inputs will be ignored in the water use and water infrastructure models.

Assumption 4.17 (for the wastewater generation rate of the reproduction process) The wastewater generation rate, ρ, of the reproduction process, normally in the range $0 < \rho \leq 1$, is set to $\rho = 1$.

Assumption 4.16 means that the welfare or consumption level that is achieved does not lead to an increase in available human labour amounts for the production and reproduction activities. In practise, Assumption 4.17 is not realistic because it means that no portion of the used water amount is lost as wastewater. Equation (4.26) is a simple correlation between the consumption good and water amounts that can be derived for the water use model.

$$\begin{aligned} W_Q^{RP}(\tau) &= \psi_Q^{RP}\, Y_Q^{RP}(\tau) = \psi_Q^{RP}\, Q(\tau) \\ \textit{with}\ \psi_Q^{RP} &> 0 \wedge \left\{ Q(\tau) := \omega\, Y_Q^{RP}(\tau) \wedge \omega = 1 \right\}. \end{aligned} \tag{4.26}$$

To distinguish more clearly between the activity level of the reproduction process $Y_Q^{RP}(\tau)$ and the consumption good amount $Q(\tau)$ we use the relation $Q(\tau) = \omega\, Y_Q^{RP}(\tau) \wedge \omega = 1$. This is necessary because the consumption amounts are components of the welfare function to be optimized and the activity level is used as a variable on the constraints for deriving the optimal conditions.

If the domestic wastewater is not mixed with the wastewater of production sector, the concentration of the domestic wastewater has the following value, or approximation, respectively, in the case where high dilution effects exist

$$\begin{aligned} c_Q^{RP}(\tau+1) &= \frac{Y_Q^{RP}(\tau)}{Y_Q^{RP}(\tau) + W_Q^{RP}(\tau)} \\ &= \frac{Y_Q^{RP}(\tau)}{Y_Q^{RP}(\tau)\,(1 + \psi_Q^{RP})} = \frac{1}{1 + \psi_Q^{RP}} \approx \frac{1}{\psi_Q^{RP}}\,,\ \mathit{if}\ \psi_Q^{RP} >> 1\,. \end{aligned} \tag{4.27}$$

The concentration of the domestic wastewater would be approximately equivalent to the reciprocal value of its specific water coefficient, which can be used as a dilution factor.

4.2.3 The Wastewater Treatment Coefficients

All residuals of the production and reproduction activities are ultimately transported to the environmental protection sector, which could contain different wastewater treatment processes. Such treatment processes could be for the unavoidable by-products generated by the extraction of the raw materials, for the deterioration residuals of capital stock, or for the residuals of the consumption good. Instead of having different treatment processes, we assume

Assumption 4.18 (for wastewater treatment) Only one joint wastewater treatment process exists, which has the flexibility and capability to reduce the concentrations of all water pollution components to exogenously given specific target concentrations (or environmental standards).

Linear processes are normally not so flexible. However, for simplification reasons, we assume

Assumption 4.19 (for human labour, capital, energy, and water inputs for wastewater treatment) Human labour and capital inputs for the wastewater treatment process are determined by the wastewater amounts independent of the wastewater pollution that is caused by the residuals of the production and reproduction activities. The required energy inputs are determined by the relation between the energy (or exergy) and the change of pollutant concentration to the exogenously given target concentration or environmental standards to be achieved.

Apart from the pollution caused by the discharge of consumption goods, we only have one joint unavoidable by-product, $\hat{j} = 1$, generated by the extraction processes for producing the consumption and the capital goods. Based on the general Eq. (3.9) in Chap. 3, the relation $c^Q_{i\hat{j}} = 1 - c^Q_i \wedge c^K_{\mu\hat{j}} = 1 - c^K_\mu$ leads us to the approximation of the wastewater concentration of the joint by-product

$$c_{\hat{j}} \approx \frac{\sum_{i=1}^{2} \frac{1-c_i^Q}{c_i^Q} \kappa_i^Q Y^Q + \sum_{\mu=1}^{2} \frac{1-c_\mu^K}{c_\mu^K} \kappa_\mu^K Y^K}{\psi_Q^{RP} Y_Q^{RP} + \sum_{i=1}^{2} \frac{\tilde{\psi}_i^Q}{c_i^Q} \kappa_i^Q Y^Q + \sum_{\mu=1}^{2} \frac{\tilde{\psi}_\mu^K}{c_\mu^K} \kappa_\mu^K Y^K}. \tag{4.28}$$

When assuming

Assumption 4.20 (for the difference in consumption goods amounts between two activity intervals) The difference in consumption good amounts for two activity intervals that follow each other is low. In addition, the consumption level of the next activity interval can be used to approximate the consumption level of the previous activity interval.

The approximation of the water concentration of the unavoidable by-product in Eq. (4.28) can therefore be simplified to

$$c_{\hat{j}} \approx \frac{\sum_{i=1}^{2} \frac{1-c_i^Q}{c_i^Q} \kappa_i^Q Y^Q + \sum_{\mu=1}^{2} \frac{1-c_\mu^K}{c_\mu^K} \kappa_\mu^K Y^K}{Y^Q \left\{ \psi_Q^{RP} + \sum_{i=1}^{2} \frac{\tilde{\psi}_i^Q}{c_i^Q} \kappa_i^Q \right\} + \sum_{\mu=1}^{2} \frac{\tilde{\psi}_\mu^K}{c_\mu^K} \kappa_\mu^K Y^K}. \tag{4.29}$$

The water pollution (or concentration) of the unavoidable by-product is a function of the extracted raw materials. These raw materials are required to produce the consumption good amounts that satisfy the welfare level over time because the changes in concentrations of the raw materials are declining functions of past activities. The wastewater concentration of the by-product depends on the dilution for production plus the relative dilution for reproduction, where the relation between the amounts of the consumption good and the material from the natural environment for extracting the desired raw materials must be considered.

For the water pollution measured as concentration and generated by the consumption of the consumption good in the reproduction sector, the approximation

$$c_Q^{RP} \approx \frac{Y_Q^{RP}}{\psi_Q^{RP} Y_Q^{RP} + \sum_{i=1}^{2} \frac{\tilde{\psi}_i^Q}{c_i^Q} \kappa_i^Q Y^Q + \sum_{\mu=1}^{2} \frac{\tilde{\psi}_\mu^K}{c_\mu^K} \kappa_\mu^K Y^K} \tag{4.30}$$

Can also be simplified to

$$c_Q^{RP} \approx \frac{Y^Q}{Y^Q \left\{ \psi_Q^{RP} + \sum_{i=1}^{2} \frac{\tilde{\psi}_i^Q}{c_i^Q} \kappa_i^Q \right\} + \sum_{\mu=1}^{2} \frac{\tilde{\psi}_\mu^K}{c_\mu^K} \kappa_\mu^K Y^K}. \tag{4.31}$$

When forming the quotient

$$\begin{aligned} \frac{c_{\hat{j}}}{c_Q^{RP}} &\approx \frac{\sum_{i=1}^{2} \frac{1-c_i^Q}{c_i^Q} \kappa_i^Q Y^Q + \sum_{\mu=1}^{2} \frac{1-c_\mu^K}{c_\mu^K} \kappa_\mu^K Y^K}{Y^Q} \\ &= \sum_{i=1}^{2} \frac{1-c_i^Q}{c_i^Q} \kappa_i^Q + \sum_{\mu=1}^{2} \frac{1-c_\mu^K}{c_\mu^K} \kappa_\mu^K \frac{Y^K}{Y^Q} , \end{aligned} \tag{4.32}$$

We get that the relation between the concentrations of these two pollution components depends on the structural change between the water amounts for producing the consumption good and the required capital good.

For the estimation of the water and wastewater amounts, we must consider that the activity level of the wastewater treatment process is equal to the sum of the water amounts of the production and reproduction sectors

$$\begin{aligned} W^{WWT} &:= W^{RP} + W^P \\ &= \psi_Q^{RP} Y_Q^{RP} + \sum_{i=1}^{2} \frac{\tilde{\psi}_i^Q}{c_i^Q} \kappa_i^Q Y^Q + \sum_{\mu=1}^{2} \frac{\tilde{\psi}_\mu^K}{c_\mu^K} \kappa_\mu^K Y^K . \end{aligned} \tag{4.33}$$

For this expression, we can use an approximation if a minor difference between $Y_Q^{RP} \wedge Y^Q$ exists

$$W^{WWT} \approx Y^Q \left\{ \psi_Q^{RP} + \sum_{i=1}^{2} \frac{\tilde{\psi}_i^Q}{c_i^Q} \kappa_i^Q \right\} + \sum_{\mu=1}^{2} \frac{\tilde{\psi}_\mu^K}{c_\mu^K} \kappa_\mu^K Y^K . \tag{4.34}$$

The labour and water inputs of the wastewater treatment process are linear functions of the wastewater amount to be treated. This means that the wastewater pollution components do not influence these input coefficients. The energy demand, however, for the treatment of the two pollution components, the pollution by the consumption good Q and the unavoidable by-product $\hat{j}$, is a function of the wastewater amount, the wastewater concentrations of these two pollution components, and their exogenously given target concentrations (or emission standards). Therefore, the reduction of the pollution components from the wastewater can be considered as an extraction process and we can use the same calculation steps that we applied for the energy demand of the extraction processes. If the concentrations of pollution components are comparably low, the approximation

$$\left. \frac{\Delta Ex_i}{\Delta c_i} \right|_{c_i(t)}^{c_i^*(t+1)} \approx RT_0 \left\{ \Delta N_i(t) \ \ln \frac{c_i^*(t+1)}{c_i(t)} \right\}$$

Also allows us to estimate the exergy change for reducing the wastewater concentrations $c_{Q,\hat{j}}(t)$ of the two pollution components $Q,\hat{j}$. The concentrations of $Q,\hat{j}$ will be reduced to the exogenously given target concentration $c^*_{Q,\hat{j}}$ by extracting the pollution amounts $\Delta N_{Q,\hat{j}}(t)$ from the wastewater amount $W^{RP+P}(t)$ as follows

$$\left.\frac{\Delta Ex_{Q,\hat{j}}}{\Delta c_{Q,\hat{j}}}\right|_{c_{Q,\hat{j}}(t)}^{c^*_{Q,\hat{j}}(t+1)} = \varepsilon\,\Delta N_{Q,\hat{j}}(t)\,\ln\frac{c^*_{Q,\hat{j}}(t+1)}{c_{Q,\hat{j}}(t)}$$

$$= \varepsilon\, W^{RP+P}(t)\,\Delta\, c_{Q,\hat{j}}(t)\,\ln\frac{c^*_{Q,\hat{j}}(t+1)}{c_{Q,\hat{j}}(t)}$$

with $\varepsilon := RT_0$ and

$$N_{Q,\hat{j}} := W^{RP+P}(t)\,\Delta\, c_{Q,\hat{j}}(t) = W^{RP+P}(t)\left[c_{Q,\hat{j}}(t) - c^*_{Q,\hat{j}}(t+1)\right]. \tag{4.35}$$

With the time-invariant energy efficiency coefficient for the wastewater treatment process, η^{WT}, in accordance with Assumption 4.15, the energy demand for reducing the specific water pollution components $Q,\hat{j}$ is equal to

$$\left.\frac{\Delta E_{Q,\hat{j}}}{\Delta c_{Q,\hat{j}}}\right|_{c_{Q,\hat{j}}(t)}^{c^*_{Q,\hat{j}}(t+1)} = \frac{\varepsilon}{\eta^{WT}}\, W^{RP+P}(t)\,\Delta\, c_{Q,\hat{j}}(t)\,\ln\frac{c^*_{Q,\hat{j}}(t+1)}{c_{Q,\hat{j}}(t)}\,. \tag{4.36}$$

The mole-specific energy coefficient for reducing the wastewater concentration of the pollution component of one unit of the wastewater amount can be approximated using Eq. (4.37).

$$e^{WWT}_{Q,\hat{j}}(t) := \frac{1}{W^{RP+P}(t)\,\Delta\, c_{Q,\hat{j}}(t)}\left.\frac{\Delta E_{Q,\hat{j}}}{\Delta c_{Q,\hat{j}}}\right|_{c_{Q,\hat{j}}(t)}^{c^*_{Q,\hat{j}}(t+1)} = \frac{\varepsilon}{\eta^{WT}}\ \ln\frac{c^*_{Q,\hat{j}}(t+1)}{c_{Q,\hat{j}}(t)}\,. \tag{4.37}$$

The concentrations are considered to be low for this approximation, so structural changes between the pollution concentrations can be ignored. The two energy coefficients can therefore be added to the sum. Later, we will ignore the precise timing by considering an activity interval within which the wastewater treatment process is taking place

$$e^{WWT}_{Q,\hat{j}}(\tau) = e^{WT}_{Q}(\tau) + e^{WT}_{\hat{j}}(\tau) = \frac{\varepsilon}{\eta^{WT}}\left\{\ln\frac{c^*_Q(\tau)}{c_Q(\tau)} + \ln\frac{c^*_{\hat{j}}(\tau)}{c_{\hat{j}}(\tau)}\right\}. \tag{4.38}$$

This approximation implies that we do not require inputs for transporting wastewater to the wastewater treatment sector or for disposing of the extracted amounts of the pollution components into the natural environment.

The coefficients for the wastewater treatment process can finally be summarized as follows

Coefficients for the wastewater treatment process:

$e(c^{WT}_{Q\hat{j}}(t))$ units of energy required to achieve the target concentrations $c^*_{Q\hat{j}}$

$\oplus\, l^{WT}$ units of human labour for 1 unit of polluted water
$\oplus\, psi^{WWT} = 1$ units of polluted water

$\mapsto \psi^{EP} = 1$ unit of the treated water amount with the target concentration $c^*_{Q\hat{j}}$
$\oplus$ residuals not considered.

Chapter 5
Constraints of the Water Use Model

Abstract The model constraints for the flow variables are introduced for the consumption good amounts and the required raw materials. The constraints for the water, wastewater, and energy amounts are formulated as well as those for sustaining and developing the capital stock variable. The relation between the human labour inputs and the capital stock reflects the technological progress within the model. The development of the capital stock depends on the path for the consumption good and the time preference of the target, or inter-temporal, welfare function. The number of variables in the system of constraints will be reduced by substituting the capital good amounts and the accumulated capital stock with the consumption good amounts. This substitution will be done using the so-called no-storage assumption. This assumption also allows the aggregation of process inputs to sector inputs, which simplifies the determination of the optimality conditions.

Before we formulate the constraints of the water use model, we will reduce the number of variables within the different activity intervals using

Assumption 5.1 (for positive quantities and no storage of the natural resources or goods produced) The natural resources appropriated from the natural environment and the goods that are produced should always have positive quantities. They will not be stored and therefore must be used completely within the activity periods and intervals.

Under this condition, the activity levels of the different processes for extracting the raw materials and for generating the free energy and water required to produce the consumption and the capital good are determined by the activity level of the production process for the consumption good. The activity level of the wastewater treatment process is also determined by the activity levels of the consumption good production processes since the wastewater will not be stored either. The input and output processes for extracting the raw materials and generating free energy, water, and wastewater can therefore be considered an integrated part of the production processes for the consumption and capital goods. This consideration also means that the precise timing for the different processes within an activity interval can be

H. Niemes, M. Schirmer, *Entropy, Water and Resources*,
DOI 10.1007/978-3-7908-2416-2_5, © Springer-Verlag Berlin Heidelberg 2010

ignored. Nevertheless, when we derive the constraints on the different inputs and outputs, we have to follow the different sequences for the transformation of inputs into outputs.

5.1 The Constraints for the Consumption Good Amounts

The first essential constraint for our basic model (identified as BM with sequential numbering) is that the consumption good, Y^Q, being produced within the activity interval τ and available for consumption in the next activity interval $\tau + 1$ is

$$H_Q(\tau) := Y^Q(\tau) - Q(\tau+1) \geq 0 \tag{BM1}$$

5.2 The Constraints for Extracting Raw Materials

The following constraints for the raw materials that are required for producing the consumption or capital good must be fulfilled when using the time-invariant, specific input coefficients, $\kappa_{i,\mu}^{Q,K}$, of the raw materials for producing one unit of the consumption or the capital good. The indices $_{ii}^{Q} \wedge {}_{\mu\mu}^{K}$ stand for the pure raw materials received after the separation that are then combined into the consumption or capital goods

$$H_{N_{ii}^Q}(\tau) = N_{ii}^Q(\tau) - \kappa_i^Q Y^Q(\tau) \geq 0 \wedge i = 1,2. \tag{BM2}$$

$$H_{N_{\mu\mu}^K}(\tau) = N_{\mu\mu}^K(\tau) - \kappa_\mu^K Y^K(\tau) \geq 0 \wedge \mu = 1,2. \tag{BM3}$$

The activity levels of the extraction processes also depend on the concentrations of the raw materials. By taking into account the no-storage Assumption 5.1, we get the relationship

$$N_{i=i,2}^Q(\tau) = \frac{\kappa_i^Q}{c\left(\overline{N}_i^{Q,K}(\tau)\right)} Y_i^Q(\tau) \wedge \overline{N}_{i=1,2}^Q = \sum_{t=1}^{\tau-1} N_i^Q(t) = \sum_{t=1}^{\tau-1} \frac{\kappa_i^Q}{c\left(\overline{N}_i^{Q,K}(t)\right)} Y_i^Q(t) \tag{5.1}$$

$$\begin{aligned} N_{\mu=1,2}^K(\tau) &= \frac{\kappa_\mu^K}{c\left(\overline{N}_\mu^{Q,K}(\tau)\right)} Y_\mu^K(\tau) \wedge \overline{N}_{\mu=1,2}^K(\tau) = \sum_{t=1}^{\tau-1} N_\mu^K(t) \\ &= \sum_{t=1}^{\tau-1} \frac{\kappa_{i=i,2}^K}{c\left(\overline{N}_\mu^{Q,K}(t)\right)} Y_\mu^K(t) \end{aligned} \tag{5.2}$$

The concentrations of the raw materials are given exogenously by declining functions. We therefore obtain the constraints

$$H_{N^Q_{i=1,2}}(\tau) = N^Q_i(\tau) - Y^Q(\tau) \frac{\kappa^Q_i}{c(\overline{N}^Q_i(0))} e^{\varphi^Q_i \kappa^Q_i \sum_{t=1}^{\tau-1} Y^Q(t)} \geq 0 \tag{BM4}$$

$$H_{\overline{N}^Q_{i=1,2}}(\tau) = \overline{N}^Q_i(\tau) - \frac{\kappa^Q_i}{c(\overline{N}^Q_i(0))} \sum_{t=1}^{\tau-1} Y^Q(t)\, e^{\varphi^Q_i \kappa^Q_i \sum_{t=1}^{\tau-1} Y^Q(t)} \geq 0 \tag{BM5}$$

$$H_{N^K_{\mu=1,2}}(\tau) = N^K_\mu(\tau) - Y^K(\tau) \frac{\kappa^K_\mu}{c(\overline{N}^K_\mu(0))} e^{\varphi^K_\mu \kappa^K_\mu \sum_{t=1}^{\tau-1} Y^K(t)} \geq 0 \tag{BM6}$$

$$H_{\overline{N}^K_{\mu=1,2}}(\tau) = \overline{N}^K_\mu(\tau) - \frac{\kappa^K_\mu}{c(\overline{N}^K_\mu(0))} \sum_{t=1}^{\tau-1} Y^K(t)\, e^{\varphi^K_\mu \kappa^K_\mu \sum_{t=1}^{\tau-1} Y^K(t)} \geq 0 \tag{BM7}$$

While the constraints BM4 and BM6 stand for the flow of the raw materials, the constraints BM5 and BM7 characterize the raw materials stocks. The flow constraints can also be embedded into the stock constraints. Therefore, in the context of deriving later short- and long-term shadow prices for the use of the natural resources, it is important to distinguish between the flow and stock variables. This also confirms the analogy of the dynamic characteristics of the use of capital goods and natural resources.

5.3 Constraints for Water and Wastewater Amounts

The activity level $W^{WWT}(\tau)$ of the wastewater treatment process is already given in Eq. (4.33) or its approximation (4.34). Since the wastewater generation rate for the production and reproduction sectors are assumed to have values of 1, the activity level $W^W(\tau) = W^{WWT}(\tau)$ of the water production process is also given by Eq. (4.34). When inserting the change in concentration as a declining function of past activities, we get the following detailed constraints for the water production and wastewater treatment amounts

$$H_W(\tau) = W^W(\tau) - \left\{ \begin{array}{l} \psi^{RP}_Q\, Y(^Q\tau) + Y^Q(\tau) \sum_{i=1}^{2} \frac{\tilde{\psi}^Q_i \kappa^Q_i}{c(\bar{N}^Q_i(0))} e^{\varphi^Q_i \kappa^Q_i \sum_{t=1}^{\tau-1} Y^Q(t)} + \\ + Y^K(\tau) \sum_{\mu=1}^{2} \frac{\tilde{\psi}^K_\mu \kappa^K_\mu}{c(\bar{N}^K_\mu(0))} e^{\varphi^K_\mu \kappa^K_\mu \sum_{t=1}^{\tau-1} Y^K(t)} \end{array} \right\} \geq 0. \tag{BM8}$$

$$H_{WWT}(\tau) = H_W(\tau). \tag{BM9}$$

If the wastewater generation rate is less than 1, then the wastewater amounts would be lower than the water amounts. This case will be covered in Part II in the context of water and wastewater saving strategies.

5.4 Constraints for Free Energy

Determining the constraints for the free energy input is a more complicated procedure. We will therefore include more intermediate steps. In the case where the activity levels of the other processes are already known, the direct and indirect amounts of energy (or exergy) required for producing the consumption and capital goods are also known. The direct energy inputs for combining the raw materials to the desired consumption and capital good amounts $Y^Q(\tau)$ and $Y^K(\tau)$ are

$$E^Q(\tau) = e^Q\, Y^Q(\tau) = \frac{ex^Q}{\eta^Q} Y^Q(\tau), \tag{5.3}$$

$$E^K(\tau) = e^K\, Y^K(\tau) = \frac{ex^K}{\eta^K} Y^K(\tau). \tag{5.4}$$

These direct energy inputs must be added to the indirect energy inputs for producing the raw materials

$$E^Q_{i=1+2}(\tau) := \sum_{i=1}^{2} E^Q_i(\tau) = Y^Q(\tau) \sum_{i=1}^{2} \frac{ex^Q_i(\tau)}{\eta^Q_i} \frac{\kappa^Q_i}{c(\bar{N}^Q_i(\tau))}. \tag{5.5}$$

$$E^K_{\mu=1+2}(\tau) := \sum_{\mu=1}^{2} E^K_\mu(\tau) = Y^K(\tau) \sum_{\mu=1}^{2} \frac{ex^K_\mu(\tau)}{\eta^K_\mu} \frac{\kappa^K_i}{c(\bar{N}^K_\mu(\tau))}. \tag{5.6}$$

Equation (4.19) for the energy input coefficients of the different extraction processes of raw materials, which are divided into appropriation and separation steps, we obtain the indirect energy inputs for producing the raw materials in detail

$$E^Q_{i=1+2}(\tau) = Y^Q(\tau) \sum_{i=1}^{2} \frac{\kappa^Q_i}{\eta^Q_i} \left\{ \begin{array}{l} \frac{ex^{app}_i}{c(\bar{N}^Q_i(0))} e^{\varphi^Q_i \bar{N}^Q_i(\tau)} \\ + \varepsilon \left(- \ln c(\bar{N}^Q_i(0)) + \varphi^Q_i \bar{N}^Q_i(\tau) \right) \end{array} \right\}. \tag{5.7}$$

$$E^K_{\mu=1+2}(\tau) = k^Q Y^Q(\tau) \sum_{\mu=1}^{2} \frac{\kappa^K_\mu}{\eta^K_\mu} \left\{ \begin{array}{l} \frac{ex^{app}_i}{c(\bar{N}^K_\mu(0))} e^{\varphi^K_\mu \bar{N}^K_\mu(\tau)} \\ + \varepsilon \left(- \ln c(\bar{N}^K_\mu(0)) + \varphi^K_\mu \bar{N}^K_\mu(\tau) \right) \end{array} \right\}. \tag{5.8}$$

The raw materials will not be stored and the extracted raw material amounts are determined by the activity levels of the consumption and capital good processes in the past, the detailed version of these equations are therefore

$$E^Q_{i=1+2}(\tau) = Y^Q(\tau) \sum_{i=1}^{2} \frac{\kappa^Q_i}{\eta^Q_i} \left\{ \begin{array}{l} \frac{ex^{app}_i}{c(\bar{N}^Q_i(0))} e^{\varphi^Q_i \kappa^Q_i \sum_{t=1}^{\tau-1} Y^Q(t)} \\ + \varepsilon \left(- \ln c(\bar{N}^Q_i(0)) + \varphi^Q_i \kappa^Q_i \sum_{t=1}^{\tau-1} Y^Q(t) \right) \end{array} \right\}. \tag{5.9}$$

$$E^K_{\mu=1+2}(\tau) = Y^K(\tau)\sum_{\mu=1}^{2}\frac{\kappa^K_\mu}{\eta^K_\mu}\left\{\begin{array}{l}\frac{ex^{app}_\mu}{c(\bar{N}^K_\mu(0))}\,e^{\varphi^K_\mu\kappa^K_\mu\sum\limits_{t=1}^{\tau-1}Y^K(t)}\\ +\,\varepsilon\left(-\ln c(\bar{N}^K_\mu(0)) + \varphi^K_\mu\kappa^K_\mu\sum\limits_{t=1}^{\tau-1}Y^K(t)\right)\end{array}\right\}. \tag{5.10}$$

The indirect energy input amounts of the raw materials for producing the consumption and capital goods depend on current and past activities. Therefore, over time, concentrations are decreasing and indirect energy inputs for the appropriation and separation steps of the extraction processes are increasing.

The indirect energy inputs for producing the water amounts are equal to

$$E^W(\tau) = \frac{ex^W}{\eta^W}\left\{Y^Q(\tau)\left\{\psi^{RP}_Q + \sum_{i=1}^{2}\frac{\tilde{\psi}^Q_i\,\kappa^Q_i}{c^Q_i(\bar{N}^Q_i(\tau))}\right\} + Y^K(\tau)\sum_{\mu=1}^{2}\frac{\tilde{\psi}^K_\mu\,\kappa^K_\mu}{c^K_\mu(\bar{N}^K_\mu(\tau))}\right\}, \tag{5.11}$$

Or, with the concentration changes of the raw material amounts extracted in the past, equal to

$$E^W(\tau) = \frac{ex^W}{\eta^W}\left\{\begin{array}{l}Y^Q(\tau)\left\{\psi^{RP}_Q + \sum\limits_{i=1}^{2}\frac{\tilde{\psi}^Q_i\,\kappa^Q_i}{c(\bar{N}^Q_i(0))}\,e^{\varphi^Q_i\kappa^Q_i\sum\limits_{t=1}^{\tau-1}Y^Q(t)}\right\}\\ +\,Y^K(\tau)\sum\limits_{\mu=1}^{2}\frac{\tilde{\psi}^K_\mu\,\kappa^K_\mu}{c(\bar{N}^K_\mu(0))}\,e^{\varphi^K_\mu\kappa^K_\mu\sum\limits_{t=1}^{\tau-1}Y^K(t)}\end{array}\right\}. \tag{5.12}$$

The indirect energy inputs for shifting the wastewater concentration down to the exogenously given emission standard must be integrated because water becomes wastewater. With the energy coefficient from Eq. (4.38)

$$e^{WWT}_{Q+\hat{j}}(\tau) = e^{WWT}_Q(\tau) + e^{WWT}_{\hat{j}}(\tau) = \frac{\varepsilon}{\eta^{WWT}}\left\{\ln\frac{c^*_Q(\tau)}{c_Q(\tau)} + \ln\frac{c^*_{\hat{j}}(\tau)}{c_{\hat{j}}(\tau)}\right\}$$

The indirect energy input amounts for wastewater treatment are

$$\begin{aligned}E^{WWT}(\tau) &:= e^{WWT}_{Q+\hat{j}}(\tau)\,W^{RP+P}(\tau) = (e^{WWT}_Q(\tau) + e^{WWT}_{\hat{j}}(\tau))\,W^{RP+P}(\tau)\\ &= \frac{\varepsilon}{\eta^{WWT}}\left\{\ln\frac{c^*_Q(\tau)}{c_Q(\tau)} + \ln\frac{c^*_{\hat{j}}(\tau)}{c_{\hat{j}}(\tau)}\right\}W^{RP+P}(\tau)\\ &= \frac{\varepsilon}{\eta^{WWT}}\left\{\ln\frac{c^*_Q(\tau)}{c_Q(\tau)} + \ln\frac{c^*_{\hat{j}}(\tau)}{c_{\hat{j}}(\tau)}\right\}\left\{\begin{array}{l}Y^Q(\tau)\left\{\begin{array}{l}\psi^{RP}_Q\\ +\sum\limits_{i=1}^{2}\frac{\tilde{\psi}^Q_i\,\kappa^Q_i}{c(\bar{N}^Q_i(0))}\,e^{\varphi^Q_i\kappa^Q_i\sum\limits_{t=1}^{\tau-1}Y^Q(t)}\end{array}\right\}\\ +\,Y^K\sum\limits_{\mu=1}^{2}\frac{\tilde{\psi}^K_\mu\,\kappa^K_\mu}{c(\bar{N}^K_\mu(0))}\,e^{\varphi^K_\mu\kappa^K_\mu\sum\limits_{t=1}^{\tau-1}Y^K(t)}\end{array}\right\}.\end{aligned} \tag{5.13}$$

These concentration shifting factors are quite complicated expressions

$$f_Q^{WWT}(\tau) := c_Q^*(\tau)\big/c_Q(\tau) \wedge f_{\hat{j}}^{WWT}(\tau) := c_{\hat{j}}^*(\tau)\big/c_{\hat{j}}(\tau).$$

In practise, however, it is usual to consider both the target water concentration and these concentration shifting factors as exogenously given by taking into account the existing wastewater treatment technologies and to base the wastewater treatment requirements on measured wastewater concentrations of the different pollution components. In order to avoid handling such complicated expressions in our model, we assume

Assumption 5.2 (for the wastewater treatment technology achieving the exogenously given target standards of the treated wastewater) Apart from the target standards for the treated wastewater, the wastewater concentration shifting factors and the wastewater treatment technology are exogenously given to such an extent that the change of the wastewater concentration shifting factor can be ignored.

Now that the different energy components have been determined, the sum of the direct and indirect energy inputs can be understood as the total net free energy amount, $E^n(\tau)$, because the internal energy demand of the energy process is not included in this sum

$$E^{nE}(\tau) := E^Q(\tau) + E^K(\tau) + \sum_{i=1}^{2} E_i^Q(\tau) + \sum_{i=1}^{2} E_i^K(\tau) + E^W(\tau) + E^{WWT}(\tau). \tag{5.14}$$

The net free energy amount must be generated in the energy process with a time-invariant energy efficiency coefficient η^E. The material extracted from the natural environment has an exogenously given declining function for the concentration of the energy-specific material used for generating free energy with energy content σ^E

$$c^E(\tau) = c^E(0)\, e^{-\varphi^E \bar{N}^E(\tau)} = c^E(0)\, e^{-\varphi^E \sum\limits_{t=1}^{\tau-1} N^E(t)} \quad with\ \bar{N}^E(\tau) := \sum\nolimits_{t=1}^{\tau-1} N^E(t), \tag{5.15}$$

The material amounts to be appropriated from the natural environment for energy production are therefore equal to

$$N^E(\tau) = \frac{E^n(\tau)}{c^E(\tau)\,\eta^E\sigma^E} = \frac{E^n(\tau)}{c^E(0)\,\eta^E\sigma^E} e^{\varphi^E \sum\limits_{t=1}^{\tau-1} E^n(t)} \tag{5.16}$$
$$with\, c^E := N_E^E(\tau)\big/N^E(\tau)\,,\ \eta^E := E^n(\tau)\big/E^g(\tau),\ \sigma^E := E^n(\tau)\big/N_E^E(\tau)\ .$$

When using the terms for the gross and net free energy contents of the extracted material, Eq. (5.16) can also be written as

$$E^{gE} = \frac{E^n(\tau)}{\eta^E} = N^E(\tau)\, c^E(\tau)\,\sigma^E. \tag{5.17}$$

The constraint for the natural resources used for generating free energy is obtained as

$$H_{N^E}(\tau) = N^E(\tau) - \frac{e^{\varphi^E \sum_{t=1}^{\tau-1} \left\{ E^Q(t) + E^K(t) + \sum_{i=1}^{2} E_i^Q(t) + \sum_{\mu=1}^{2} E_\tau^K(t) + E^{WT}(t) + E^{WD}(t) + E^{WWC}(t) + E^{WWT}(t) \right\}}}{c^E(0)\,\eta^E \sigma^E} * \ldots$$

$$\ldots * \left\{ E^Q(\tau) + E^K(\tau) + \sum_{i=1}^{2} E_i^Q(\tau) + \sum_{\mu=1}^{2} E_\mu^K(\tau) + E^W(\tau) + E^{WWT}(\tau) \right\} \geq 0. \qquad \text{(BM10*)}$$

The problem of generating free energy from natural resources will be excluded in the model by

Assumption 5.3 (for the indirect labour input for generating the required free energy) The material used for generating free energy is considered temporarily as a renewable natural resource. This means that the degradation rate of the concentration is temporarily set to be $\varphi^E = 0$.

The constraint (BM10*) can therefore be simplified to

$$H_{N^E}(\tau) = N^E(\tau) - \frac{E^Q(\tau) + E^K(\tau) + \sum_{i=1}^{2} E_i^Q(\tau) + \sum_{\mu=1}^{2} E_\mu^K(\tau) + E^{WT}(\tau) + E^{WWT}(\tau)}{c^E(0)\,\eta^E \sigma^E} \geq 0 \qquad \text{(BM10)}$$

With the following detailed expressions for the energy inputs of the different processes

$$E^Q(\tau) = \frac{ex^Q}{\eta^Q} Y^Q(\tau)$$

$$E^K(\tau) = \frac{ex^K}{\eta^K} Y^K(\tau).$$

$$E_{i=1+2}^Q(\tau) = Y^Q(\tau) \sum_{i=1}^{2} \frac{\kappa_i^Q}{\eta_i^Q} \left\{ \begin{array}{l} \dfrac{ex_i^{app}}{c(\bar{N}_i^Q(0))} e^{\varphi_i^Q \kappa_i^Q \sum_{t=1}^{\tau-1} Y^Q(t)} \\ + \varepsilon \left(-\ln c(\bar{N}_i^Q(0)) + \varphi_i^Q \kappa_i^Q \sum_{t=1}^{\tau-1} Y^Q(t) \right) \end{array} \right\}$$

$$E_{\mu=1+2}^K(\tau) = Y^K(\tau) \sum_{\mu=1}^{2} \frac{\kappa_\mu^K}{\eta_\mu^K} \left\{ \begin{array}{l} \dfrac{ex_\mu^{app}}{c(\bar{N}_\mu^K(0))} e^{\varphi_\mu^K \kappa_\mu^K \sum_{t=1}^{\tau-1} Y^K(t)} \\ + \varepsilon \left(-\ln c(\bar{N}_\mu^K(0)) + \varphi_\mu^K \kappa_\mu^K \sum_{t=1}^{\tau-1} Y^K(t) \right) \end{array} \right\}.$$

$$
\begin{aligned}
&E^{W}(\tau) \\
&\approx \frac{ex^{W}}{\eta^{W}} \left\{ \begin{array}{l} Y^{Q}(\tau) \left\{ \psi_{Q}^{RP} + \sum\limits_{i=1}^{2} \frac{\tilde{\psi}_i^{Q}\, \kappa_i^{Q}}{c(\overline{N}_i^{Q}(0))}\, e^{\phi_i^{Q} \kappa_i^{Q} \sum\limits_{t=1}^{\tau-1} Y^{Q}(t)} \right\} \\ + Y^{K}(\tau) \sum\limits_{\mu=1}^{2} \frac{\tilde{\psi}_{\mu}^{K}\, \kappa_{\mu}^{K}}{c(\overline{N}_{\mu}^{K}(0))}\, e^{\varphi_{\mu}^{K} \kappa_{\mu}^{K} \sum\limits_{t=1}^{\tau-1} Y^{K}(t)} \end{array} \right\} \\
&E^{WWT}(\tau) \\
&\approx \frac{\varepsilon}{\eta^{WT}} \left\{ \ln \frac{c_Q^{*}(\tau)}{c_Q(\tau)} + \ln \frac{c_j^{*}(\tau)}{c_j(\tau)} \right\} \left\{ \begin{array}{l} Y^{Q}(\tau) \left\{ \psi_{Q}^{RP} + \sum\limits_{i=1}^{2} \frac{\tilde{\psi}_i^{Q}\, \kappa_i^{Q}}{c(\overline{N}_i^{Q}(0))}\, e^{\varphi_i^{Q} \kappa_i^{Q} \sum\limits_{t=1}^{\tau-1} Y^{Q}(t)} \right\} \\ + Y^{K}(\tau) \sum\limits_{\mu=1}^{2} \frac{\tilde{\psi}_{\mu}^{K}\, \kappa_{\mu}^{K}}{c(\overline{N}_{\mu}^{K}(0))}\, e^{\varphi_{\mu}^{K} \kappa_{\mu}^{K} \sum\limits_{t=1}^{\tau-1} Y^{K}(t)} \end{array} \right\}.
\end{aligned}
$$

5.5 Constraints for Human Labour Inputs

In accordance with the assumptions made for the human labour inputs with the time-invariant total labour input amounts $L^P(\tau) = L^P > 0$ for operating the different production processes $g \in \left\{ {}^{Q}, {}^{K}, {}^{Q}_{i=1,2}, {}^{K}_{\mu=1,2}, {}^{W}, {}^{WWT}, {}^{E} \right\}$, we obtain the general constraint for labour input amounts. These are defined as the product of the input coefficients and activity levels for the different processes

$$
H_L(\tau) = L^P - \sum_{g} l^{g}(\tau)\, Y^{g}(\tau) \geq 0 \wedge L^{g}(\tau) := l^{g}(\tau)\, Y^{g}(\tau).
$$

The direct human labour input amounts for combining the raw materials to the desired consumption and capital good amounts are equal to

$$
\begin{aligned}
L^{Q}(\tau) &= Y^{Q}(\tau)\, l^{Q}(\bar{K}(0))\, e^{-\varphi_l \bar{Y}_K^{Q}(\tau)} > 0 \text{ with } \bar{Y}_K^{Q}(\tau) := \sum_{t=1}^{\tau-1} Y_K^{Q}(t) \\
L^{K}(\tau) &:= Y^{K}(\tau)\, l^{K}.
\end{aligned}
\tag{5.18}
$$

The indirect human labour inputs for producing the raw materials are given by the equation

$$
\begin{aligned}
L^{Q}_{i=1+2}(\tau) &= \sum_{i=1}^{2} L_i^{Q}(\tau) = Y^{Q}(\tau) \sum_{i=1}^{2} l_i^{Q} \frac{\kappa_i^{Q}}{c(\bar{N}_i^{Q}(\tau))} \\
L^{K}_{\mu=1+2}(\tau) &= \sum_{\mu=1}^{2} L_i^{K}(\tau) = Y^{K}(\tau) \sum_{i=1}^{2} l_i^{K} \frac{\kappa_i^{K}}{c(\bar{N}_i^{K}(\tau))}.
\end{aligned}
\tag{5.19}
$$

Since the concentrations of the raw materials are changing over time, the detailed versions of these equations are

$$L^Q_{i=1+2}(\tau) = Y^Q(\tau)\sum_{i=1}^{2} l^Q_i \frac{\kappa^Q_i e^{\varphi^Q_i \bar{N}^K_i(\tau)}}{c(\bar{N}^K_i(0))} = Y^Q(\tau)\sum_{i=1}^{2} \frac{l^Q_i \kappa^Q_i e^{\varphi^Q_i \kappa^Q_i \sum_{t=1}^{\tau-1} Y^Q(t)}}{c(\bar{N}^K_i(0))}.$$

$$L^K_{\mu=1+2}(\tau) = Y^K(\tau)\sum_{\mu=1}^{2} l^K_\mu \frac{\kappa^K_\mu e^{\varphi^K_\mu \bar{N}^K_\mu(\tau)}}{c(\bar{N}^K_\mu(0))} = Y^K(\tau)\sum_{\mu=1}^{2} \frac{l^K_\mu \kappa^K_\mu e^{\varphi^K_\mu \kappa^K_i \sum_{t=1}^{\tau-1} Y^K(t)}}{c(\bar{N}^K_\mu(0))}.$$

(5.20)

The indirect labour inputs that is required for water production and waste water treatment are simply the product of the water amounts and the labour input coefficient of these two processes

$$L^W(\tau) = l^W \left\{ \begin{array}{l} Y^Q(\tau)\left\{ \psi^{RP}_Q + \sum_{i=1}^{2} \frac{\tilde{\psi}^Q_i \kappa^Q_i}{c(\bar{N}^Q_i(0))} e^{\varphi^Q_i \kappa^Q_i \sum_{t=1}^{\tau-1} Y^Q(t)} \right\} \\ + Y^K(\tau) \sum_{\mu=1}^{2} \frac{\tilde{\psi}^K_\mu \kappa^K_\mu}{c(\bar{N}^K_\mu(0))} e^{\varphi^K_\mu \kappa^K_\mu \sum_{t=1}^{\tau-1} Y^K(t)} \end{array} \right\}. \quad (5.21)$$

$$L^{WWT}(\tau) = l^{WWT} \left\{ \begin{array}{l} Y^Q(\tau)\left\{ \psi^{RP}_Q + \sum_{i=1}^{2} \frac{\tilde{\psi}^Q_i \kappa^Q_i}{c(\bar{N}^Q_i(0))} e^{\varphi^Q_i \kappa^Q_i \sum_{t=1}^{\tau-1} Y^Q(t)} \right\} \\ + Y^K(\tau) \sum_{\mu=1}^{2} \frac{\tilde{\psi}^K_\mu \kappa^K_\mu}{c(\bar{N}^K_\mu(0))} e^{\varphi^K_\mu \kappa^K_\mu \sum_{t=1}^{\tau-1} Y^K(t)} \end{array} \right\}. \quad (5.22)$$

We must also take into account the indirect labour input for the energy production process. The indirect labour input of the energy production process is a function of the material appropriated from the natural environment for generating the free energy and the labour input coefficient of this process with a certain number of indirect labour input components

$$L^E(\tau) := l^E N^E(\tau) = \frac{l^E}{c^E(\tau)\sigma^E \eta^E} E^n(\tau).$$

The portion of free energy that is the human labour input is minor when compared with the energy content of the material that is used to generate the free energy for the economic transformation processes. If we argue that the human labour input is therefore a special form of useable free energy, then it seems acceptable to ignore the indirect labour input for energy production. Since we have already excluded the change of energy efficiency of the different processes, we can assume

Assumption 5.4 (for the indirect labour input for generating the required free energy) The human labour inputs for generating free energy required by the other economic transformation processes will not be taken into consideration in our basic model. If we wanted to analyse the energy sector, this assumption would have to be relaxed.

Under our model conditions, the total sum of direct and indirect energy inputs, the total net free energy amount $E^n(\tau)$, has the components

$$E^{nE}(\tau) := E^Q(\tau) + E^K(\tau) + \sum_1^2 E_i^Q(\tau) + \sum_1^2 E_\mu^K(\tau) + E^W(\tau) + E^{WWT}(\tau).$$

When the details about these components are inserted, we must integrate the expression for the indirect labour input for the energy production process with the detailed expressions of the different energy components that are summarized in the context of energy constraint BM10.

$$L^E(\tau) = \frac{l^E}{c^E \sigma^E \eta^E} \left\{ E^Q(\tau) + E^K(\tau) + \sum_{i=1}^{2} E_i^Q(\tau) + \sum_{\mu=1}^{2} E_\mu^K(\tau) + E^W(\tau) + E^{WWT}(\tau) \right\}. \tag{5.23}$$

We finally obtain the general labour constraint for the labour input amounts

$$H_L(\tau) = L^P - \left\{ L^Q(\tau) + L^K(\tau) + L^Q_{i=1+2}(\tau) + L^K_{\mu=1+2}(\tau) + L^W(\tau) + L^{WWT}(\tau) \right\} \geq 0 \tag{B11}$$

With the following detailed expressions for different human labour inputs

$$L^Q(\tau) = \tilde{l}^Q Y^Q(\tau) \bar{K}(0) e^{-\lambda^Q k^Q \sum_{t=1}^{\tau-1} Y^Q(t)}$$
$$L^K(\tau) = l^K Y^K(\tau).$$

$$L^Q_{i=1+2}(\tau) = Y^Q(\tau) \sum_{i=1}^{2} \frac{l_i^Q \kappa_i^Q e^{+\varphi_i^Q \kappa_i^Q \sum_{t=1}^{\tau-1} Y^Q(t)}}{c(\bar{N}_i^K(0))}$$

$$L^K_{\mu=1+2}(\tau) = Y^K(\tau) \sum_{\mu=1}^{2} \frac{l_\mu^K \kappa_\mu^K}{c(\bar{N}_\mu^K(0))} e^{+\varphi_\mu^K \kappa_\mu^K \sum_{t=1}^{\tau-1} Y^K(t)}.$$

$$L^W(\tau) \approx l^W \left\{ \begin{array}{l} Y^Q(\tau) \left\{ \psi_Q^{RP} + \sum_{i=1}^{2} \frac{\tilde{\psi}_i^Q \kappa_i^Q}{c(\bar{N}_i^Q(0))} e^{\varphi_i^Q \kappa_i^Q \sum_{t=1}^{\tau-1} Y^Q(\tau)} \right\} \\ + Y^K(\tau) \sum_{\mu=1}^{2} \frac{\tilde{\psi}_\mu^K \kappa_\mu^K}{c(\bar{N}_\mu^K(0))} e^{\varphi_\mu^K \kappa_\mu^K \sum_{t=1}^{\tau-1} Y^K(t)} \end{array} \right\}$$

$$L^{WWT}(\tau) = l^{WWT} \left\{ \begin{array}{l} Y^Q(\tau) \left\{ \psi_Q^{RP} + \sum_{i=1}^{2} \frac{\tilde{\psi}_i^Q \kappa_i^Q}{c(\bar{N}_i^Q(0))} e^{\varphi_i^Q \kappa_i^Q \sum_{t=1}^{\tau-1} Y^Q(t)} \right\} \\ + Y^K(\tau) \sum_{\mu=1}^{2} \frac{\tilde{\psi}_\mu^K \kappa_\mu^K}{c(\bar{N}_\mu^K(0))} e^{\varphi_\mu^K \kappa_\mu^K \sum_{t=1}^{\tau-1} Y^K(t)} \end{array} \right\}.$$

$$L^E(\tau) = \frac{l^E}{c^E(\tau)\,\sigma^E \eta^E}\left\{E^Q(\tau) + E^K(\tau) + \sum_{i=1}^{2} E_i^Q(\tau) + \sum_{\mu=1}^{2} E_\mu^K(\tau) + E^W(\tau) + E^{WWT}(\tau)\right\}$$

with the details being summarized in the conntext of energy constraint BM10

5.6 Constraints for Sustaining and Developing the Capital Stock

The general constraint for building up capital stocks is

$$H_{\bar{K}}(\tau) := \sum_g K^g(0) + \sum_{t=1}^{\tau-1} Y^K(t) - d\sum_{t=1}^{\tau-1}\sum_g k^g Y^g(t) - \sum_g k^g Y^g(\tau) \geq 0.$$

This constraint means that an initial capital stock $\bar{K}(0) = \sum_g K^g(0)$ that is distributed among the different economic processes at the beginning of the planning horizon will be increased by the capital good amounts $\sum_{t=1}^{\tau-1} Y^{\bar{g}=K}(t)$. The capital good amounts are increased by the capital good process $\bar{g} = K$ in past intervals $1,\dots,\tau - 1$ and decreased by the depreciation amounts $d\sum_{t=1}^{\tau-1}\sum_g k^g Y^g(t)$. The absolute value of the surplus capital good amounts $|-\sum_g k^g Y^g(\tau)|$ will be available for the production activities of activity interval τ.

In our basic model, capital good inputs are only required for producing the consumption good amounts. This means that the constraint of capital good can be simplified to

$$H_{\bar{K}}(\tau) := K^Q(0) + \sum_{t=1}^{\tau-1} Y^K(t) - dk^Q \sum_{t=1}^{\tau-1} Y^Q(t) - k^Q Y^Q(\tau) \geq 0. \quad (B12)$$

Constraint B12 shows that the following relations that are derived from the capital stock development between two adjacent time intervals closely connect the two variables on the production side.

$$\begin{aligned} K^Q(0) + \sum_{t=1}^{\tau} Y^K(t) &= dk^Q \sum_{t=1}^{\tau} Y^Q(t) + k^Q Y^Q(\tau+1) \\ -K^Q(0) + \sum_{t=1}^{\tau-1} Y^K(t) &= dk^Q \sum_{t=1}^{\tau-1} Y^Q(t) + k^Q Y^Q(\tau) \\ = Y^K(\tau) &= dk^Q Y^Q(\tau) + k^Q\left\{Y^Q(\tau+1) - Y^Q(\tau)\right\} \qquad (5.24) \\ &= dk^Q Y^Q(\tau) + k^Q Y^Q(\bar{\tau})\,\gamma^\tau \\ &= k^Q Y^Q(\tau)\,\{d + \gamma^\tau\} \ \textit{with}\ \gamma^\tau := \frac{Y^Q(\tau+1) - Y^Q(\tau)}{Y^Q(\tau)}. \end{aligned}$$

As a result, activity levels of the consumption good processes can substitute the activity level of the capital good process. The sum of these expressions for future time intervals $\tau^f \in \{\bar{\tau}+1,...,T\}$ is interesting from the economic point of view

$$\begin{aligned}
\sum_{\tau^f=\bar{\tau}+1}^{T} Y^K(\tau^f) &= dk^Q \sum_{\tau^f=\bar{\tau}+1}^{T} Y^Q(\tau^f) + k^Q \sum_{\tau^f=\bar{\tau}+1}^{T} \left\{ Y^Q(\tau^f+1) - Y^Q(\tau^f) \right\} \\
&= dk^Q \sum_{\tau^f=\bar{\tau}+1}^{T} Y^Q(\tau^f) + k^Q \sum_{\tau^f=\bar{\tau}+1}^{T} Y^Q(\tau^f) \left\{ \frac{Y^Q(\tau^f+1) - Y^Q(\tau^f)}{Y^Q(\tau^f)} \right\} \\
&= dk^Q \sum_{\tau^f=\bar{\tau}+1}^{T} Y^Q(\tau^f) + k^Q \sum_{\tau^f=\bar{\tau}+1}^{T} Y^Q(\tau^f)\gamma^{\tau^f} \text{ with } \gamma^{\tau^f} \coloneqq \frac{Y^Q(\tau^f+1) - Y^Q(\tau^f)}{Y^Q(\tau^f)}.
\end{aligned} \tag{5.25}$$

Equation (5.25) means that the development of the capital stock is clearly determined by the time development path of the consumption good amounts and also by the inter-temporal welfare function with the exogenously given time preferences. In the case where there is no increase in consumption, $\gamma^f = 0$, the produced capital good amounts are required to cover at least the depreciation of the capital stock. When the capital stock is completely used and sustained, but not developed, then $d + \gamma^f = 0$. The weaker condition, when the capital stock is completely used, means that growth rates can be negative, but are limited by $\gamma^f \geq -d$.

For the special case of a unique growth rate in consumption good amounts for future time intervals, we obtain

$$\begin{aligned}
\sum_{\tau^f=\bar{\tau}+1}^{T} Y^K(\tau^f) &= k^Q\{d+\gamma^\tau\} \sum_{\tau^f=\bar{\tau}+1}^{T} Y^Q(\tau^f) \\
\mathit{with}\, \gamma^\tau &:= \frac{Y^Q(\tau^f+1) - Y^Q(\tau^f)}{Y^Q(\tau^f)} \mathit{for}\, \forall \tau^f .
\end{aligned} \tag{5.26}$$

5.7 Aggregation of Processes to Sectors

Later, we will have a large number of expressions for deriving the optimal conditions. It is therefore helpful to aggregate in advance the water treatment, water distribution, wastewater, and wastewater treatment processes to the water and wastewater sector, and the other processes to another sector.

5.7.1 Energy and Human Labour Inputs for the Production Sector

The set of processes $g \in \left\{ Q, K, \underset{i=1,2}{Q}, \underset{\upsilon=1,2}{K} \right\}$ represents the consumption and capital good processes and corresponding extraction processes of raw materials form the so-called production sector. The total energy and human labour input amounts are equal to sums

$$E^P_{\text{sector}} = E^Q(\tau) + E^K(\tau) + E^Q_{i=1+2}(\tau) + E^K_{\mu=1+2}(\tau)$$
$$= Y^Q(\tau)\left\{\begin{array}{l} \frac{ex^Q}{\eta^Q} \\ +\frac{ex^K}{\eta^K}k^Q\left\{d^Q+\gamma^Q_\tau\right\} \\ +\sum\limits_{i=1}^{2}\frac{\kappa^Q_i}{\eta^Q_i}\left\{\begin{array}{l}\frac{ex^{app}_i}{c(\bar{N}^Q_i(0))}e^{\varphi^Q_i\kappa^Q_i\sum\limits_{t=1}^{\tau-1}Y^Q(t)}+ \\ \varepsilon\left(-\ln c(\bar{N}^Q_i(0))+\varphi^Q_i\kappa^Q_i\sum\limits_{t=1}^{\tau-1}Y^Q(t)\right)\end{array}\right\} \\ +k^Q\left\{d^Q+\gamma^Q_\tau\right\}\sum\limits_{\mu=1}^{2}\frac{\kappa^K_\mu}{\eta^K_\mu}\left\{\begin{array}{l}\frac{ex^{app}_\mu}{c(\bar{N}^K_\mu(0))}e^{\varphi^K_\mu\kappa^K_\mu k^Q\left\{\sum\limits_{t=1}^{\tau-1}Y^Q(t)\left\{d^Q+\gamma^Q_t\right\}\right\}}+ \\ \varepsilon\left(\begin{array}{l}-\ln c(\bar{N}^K_\mu(0)) \\ +\varphi^K_\mu\kappa^K_\mu k^Q\left\{\sum\limits_{t=1}^{\tau-1}Y^Q(t)\left\{d^Q+\gamma^Q_t\right\}\right\}\end{array}\right)\end{array}\right\}\end{array}\right\}. \tag{5.27}$$

$$L^P_{sector} = L^Q(\tau) + L^K(\tau) + L^Q_{i=1+2}(\tau) + L^K_{\mu=1+2}(\tau)$$
$$= Y^Q(\tau)\left\{\begin{array}{l} \tilde{l}^Q\bar{K}(0)\,e^{-\lambda^Q k^Q\left\{\sum\limits_{t=1}^{\tau-1}Y^Q(t)\left\{d^Q+\gamma^Q_t\right\}\right\}} \\ +l^K k^Q\left\{d^Q+\gamma^Q_\tau\right\} \\ +\sum\limits_{i=1}^{2}\frac{l^Q_i\kappa^Q_i e^{+\varphi^Q_i\kappa^Q_i\sum\limits_{t=1}^{\tau-1}Y^Q(t)}}{c(\bar{N}^Q_i(0))} \\ +k^Q\left\{d^Q+\gamma^Q_\tau\right\}\sum\limits_{\mu=1}^{2}\frac{l^K_\mu\kappa^K_\mu}{c(\bar{N}^K_\mu(0))}e^{+\varphi^K_\mu\kappa^K_\mu k^Q\left\{\sum\limits_{t=1}^{\tau-1}Y^Q(t)\left\{d^Q+\gamma^Q_t\right\}\right\}}\end{array}\right\}. \tag{5.28}$$

5.7.2 Energy and Human Labour Inputs for the Water Sectors

The set of processes $\hat{g} \in \left\{^{WT}, ^{WD}, ^{WWC}, ^{WWT}\right\}$ represents the water and wastewater sectors. The total energy and human labour input amounts are equal to sums where the capital good amounts are substituted by the relation $Y^K(\tau) = k^Q Y^Q(\tau)\{d+\gamma^\tau\}$. The energy and human labour input amounts for water production and wastewater treatment are equal to

$$E^W_{sector}(\tau) = W^W(\tau)\left\{\tilde{e}^{WT}\sum_{\omega}\ln\frac{c^*_W(0)}{c_W(\tau)}\right\} \tag{5.29}$$

$$L^W_{sector}(\tau) = W^{WT}(\tau)\,l^{WT} \tag{5.30}$$

with

$$W^{W}(\tau) = \left\{ \begin{array}{l} Y^{Q}(\tau) \left\{ \psi_{Q}^{RP} + \sum\limits_{i=1}^{2} \frac{\tilde{\psi}_{i}^{Q} \kappa_{i}^{Q}}{c(\bar{N}_{i}^{Q}(0))} e^{\varphi_{i}^{Q} \kappa_{i}^{Q} \sum\limits_{t=1}^{\tau-1} Y^{Q}(\tau)} \right\} \\ + Y^{K}(\tau) \sum\limits_{\mu=1}^{2} \frac{\tilde{\psi}_{\mu}^{K} \kappa_{\mu}^{K}}{c(\bar{N}_{\mu}^{K}(0))} e^{\varphi_{\mu}^{K} \kappa_{\mu}^{K} \sum\limits_{t=1}^{\tau-1} Y^{K}(t)} \end{array} \right\}$$

$$= Y^{Q}(\tau) \left\{ \begin{array}{l} \left\{ \psi_{Q}^{RP} + \sum\limits_{i=1}^{2} \frac{\tilde{\psi}_{i}^{Q} \kappa_{i}^{Q}}{c(\bar{N}_{i}^{Q}(0))} e^{\varphi_{i}^{Q} \kappa_{i}^{Q} \sum\limits_{t=1}^{\tau-1} Y^{Q}(\tau)} \right\} \\ + k^{Q} \{ d + \gamma^{\tau} \} \sum\limits_{\mu=1}^{2} \frac{\tilde{\psi}_{\mu}^{K} \kappa_{\mu}^{K}}{c(\bar{N}_{\mu}^{K}(0))} e^{\varphi_{\mu}^{K} \kappa_{\mu}^{K} \sum\limits_{t=1}^{\tau-1} k^{Q} Y^{Q}(t) \{ d+\gamma^{t} \})} \end{array} \right\} \tag{5.31}$$

$$E_{sector}^{WWT}(\tau) = W^{WW}(\tau)\, \tilde{\varepsilon}^{WWT} \left\{ \ln \frac{c_{Q}^{*}(0)}{c_{Q}(\tau)} + \ln \frac{c_{\hat{j}}^{*}(0)}{c_{\hat{j}}(\tau)} \right\}. \tag{5.32}$$

$$L_{sector}^{WWT}(\tau) = W^{WW}(\tau)\, l^{WWT} \tag{5.33}$$

with

$$W^{WWT}(\tau) = \left\{ \begin{array}{l} Y^{Q}(\tau) \left\{ \psi_{Q}^{RP} + \sum\limits_{i=1}^{2} \frac{\tilde{\psi}_{i}^{Q} \kappa_{i}^{Q}}{c(\bar{N}_{i}^{Q}(0))} e^{\varphi_{i}^{Q} \kappa_{i}^{Q} \sum\limits_{t=1}^{\tau-1} Y^{Q}(t)} \right\} \\ + Y^{K}(\tau) \sum\limits_{\mu=1}^{2} \frac{\tilde{\psi}_{\mu}^{K} \kappa_{\mu}^{K}}{c(\bar{N}_{\mu}^{K}(0))} e^{\varphi_{\mu}^{K} \kappa_{\mu}^{K} \sum\limits_{t=1}^{\tau-1} Y^{K}(t)} \end{array} \right\}$$

$$= Y^{Q}(\tau) \left\{ \begin{array}{l} \left\{ \psi_{Q}^{RP} + \sum\limits_{i=1}^{2} \frac{\tilde{\psi}_{i}^{Q} \kappa_{i}^{Q}}{c(\bar{N}_{i}^{Q}(0))} e^{\varphi_{i}^{Q} \kappa_{i}^{Q} \sum\limits_{t=1}^{\tau-1} Y^{Q}(t)} \right\} \\ + k^{Q} \{ d + \gamma^{\tau} \} \sum\limits_{\mu=1}^{2} \frac{\tilde{\psi}_{\mu}^{K} \kappa_{\mu}^{K}}{c(\bar{N}_{\mu}^{K}(0))} e^{\varphi_{\mu}^{K} \kappa_{\mu}^{K} \sum\limits_{t=1}^{\tau-1} k^{Q} Y^{Q}(t) \{ d+\gamma^{t} \})} \end{array} \right\}. \tag{5.34}$$

Chapter 6
Optimality Conditions of the Water Use Model

Abstract The optimality concept is focused on deriving the so-called non-profit conditions. These non-profit conditions are expressed in the form of actual and inter-temporal marginal costs for producing the desired quantity of the consumption good and the associated water and generated wastewater amounts. These marginal costs are based on the shadow prices for the two essential input factors, the human labour inputs and the energy inputs. The concentrations of the extracted raw materials are given exogenously. Therefore, there are declining functions for the production and reproduction activities in the past, which cause actual and inter-temporal marginal costs for using non-renewable natural resources in the future. The decrease of the raw materials concentrations is combined with an increase in un-avoidable residuals, which must be transported by water to the wastewater treatment sector. As a result, the use of non-renewable resources also leads to actual and inter-temporal marginal costs for water production and wastewater treatment.

6.1 The Optimization Concept

Based on the theoretical foundations, the general design, the specifications, and the model constraints of the previous chapters, we are now in the position to derive the optimal conditions for our basic model. The objective is to maximize the quasi-concave, inter-temporal welfare function

$$W(\vec{Q}(1),...,\vec{Q}(T)) = \sum\nolimits_{\tau=1}^{T} (1+\delta)^{1-\tau} W_\tau(\vec{Q}(\tau)), \tag{6.1}$$

Where the consumption good vector, $\vec{Q}(1)$, for the first activity interval determines the exogenously given initial conditions for our model.

Since the consumption good vector has only one component in our basic model, the following welfare function must be maximized

$$W(\bar{Q}(1), Q(2),...,Q(T)) = W_1(\bar{Q}(1) + \sum_{\tau=2}^{T} (1+\delta)^{1-\tau} W_\tau(Q(\tau)). \tag{6.2}$$

H. Niemes, M. Schirmer, *Entropy, Water and Resources*,
DOI 10.1007/978-3-7908-2416-2_6, © Springer-Verlag Berlin Heidelberg 2010

When compared to the welfare function used by Faber et al. (1995:156), this welfare function includes the initial welfare level achieved by the initial consumption good amount at the beginning of the planning horizon although the initial conditions are not part of the maximization process.

A feasible production program is called optimal if it maximizes this welfare function. According to the procedure of Kuhn-Tucker (for more details see Hadley 1964:185–212) this means to form the LAGRANGE-function (Eq. 6.3) from the welfare function and the system of constraints.

$$V = W_I(\bar{Q}(1) + \sum_{\tau=2}^{T} (1+\delta)^{1-\tau} W_\tau(Q(\tau)) + \sum_{\tau=1}^{T} \left\{ \sum_g p_g(\tau) H_g(\tau) \right\} \Leftrightarrow$$
$$\tilde{V} := V - W_1(\bar{Q}(1) = \sum_{\tau=2}^{T} (1+\delta)^{1-\tau} W_\tau(Q(\tau)) \qquad (6.3)$$
$$+ \sum_{\tau=1}^{T} \left\{ \sum_g p_g(\tau) H_g(\tau) \right\}$$

With the indices g as notations for the model constraints

$$H_Q(\tau+1),\ H_{N_{ii}^Q}(\tau), H_{N_{ii}^K}(\tau), H_{\overline{N}_{ii}^Q}(\tau), H_{\overline{N}_{ii}^K}(\tau),$$
$$H_W(\tau), H_{WWT}(\tau), H_{N^E}(\tau), H_L(\tau), H_{\bar{K}}(\tau).$$

Using the notation $Y_g(\bar{\tau})$ for the variables of our model, an optimal feasible production program must fulfil the general conditions for each of the activity intervals. For a selected activity interval $\bar{\tau} \in \{1, ..., T\}$, for example,

$$\left\{ \frac{\vartheta \tilde{V}}{\vartheta Y_g(\bar{\tau})} = 0 \; for \; Y_g(\bar{\tau}) > 0 \right\} \vee \left\{ \frac{\vartheta \tilde{V}}{\vartheta Y_g(\bar{\tau})} \leq 0 \; for \; Y_g(\bar{\tau}) = 0 \right\}$$
$$with \qquad (6.4)$$
$$\{ H_g(\bar{\tau}) = 0 \; for \; p_g(\bar{\tau}) \geq 0 \} \vee \{ H_g(\bar{\tau}) > 0 \; for \; p_g(\bar{\tau}) = 0 \}$$
$$\wedge \; p_g(\bar{\tau}+1) \geq 0,$$

The LAGRANGE-multipliers $p_g(\bar{\tau})$ in Eq. (6.3) can be interpreted as shadow prices of the variables $Y_g(\bar{\tau})$.

The no-storage assumption of the different process input and output amounts allows us to base the different variables, $Y_g(\bar{\tau})$, solely on one or a limited number of variables for the production side, more specifically, the activity levels of the production processes for the consumption good $Y^Q(\bar{\tau})$ and the capital good $Y^K(\bar{\tau})$. The consumption good amount $Q(\bar{\tau})$ is the variable for the demand side. These simplifications imply that all production processes are always in operation and there is no need to carry out annoying case distinctions for the interpretation of the conditions of optimal feasible production programs. The following parts of this general condition will be the subject of our analysis

$$\left\{\frac{\vartheta \tilde{V}}{\vartheta Y_g(\bar{\tau})} = 0 \, for \, Y_g(\bar{\tau}) > 0\right\} \, with \, \left\{H_g(\bar{\tau}) = 0 \, for \, p_g(\bar{\tau}) \geq 0\right\} \wedge p_g(\bar{\tau}) > 0. \tag{6.5}$$

A selected activity interval, $\bar{\tau}$, divides the planning span into activity intervals in the past and the future by using the time indices $\tau^p \in \{1, ..., \bar{\tau} - 1\}$ and $\tau^f \in \{\bar{\tau} + 1, ..., T\}$ for the separation of activity intervals. The differentiation of this function by variables of a selected activity interval $\bar{\tau}$ has no effect on the past, but is addressing the actual and future activity intervals in case there are inter-temporal connections between some model constraints.

One example of such an inter-temporal connection in our model is the exogenously given declining functions for the concentration of the extracted raw materials and the creation and use of the capital stock, which influence the labour input for the production process of the consumption good. The activities in the selected interval determine not only the initial condition of the next interval, but also for all future intervals up to the end of the planning horizon. Therefore, the differentiation of the LAGRANGE-function will lead to shadow price expressions with additional components for inter-temporal negative and positive externalities.

From the economic point of view, it is useful to distinguish between the demand and production side. We can also say that every optimal production program in our model must fulfil the general conditions for the demand side and the production side. The variables for the demand side are the consumption good amounts that are consumed. The variables for the production side are the activity levels of the production processes, which are equal to the consumption good amount $Y^Q(\bar{\tau})$ and the capital good $Y^K(\bar{\tau})$ since the corresponding output coefficient are equal to 1.

6.2 Optimality Conditions for the Demand Side

The consumption good constraint describes how the demand side is connected with the production side. In our case with no durable consumption goods, we use the equivalent constraints of how both sides are connected

$$H_Q(\tau) = Y^Q(\tau) - Q(\tau + 1) \geq 0 \Leftrightarrow H_{YQ}(\tau) = Y^Q(\tau) - Q(\tau + 1) \geq 0\,. \tag{BM12}$$

If we had durable consumption goods, which are not completely consumed within one activity interval, another inter-temporal connection on the demand side would have to be used.

We obtain the following identity relation between the shadow prices for consumption and production goods

$$p_Q(\bar{\tau} + 1) = (1 + \delta)^{-\bar{\tau}} \frac{\vartheta W_{\bar{\tau}+1}(Q(\bar{\tau} + 1)}{\vartheta Q(\bar{\tau} + 1)} = p_{YQ}(\bar{\tau}). \tag{6.6}$$

This equation means (for more details see Faber et al. 1995:159) that, at the optimum, the shadow price for the consumption good must correspond to its social marginal cost, which is determined by the differentiation of the inter-temporal welfare function. Two different effects are thus taken into consideration. The term $\vartheta W_{\bar{\tau}+1}(Q(\bar{\tau}+1)/\vartheta Q(\bar{\tau}+1)$ expresses the preferences for consumption with respect to interval period $\bar{\tau}+1$, while $(1+\delta)^{-\bar{\tau}}$ represents the time preferences for consumption.

6.3 Optimality Conditions for the Production Side

In the case where optimal conditions are fulfilled, we also achieve the so-called non-profit condition where the shadow price of the consumption good $p_{YQ}(\bar{\tau})$ is equal to the sum of the actual and inter-temporal marginal costs of the different cost components for the production of one unit of the consumption good. The marginal costs are equal to the average costs as a result of the linearity of the technology used in our model. As described by Faber et al. (1995:155–169), the use of the non-profit condition reduces the number of substitution procedures for different shadow prices of intermediary goods.

The essential input factors for operating the different production processes in our model are the natural resources, human labour, free energy, and the capital good input. Shadow prices are determined by the marginal costs of these essential input factors. In this context, we should keep in mind that the amounts of the natural resources appropriated from the natural environment for extracting the raw materials and the water for producing the consumption and capital good are less relevant because of the law of mass conservation. Their quality changes are measured in the form of concentrations of the natural resources. These concentrations are the dominant characteristics because of the relations between the concentration changes and the energy and labour input requirements for extracting these resources from the natural environment.

The capital good is a durable good that is available for the current and future activity intervals, whereas the consumption good is used completely for reproduction. The shadow price of the capital good $\bar{p}_{YK}(\bar{\tau}\mapsto T)$ must therefore be treated differently than the shadow price of the consumption good $p_{YQ}(\bar{\tau})$.

6.3.1 Non-profit Conditions for the Production Sector

When differentiating the LAGRANGE function, we will apply the Leibniz rule for differentiating a function $f(x) = u(x)\,v(x)$ or a collection of functions $f_1,\ldots,f_n$

$$\left\{\frac{d}{dx}f(x) = \frac{du(x)}{dx}\,v(x) + u(x)\,\frac{dv(x)}{dx}\right\} \vee \left\{\frac{d}{dx}\prod_{i=1}^{k} f_i(x) = \left(\sum_{i=1}^{k}\frac{\frac{d}{dx}f_i(x)}{f_i(x)}\right)\prod_{i=1}^{k} f_i(x)\right\}. \tag{6.7}$$

Our constraints have the following type of inter-temporal functions to be differentiated by the variable $Y(\bar{\tau})$ for a selected interval $\bar{\tau} \in (1,...,T)$

$$F(p(\tau),Y(\tau)) := \sum_{\tau=1}^{T} p(\tau)Y(\tau) \sum_{\mu=1}^{2} \frac{A_\mu e^{B_\mu \bar{Y}(\tau)}}{C_\mu(0)}$$
$$\text{with } C_\mu(\tau) = C_\mu(0)\, e^{-B_\mu \bar{Y}(\tau)} \wedge \bar{Y}(\tau) = \sum_{t=1}^{\tau-1} Y(t). \tag{6.8}$$

The application of the product rule for differentiating leads to an expression with two terms

$$\frac{\vartheta}{\vartheta Y(\bar{\tau})} F(p(\tau),Y(\tau))$$
$$= p(\bar{\tau})Y(\bar{\tau}) \sum_{\mu=1}^{2} \frac{A_\mu e^{B_\mu \sum_{t=1}^{\bar{\tau}-1} Y(t)}}{C_\mu(0)} + \sum_{\tau^f=\bar{\tau}+1}^{T} p(\tau^f)Y(\tau^f) \sum_{\mu=1}^{2} \frac{A_\mu e^{B_\mu \sum_{t=1}^{\tau^f-1} Y(t)}}{C_\mu(0)} \tag{6.9}$$
$$= p(\bar{\tau}) \sum_{\mu=1}^{2} \frac{A_\mu}{C_\mu(\bar{\tau})} + \sum_{\tau^f=\bar{\tau}+1}^{T} p(\tau^f)Y(\tau^f) \sum_{\mu=1}^{2} \frac{A_\mu B_\mu}{C_\mu(\tau^f)}$$

If $p(\tau)$ is the economic definition of a shadow price for producing the good amounts $Y(\tau)$, the first term is the actual marginal costs for producing this good within the selected interval $\bar{\tau}$. The second term is the inter-temporal marginal costs of production during future intervals $\bar{\tau}^f = \bar{\tau} + 1,..., T$, for which the following relations will be used later on

$$MC^Y(\bar{\tau} \mapsto T) := \frac{\vartheta}{\vartheta Y(\bar{\tau})} F(p(\tau),Y(\tau)) = MC^Y(\bar{\tau}) + MC^Y(\bar{\tau} \to T) \text{ with}$$
$$MC^Y(\bar{\tau}) := p(\bar{\tau}) \sum_{\mu=1}^{2} \frac{A_\mu}{C_\mu(\bar{\tau})}, \; MC^Y(\bar{\tau} \to T) := \sum_{\tau^f=\bar{\tau}+1}^{T} p(\tau^f)Y(\tau^f) \sum_{\mu=1}^{2} \frac{A_\mu B_\mu}{C_\mu(\tau^f)}. \tag{6.10}$$

The optimality conditions are obtained by differentiating the LAGRANGE function by the variable $Y^Q(\bar{\tau})$ with the given constraints of the basic model for a selected time interval $\bar{\tau} \in \{1,..., T\}$. The optimality conditions are formulated as non-profit conditions. Therefore, the shadow price for producing one unit of consumption good must be equal to the sum of the marginal costs for the different cost components of the processes of the production sector as well as water and wastewater sectors.

6.3.1.1 Marginal Human Labour Costs for the Production Sector

The total marginal costs for human labour input in the production sector are equal to the sum

$$MCL^P_{sector}(\bar{\tau} \mapsto T)$$
$$= MCL^Q_Q(\bar{\tau} \mapsto T) + MCL^Q_K(\bar{\tau} \mapsto T) + MCL^Q_{Q_{N_{i=1+2}}}(\bar{\tau} \mapsto T) + MCL^Q_{K_{N_{\mu=1+2}}}(\bar{\tau} \mapsto T).$$

The different expressions of this sum are obtained by differentiating the following part of the production side of the LAGRANGE function. The production side of the LAGRANGE function is formed from the transformed system of constraints in Eq. (5.64) and the shadow prices (or LANGRANGE multipliers) of the human labour inputs for producing one unit of the consumption good during the whole planning horizon, where only the non-profit conditions are taken into account

$$MCL^{P}_{sector}(\bar{\tau} \mapsto T) = \frac{\vartheta}{\vartheta Y^{Q}(\bar{\tau})} \sum_{\tau=1}^{T} Y^{Q}(\tau) \left\{ \begin{array}{l} \tilde{l}^{Q} e^{-\lambda^{Q} k^{Q} \left\{ \sum_{t=1}^{\tau-1} Y^{Q}(t) \right\}} \\ + l^{K} k^{Q} \left\{ d^{Q} + \gamma_{\tau}^{Q} \right\} \\ + \sum_{i=1}^{2} \frac{l_{i}^{Q} \kappa_{i}^{Q} e^{+\varphi_{i}^{Q} \kappa_{i}^{Q} \sum_{t=1}^{\tau-1} Y^{Q}(t)}}{c(\overline{N}_{i}^{Q}(0))} \\ + k^{Q} \left\{ d^{Q} + \gamma_{\tau}^{Q} \right\} \sum_{\mu=1}^{2} \frac{l_{\mu}^{K} \kappa_{\mu}^{K}}{c(\overline{N}_{\mu}^{K}(0))} e^{+\varphi_{\mu}^{K} \kappa_{\mu}^{K} k^{Q} \left\{ \sum_{t=1}^{\tau-1} Y^{Q}(t) \left\{ d^{Q} + \gamma_{t}^{Q} \right\} \right\}} \end{array} \right\}. \tag{6.11}$$

For a selected time interval, we have the following marginal human labour costs components for the production sector, which will be marked as sector A. The water and wastewater sectors will be marked as sector B and C, respectively.

A.1.1 Marginal Human Labour Costs for the Consumption Good Process

$$\begin{aligned} MCL^{Q}_{Q}(\bar{\tau} \mapsto T) &= \frac{\vartheta}{\vartheta Y^{Q}(\bar{\tau})} \left\{ \sum_{\tau=1}^{T} p_{L}(\tau)\, Y^{Q}(\tau)\, \tilde{l}^{Q}(0) e^{-\lambda^{Q} k^{Q} \left\{ \sum_{t=1}^{\tau-1} Y^{Q}(t) \right\}} \right\} \\ &= p_{L}(\bar{\tau})\, \tilde{l}^{Q}(0) e^{-\lambda^{Q} k^{Q} \left\{ \sum_{t=1}^{\tau-1} Y^{Q}(t) \right\}} \\ &\quad + (-\lambda) k^{Q} \sum_{\tau^{f}=\bar{\tau}+1}^{T} p_{L}(\tau^{f}) Y^{Q}(\tau^{f}) \tilde{l}^{Q}(0) e^{-\lambda^{Q} k^{Q} \left\{ \sum_{t=1}^{\tau^{f}-1} Y^{Q}(t)\{\} \right\}} \\ &= p_{L}(\bar{\tau})\, \tilde{l}^{Q}(\bar{\tau}) + (-\lambda) k^{Q} \sum_{\tau^{f}=\bar{\tau}+1}^{T} p_{L}(\tau^{f}) Y^{Q}(\tau^{f}) \tilde{l}^{Q}(\tau^{f}). \end{aligned}$$

Aside from the actual marginal human labour costs,

$$MCL^{Q}_{Q}(\bar{\tau}) := p_{L}(\bar{\tau})\, \tilde{l}^{Q}(\bar{\tau}),$$

The innovation effect of investments into the capital stock leads to inter-temporal marginal human labour costs

$$ITMCL^Q_{Q_K}(\bar{\tau} \to T) := (-\lambda)k^Q \sum_{\tau^f=\bar{\tau}+1}^{T} p_L(\tau^f)Y^Q(\tau^f)\tilde{l}^Q(\tau^f).$$

A.1.2 Marginal Human Labour Costs for Producing the Capital Good

$$MCL^Q_{Q_K}(\bar{\tau} \mapsto T) = \frac{\vartheta}{\vartheta Y^Q(\bar{\tau})}\left\{\sum_{\tau=1}^{T} p_L(\tau)\, l^K\, Y(\tau)\, k^Q \left\{ d^Q + \gamma^Q_\tau \right\}\right\}$$
$$= p_L(\bar{\tau})\, l^K k^Q \left\{ d^Q + \gamma^Q_{\bar{\tau}} \right\} = p_L(\bar{\tau})\, l^K k^Q d^Q + p_L(\bar{\tau})\, l^K k^Q \gamma^Q_{\bar{\tau}}.$$

The actual marginal human labour costs for compensating the depreciation (or use) of the capital stock is

$$MCL^Q_{K_{dep}}(\bar{\tau}) := p_L(\bar{\tau}) l^K k^Q d^Q .$$

The actual marginal costs for the development of the capital stock is

$$MCL^Q_{K_{dev}}(\bar{\tau}) := p_L(\bar{\tau})\, l^K k^Q \gamma^Q_{\bar{\tau}} .$$

A.1.3 Marginal Human Labour Costs for Extracting the Raw Materials for the Consumption Good Process

$$MCL^Q_{N_{i=1+2}}(\bar{\tau} \mapsto T)$$
$$= \frac{\vartheta}{\vartheta Y^Q(\bar{\tau})}\left\{\sum_{\tau=1}^{T} p_L(\tau)\, Y^Q(\tau) \sum_{i=1}^{2} \frac{l^Q_i \kappa^Q_i e^{+\varphi^Q_i \kappa^Q_i \sum_{t=1}^{\tau-1} Y^Q(t)}}{c(\overline{N}^Q_i(0))}\right\}$$
$$= p_L(\bar{\tau}) \sum_{i=1}^{2} \frac{l^Q_i \kappa^Q_i e^{+\varphi^Q_i \kappa^Q_i \sum_{t=1}^{\tau-1} Y^Q(t)}}{c(\overline{N}^Q_i(0))}$$
$$+ \sum_{\tau^f=\bar{\tau}+1}^{T} p_L(\tau^f)\, Y^Q(\tau^f) \sum_{i=1}^{2} \frac{l^Q_i \kappa^Q_i \varphi^Q_i \kappa^Q_i e^{+\varphi^Q_i \kappa^Q_i \sum_{t=1}^{\tau^f-1} Y^Q(t)}}{c(\overline{N}^Q_i(0))}$$
$$= p_L(\bar{\tau}) \sum_{i=1}^{2} \frac{l^Q_i \kappa^Q_i}{c(\overline{N}^Q_i(\bar{\tau}))} + \sum_{\tau^f=\bar{\tau}+1}^{T} p_L(\tau^f)\, Y^Q(\tau^f) \sum_{i=1}^{2} \frac{l^Q_i \kappa^Q_i \varphi^Q_i \kappa^Q_i}{c(\overline{N}^Q_i(\tau^f))}.$$

In addition to the actual marginal human labour costs

$$MCL^{Q}_{N_{i=1,2}}(\bar{\tau}) := p_L(\bar{\tau}) \sum_{i=1}^{2} \frac{l^{Q}_{i}\,\kappa^{Q}_{i}}{c(\overline{N}^{Q}_{i}(\bar{\tau}))},$$

The use of non-renewable natural resources for producing the consumption good generates inter-temporal marginal human labour costs

$$ITMCL^{Q}_{N_{i=1+2}}(\tau \to T) = \sum_{\tau^f=\bar{\tau}+1}^{T} p_L(\tau^f)\, Y^{Q}(\tau^f) \sum_{i=1}^{2} \frac{l^{Q}_{i}\,\kappa^{Q}_{i}\,\varphi^{Q}_{i}\,\kappa^{Q}_{i}}{c(\overline{N}^{Q}_{i}(\tau^f))}.$$

A.1.4 Marginal Human Labour Costs for Extracting the Raw Materials to Produce the Capital Good Amounts

$$\begin{aligned}
&MCL^{Q}_{K_{N_{\mu=1+2}}}(\bar{\tau} \mapsto T)\\
&= \frac{\vartheta}{\vartheta Y^{Q}(\bar{\tau})} \sum_{\tau=1}^{T} p_L(\tau) Y^{Q}(\tau) k^{Q} \left\{ d^{Q} + \gamma^{Q}_{\tau} \right\} \sum_{\mu=1}^{2} \frac{l^{K}_{\mu}\,\kappa^{K}_{\mu}}{c(\overline{N}^{K}_{\mu}(0))}\, e^{+\varphi^{K}_{\mu}\kappa^{K}_{\mu} k^{Q} \left\{ \sum_{t=1}^{\tau-1} Y^{Q}(t) \left\{ d^{Q} + \gamma^{Q}_{t} \right\} \right\}}\\
&= p_L(\bar{\tau})\, k^{Q} \left\{ d^{Q} + \gamma^{Q}_{\bar{\tau}} \right\} \sum_{\mu=1}^{2} \frac{l^{K}_{\mu}\,\kappa^{K}_{\mu}}{c(\overline{N}^{K}_{\mu}(\tau^f)}\\
&\quad + \sum_{\tau^f=\bar{\tau}+1}^{T} p_L(\tau^f)\, k^{Q} Y^{Q}(\tau^f) \left\{ d^{Q} + \gamma^{Q}_{\tau^f} \right\} \left\{ \sum_{\mu=1}^{2} \varphi^{K}_{\mu}\kappa^{K}_{\mu} \frac{l^{K}_{\mu}\,\kappa^{K}_{\mu}\,\varphi^{K}_{\mu}\kappa^{K}_{\mu}}{c(\overline{N}^{K}_{\mu}(\tau^f)} \right\}.
\end{aligned}$$

The actual human labour costs for using non-renewable natural resources for sustaining and developing the capital stock that is required to produce one unit of the consumption good amounts is

$$\begin{aligned}
MCL^{Q}_{K_{N_{\mu=1+2}}}(\bar{\tau}) &= p_L(\bar{\tau})\, k^{Q} \left\{ d^{Q} + \gamma^{Q}_{\bar{\tau}} \right\} \sum_{\mu=1}^{2} \frac{l^{K}_{\mu}\,\kappa^{K}_{\mu}}{c(\overline{N}^{K}_{\mu}(\tau^f)}\\
&= MCL^{Q}_{K_{dep\,N_{\mu=1+2}}}(\bar{\tau}) + MCL^{Q}_{K_{dev\,N_{\mu=1+2}}}(\bar{\tau}) \text{ with}\\
MCL^{Q}_{K_{dep\,N_{\mu=1+2}}}(\bar{\tau}) &:= p_L(\bar{\tau})\, k^{Q}\, d^{Q} \sum_{\mu=1}^{2} \frac{l^{K}_{\mu}\,\kappa^{K}_{\mu}}{c(\overline{N}^{K}_{\mu}(\tau^f)}\\
MCL^{Q}_{K_{dev\,N_{\mu=1+2}}}(\bar{\tau}) &:= p_L(\bar{\tau})\, k^{Q}\, \gamma^{Q}_{\bar{\tau}} \sum_{\mu=1}^{2} \frac{l^{K}_{\mu}\,\kappa^{K}_{\mu}}{c(\overline{N}^{K}_{\mu}(\tau^f)}.
\end{aligned}$$

The following are the inter-temporal marginal human labour costs occur

$$
\begin{aligned}
& ITMCL^{Q}_{K_{N_{\mu=1+2}}}(\bar{\tau} \to T) \\
&= \sum_{\tau^f=\bar{\tau}+1}^{T} p_L(\tau^f) k^Q Y^Q(\tau^f) \left\{ d^Q + \gamma^Q_{\tau^f} \right\} \left\{ \sum_{\mu=1}^{2} \varphi^K_\mu \kappa^K_\mu \frac{l^K_\mu \kappa^K_\mu \varphi^K_\mu \kappa^K_\mu}{c(\overline{N}^K_\mu(\tau^f))} \right\} \\
&= ITMCL^{Q}_{K_{dep\,N_{\mu=1+2}}}(\bar{\tau} \to T) + ITMCL^{Q}_{K_{dev\,N_{\mu=1+2}}}(\bar{\tau} \to T) \text{ with} \\
& ITMCL^{Q}_{K_{dep\,N_{\mu=1+2}}}(\bar{\tau} \to T) \\
&= k^Q d^Q \sum_{\tau^f=\bar{\tau}+1}^{T} p_L(\tau^f) Y^Q(\tau^f) \left\{ \sum_{\mu=1}^{2} \varphi^K_\mu \kappa^K_\mu \frac{l^K_\mu \kappa^K_\mu \varphi^K_\mu \kappa^K_\mu}{c(\overline{N}^K_\mu(\tau^f))} \right\} \\
& ITMCL^{Q}_{K_{dev\,N_{\mu=1+2}}}(\bar{\tau} \to T) \\
&= k^Q \sum_{\tau^f=\bar{\tau}+1}^{T} p_L(\tau^f) k^Q Y^Q(\tau^f) \gamma^Q_{\tau^f} \left\{ \sum_{\mu=1}^{2} \varphi^K_\mu \kappa^K_\mu \frac{l^K_\mu \kappa^K_\mu \varphi^K_\mu \kappa^K_\mu}{c(\overline{N}^K_\mu(\tau^f))} \right\}.
\end{aligned}
$$

6.3.1.2 Marginal Energy Costs for the Production Sector

The total marginal costs of the energy input for the production sector are equal to the sum

$$
\begin{aligned}
& MCE^P_{\sec tor}(\bar{\tau} \mapsto T) \\
&= MCE^Q_Q(\bar{\tau} \mapsto T) + MCE^Q_K(\bar{\tau} \mapsto T) + MCE^Q_{N_{i=1+2}}(\bar{\tau} \mapsto T) + MCE^Q_{K_{N_{\mu=1+2}}}(\bar{\tau} \mapsto T).
\end{aligned}
$$

These expressions are obtained by differentiating of the production side of the LAGRANGE function, which is formed from the transformed system of constraints and the shadow prices (or LANGRANGE multipliers) of the energy inputs for producing one unit of the consumption good during the whole planning horizon, where only the non-profit conditions are taken into account

$$
MCE^P_{\sec tor}(\bar{\tau} \mapsto T) = \frac{\vartheta \sum_{\tau=1}^{T} Y^Q(\tau) \left\{ \begin{array}{l} \frac{ex^Q}{\eta^Q} + \\ \frac{ex^K}{\eta^K} k^Q \left\{ d^Q + \gamma^Q_\tau \right\} + \\ \sum_{i=1}^{2} \frac{\kappa^Q_i}{\eta^Q_i} \left\{ \begin{array}{l} \frac{ex^{app}_i}{c(\overline{N}^Q_i(0))} e^{\varphi^Q_i \kappa^Q_i \sum_{t=1}^{\tau-1} Y^Q(t)} + \\ \varepsilon \left(-\ln c(\overline{N}^Q_i(0)) + \varphi^Q_i \kappa^Q_i \sum_{t=1}^{\tau-1} Y^Q(t) \right) \end{array} \right\} + \\ k^Q \left\{ d^Q + \gamma^Q_\tau \right\} \sum_{\mu=1}^{2} \frac{\kappa^K_\mu}{\eta^K_\mu} \left\{ \begin{array}{l} \frac{ex^{app}_\mu}{c(\overline{N}^K_\mu(0))} e^{\varphi^K_\mu \kappa^K_\mu k^Q \left\{ \sum_{t=1}^{\tau-1} Y^Q(t) \left\{ d^Q + \gamma^Q_t \right\} \right\}} \\ +\varepsilon \left(\begin{array}{l} -\ln c(\overline{N}^K_\mu(0)) \\ + \varphi^K_\mu \kappa^K_\mu k^Q \left\{ \sum_{t=1}^{\tau-1} Y^Q(t) \left\{ d^Q + \gamma^Q_t \right\} \right\} \end{array} \right) \end{array} \right\} \end{array} \right\}}{\vartheta Y^Q(\bar{\tau})}.
$$

When differentiating this expression by the consumption good amount of a selected time interval, we receive the following marginal energy costs components.

A.2.1 Marginal Energy Costs of Consumption Good Process

$$MCE_Q^Q(\bar{\tau} \mapsto T) = \frac{\vartheta}{\vartheta Y^Q(\bar{\tau})} \left\{ \sum_{\tau=1}^{T} p_E(\tau)\, Y^Q(\tau)\, \frac{ex^Q}{\eta^Q} \right\} = p_E(\bar{\tau})\, \frac{ex^Q}{\eta^Q}$$

A.2.2 Marginal Energy Costs of the Capital Good Process

$$MCE_{\underset{K}{K}}^Q(\bar{\tau} \mapsto T) = \frac{\vartheta}{\vartheta Y^Q(\tau)} \left\{ \sum_{\tau=1}^{T} p_E(\tau)\, \tfrac{ex^K}{\eta^K}\, Y(\tau)\, k^Q \left\{ d^Q + \gamma_\tau^Q \right\} \right\} =$$

$$p_E(\bar{\tau})\, \tfrac{ex^K}{\eta^K} k^Q \left\{ d^Q + \gamma_{\bar{\tau}}^Q \right\} = p_E(\bar{\tau})\, \tfrac{ex^K}{\eta^K} k^Q d^Q + p_E(\tau)\, \tfrac{ex^K}{\eta^K} k^Q \gamma_{\bar{\tau}}^Q.$$

We get two terms, namely the actual marginal energy costs for compensating the depreciation (or use) of the capital stock

$$MCE_{\underset{K}{K}use}^Q(\bar{\tau}) := \frac{ex^K}{\eta^K} k^Q d^Q,$$

And the actual marginal costs for investing into the development of the capital stock

$$MCE_{\underset{K}{K}dev}^Q(\bar{\tau}) := p_E(\bar{\tau})\, \frac{ex^K}{\eta^K} k^Q \gamma_{\bar{\tau}}^Q.$$

In order to fulfil the non-profit conditions, it is essential that only those production programs are allowed and considered feasible if the newly acquired investment covers at least the depreciation of the capital stock, which means that $d^Q + \gamma_{\bar{\tau}}^Q \geq 0 \Leftrightarrow \gamma_{\bar{\tau}}^Q \geq -d^Q$.

A.2.3 Marginal Energy Costs for the Extracting the Raw Material Required for the Consumption Good Process

$$
\begin{aligned}
&MCE^{Q}_{\underset{N_{i=1+2}}{Q}}(\bar{\tau} \mapsto T)\\
&= \frac{\vartheta}{\vartheta Y^{Q}(\bar{\tau})}\left\{\sum_{\tau=1}^{T} p_E(\tau)\, Y^{Q}(\tau) \sum_{i=1}^{2} \frac{\kappa_i^Q}{\eta_i^Q}\left\{\begin{array}{l}\dfrac{ex_i^{app}}{c(\overline{N}_i^Q(0))}\, e^{\varphi_i^Q \kappa_i^Q \sum_{t=1}^{\tau-1} Y^Q(t)} + \\ \varepsilon\left(-\ln c(\overline{N}_i^Q(0)) + \varphi_i^Q \kappa_i^Q \sum_{t=1}^{\tau-1} Y^Q(t)\right)\end{array}\right\}\right\}\\
&= p_E(\bar{\tau}) \sum_{i=1}^{2} \frac{\kappa_i^Q}{\eta_i^Q}\left\{\begin{array}{l}\dfrac{ex_i^{app}}{c(\overline{N}_i^Q(0))}\, e^{\varphi_i^Q \kappa_i^Q \sum_{t=1}^{\bar{\tau}-1} Y^Q(t)}\\ +\,\varepsilon\left(-\ln c(\overline{N}_i^Q(0)) + \varphi_i^Q \kappa_i^Q \sum_{\tau^p=1}^{\bar{\tau}-1} Y^Q(t)\right)\end{array}\right\}\\
&\quad + \sum_{\tau^f=\bar{\tau}+1}^{T}\left\{p_E(\tau^f)\, Y^Q(\tau^f)\left\{\begin{array}{l}\sum_{i=1}^{2} \dfrac{\kappa_i^Q}{\eta_i^Q} \dfrac{ex_i^{app} \varphi_i^Q \kappa_i^Q}{c(\overline{N}_i^Q(0))}\, e^{+\varphi_i^Q \kappa_i^Q \sum_{t=1}^{\tau^f-1} Y^Q(t)}\\ +\,\varepsilon\ \varphi_i^Q \kappa_i^Q \sum_{t=1}^{\tau^f-1} Y^Q(t)\end{array}\right\}\right\}\\
&= p_E(\bar{\tau}) \sum_{i=1}^{2} \frac{\kappa_i^Q}{\eta_i^Q}\left\{\frac{ex_i^{app}}{c(\overline{N}_i^Q(\bar{\tau}))} + \varepsilon\left(-\ln c(\overline{N}_i^Q(0)) + \overline{N}_i(\bar{\tau})\right)\right\}\\
&\quad + \sum_{\tau^f=\bar{\tau}+1}^{T}\left\{p_E(\tau^f)\, Y^Q(\tau^f)\left\{\sum_{i=1}^{2} \varphi_i^Q \kappa_i^Q\left\{\frac{\kappa_i^Q\, ex_i^{app}}{\eta_i^Q c(\overline{N}_i^Q(\tau^f))} + \varepsilon\,\overline{N}_i(\tau^f)\right\}\right\}\right\}.
\end{aligned}
$$

The marginal energy costs are composed of two terms: the actual marginal energy costs, which depend on the activities in the past,

$$
MCE^{Q}_{\underset{N_{i=1+2}}{Q}}(\bar{\tau}) := p_E(\bar{\tau}) \sum_{i=1}^{2} \frac{\kappa_i^Q}{\eta_i^Q}\left\{\frac{ex_i^{app}}{c(\overline{N}_i^Q(\bar{\tau}))} + \varepsilon\left(-\ln c(\overline{N}_i^Q(0)) + \overline{N}_i(\bar{\tau})\right)\right\},
$$

And the inter-temporal marginal energy costs, which are caused by the use of non-renewable natural resources for the production of the consumption good

$$
\begin{aligned}
&ITMCE^{Q}_{\underset{N=1+2}{Q}}(\bar{\tau} \to T)\\
&= \sum_{\tau^f=\bar{\tau}+1}^{T}\left\{p_E(\tau^f)\, Y^Q(\tau^f)\left\{\sum_{i=1}^{2} \varphi_i^Q \kappa_i^Q\left\{\frac{\kappa_i^Q\, ex_i^{app}}{\eta_i^Q c(\overline{N}_i^Q(\tau^f))} + \varepsilon\,\overline{N}_i(\tau^f)\right\}\right\}\right\}.
\end{aligned}
$$

A.2.4 Marginal Energy Costs for the Raw Materials Extraction Processes Required to Produce the Capital Good Amounts

The actual marginal energy costs of using the non-renewable natural resources for sustaining and developing the capital stock for the production of one unit of the consumption good are outlined below.

$$MCE^Q_{\substack{K\\ N\mu=1+2}}(\bar{\tau}) := p_E(\bar{\tau})k^Q\left\{d^Q+\gamma^Q_{\bar{\tau}}\right\}\sum_{\mu=1}^{2}\frac{\kappa^K_\mu}{\eta^K_\mu}\left\{\begin{array}{l}\frac{ex^{app}_\mu}{c(\overline{N}^K_\mu(\bar{\tau}))}\\ +\varepsilon\left(-\ \ln c(\overline{N}^K_\mu(0))\ +\overline{N}^K_\mu(\bar{\tau})\right)\end{array}\right\}$$

$$= MCE^Q_{\substack{K_{dep}\\ N\mu=1+2}}(\bar{\tau}) + MCE^Q_{\substack{K_{dev}\\ N\mu=1+2}}(\bar{\tau})\ with$$

$$MCE^Q_{\substack{K_{dep}\\ N\mu=1+2}}(\bar{\tau}) := p_E(\bar{\tau})k^Q d^Q\sum_{\mu=1}^{2}\frac{\kappa^K_\mu}{\eta^K_\mu}\left\{\frac{ex^{app}_\mu}{c(\overline{N}^K_\mu(\bar{\tau}))}+\varepsilon\left(-\ \ln c(\overline{N}^K_\mu(0))\ +\overline{N}^K_\mu(\bar{\tau})\right)\right\}$$

$$MCE^Q_{\substack{K_{dev}\\ N\mu=1+2}}(\bar{\tau}) := p_E(\bar{\tau})k^Q \gamma^Q_{\bar{\tau}}\sum_{\mu=1}^{2}\frac{\kappa^K_\mu}{\eta^K_\mu}\left\{\frac{ex^{app}_\mu}{c(\overline{N}^K_\mu(\bar{\tau}))}+\varepsilon\left(-\ \ln c(\overline{N}^K_\mu(0))\ +\overline{N}^K_\mu(\bar{\tau})\right)\right\},$$

Inter-temporal marginal energy costs are also generated for sustaining and developing the capital stock that is used for the future development path of consumption goods, which are the arguments of the inter-temporal welfare function with its exogenously given time preferences

$$ITMCE^Q_{\substack{K\\ N_{\mu=1+2}}}(\bar{\tau}\rightarrow T)$$

$$= \sum_{\tau^f=\bar{\tau}+1}^{T} p_E(\tau^f)k^Q Y^Q(\tau^f)\left\{d^Q+\gamma^Q_{\tau^f}\right\}\left\{\sum_{\mu=1}^{2}\varphi^K_\mu\kappa^K_\mu\frac{\kappa^K_\mu}{\eta^K_\mu}\left\{\frac{ex^{app}_\mu}{c(\overline{N}^K_\mu(\tau^f)}+\varepsilon\right\}\right\}$$

$$= ITMCE^Q_{\substack{K_{dep}\\ N_{\mu=1+2}}}(\bar{\tau}\rightarrow T) + ITMCE^Q_{\substack{K_{dev}\\ N_{\mu=1+2}}}(\bar{\tau}\rightarrow T)\ with$$

$$ITMCE^Q_{\substack{K_{dep}\\ N_{\mu=1+2}}}(\bar{\tau}\rightarrow T)$$

$$= k^Q d^Q\sum_{\tau^f=\bar{\tau}+1}^{T} p_E(\tau^f)Y^Q(\tau^f)\left\{\sum_{\mu=1}^{2}\varphi^K_\mu\kappa^K_\mu\frac{\kappa^K_\mu}{\eta^K_\mu}\left\{\frac{ex^{app}_\mu}{c(\overline{N}^K_\mu(\tau^f)}+\varepsilon\right\}\right\}$$

$$ITMCE^Q_{\substack{K_{dev}\\ N_{\mu=1+2}}}(\bar{\tau}\rightarrow T)$$

$$= k^Q\sum_{\tau^f=\bar{\tau}+1}^{T} p_E(\tau^f)Y^Q(\tau^f)\gamma^Q_{\tau^f}\left\{\sum_{\mu=1}^{2}\varphi^K_\mu\kappa^K_\mu\frac{\kappa^K_\mu}{\eta^K_\mu}\left\{\frac{ex^{app}_\mu}{c(\overline{N}^K_\mu(\tau^f)}+\varepsilon\right\}\right\}.$$

In economic theory, if typical short-term positive or negative externalities exist, optimal solutions can only be achieved by governmental market interventions and regulations in the form of suitable environmental instruments. If inter-temporal externalities occur, the need for governmental interventions is more complex and more important in order to balance short-term welfare interests with the interests of future generations. As a result, the optimal shadow prices for energy generated from non-renewable resources can not be arranged by short-term market processes, but by political and socio-economical processes within a society that is taking care of the interests of the next generations.

6.3.1.3 Summary of the Marginal Costs for the Production Sector

The different marginal cost components of the production sector can be summarized using the matrix that follows. The first rows show the marginal costs of the consumption good process. The second rows show the marginal costs for extracting the raw materials $N^Q_{i=1+2}$. The raw materials are combined into one unit of consumption good by using the capital stock. The raw materials $N^K_{i=1+2}$ must be converted to capital goods to sustain and to develop the capital stock in accordance with the development path of the consumption goods in the inter-temporal welfare function.

process $\hat{g}$	actual m. lab. costs	intertemp. lab costs $(\bar{\tau} \to T)$	actual m. energy costs	intertemporal m. energy costs $(\bar{\tau} \to T)$
$\begin{matrix} Q \\ Q \end{matrix}$	$MCL^Q_Q(\bar{\tau})$	$ITCL^Q_K$	$MCE^Q_Q(\bar{\tau})$	
$\begin{matrix} Q \\ N_{i=1+2} \end{matrix}$	$MCL^Q_{N_{i=1+2}}(\bar{\tau})$	$ITCL^Q_{N_{i=1+2}}$	$MCE^Q_{N_{i=1+2}}(\bar{\tau})$	$ITCE^Q_{N_{i=1+2}}$
$\begin{matrix} Q \\ K \end{matrix}$	$MCL^Q_K(\bar{\tau})$		$MCE^Q_K(\bar{\tau})$	
$\begin{matrix} Q \\ K_{dep} \\ N_{\mu=1+2} \end{matrix}$	$MCL^Q_{\substack{K_{dep} \\ N_{\mu=1+2}}}(\bar{\tau})$	$ITCE^Q_{\substack{K_{dep} \\ N_{\mu=1+2}}}$	$MCE^Q_{\substack{K_{dep} \\ N_{\mu=1+2}}}(\bar{\tau})$	$ITCE^Q_{\substack{K_{dep} \\ N_{\mu=1+2}}}$
$\begin{matrix} Q \\ K_{dev} \\ N_{\mu=1+2} \end{matrix}$	$MCL^Q_{\substack{K_{dev} \\ N_{\mu=1+2}}}(\bar{\tau})$	$ITCL^Q_{\substack{K_{dep} \\ N_{\mu=1+2}}}$	$MCE^Q_{\substack{K_{dev} \\ N_{\mu=1+2}}}(\bar{\tau})$	$ITCE^Q_{\substack{K_{dev} \\ N_{\mu=1+2}}}$
TMC^p_{sec}	$\sum_{\hat{g}} MCL^p_{\hat{g}}(\bar{\tau})$	$\sum_{\hat{g}} ITCL^p_{\hat{g}}$	$\sum_{\hat{g}} MCE^p_{\hat{g}}(\bar{\tau})$	$\sum_{\hat{g}} ITCE^p_{\hat{g}}$

6.3.2 *Non-profit Conditions for the Water and Wastewater Sectors*

In order to obtain the non-profit conditions for the water and wastewater sector, we differentiate the following LAGRANGE-function by the variable $Y^Q(\bar{\tau})$ for a selected time interval $\bar{\tau} \in (1,\dots,T)$.

6.3.2.1 Marginal Human Labour and Energy Costs for Water Production

B.1 Marginal Human Labour Costs for Water Production

Under non-profit conditions, the marginal costs of the human labour input for the water production that is needed to produce one unit of consumption good are equal to

$$MCL_W^Q(\bar{\tau} \mapsto T) = \frac{\vartheta}{\vartheta Y^Q(\bar{\tau})} \sum_{\tau=1}^{T} p_L(\tau) l^{WT} Y^Q(\tau) \left\{ \begin{array}{l} \left\{ \psi_Q^{RP} + \sum\limits_{i=1}^{2} \frac{\tilde{\psi}_i^Q \kappa_i^Q}{c(\overline{N}_i^Q(0))} e^{\varphi_i^Q \kappa_i^Q \sum\limits_{t=1}^{\tau-1} Y^Q(\tau)} \right\} \\ + k^Q \left\{ d + \gamma^\tau \right\} \sum\limits_{\mu=1}^{2} \frac{\tilde{\psi}_\mu^K \kappa_\mu^K}{c(\overline{N}_\mu^K(0))} e^{\varphi_\mu^K \kappa_\mu^K \sum\limits_{t=1}^{\tau-1} k^Q Y^Q(t)\{d+\gamma^t\})} \end{array} \right\}.$$

The following outlines the total marginal human labour cost for water production.

B.1.1 Total Actual Marginal Human Labour Costs for Water Production

$$MCL_W^Q(\bar{\tau}) = p_L(\bar{\tau}) l^{WT} \left\{ \left\{ \psi_Q^{RP} + \sum_{i=1}^{2} \frac{\tilde{\psi}_i^Q \kappa_i^Q}{c(\overline{N}_i^Q(\bar{\tau}))} \right\} + k^Q \left\{ d + \gamma^{\bar{\tau}} \right\} \sum_{\mu=1}^{2} \frac{\tilde{\psi}_\mu^K \kappa_\mu^K}{c(\overline{N}_\mu^K(\bar{\tau}))} \right\} =$$

$$MCL_W^Q(\bar{\tau}) = MCL_{\underset{Qrep}{W}}^Q(\bar{\tau}) + MCL_{\underset{N_{i=1+2}^Q}{W}}^Q(\bar{\tau}) + MCL_{\underset{N_{\mu=1+2}^K}{W}}^Q(\bar{\tau}).$$

$$\begin{aligned} & MCL_W^Q(\bar{\tau} \mapsto T) \\ & = \frac{\vartheta}{\vartheta Y^Q(\bar{\tau})} \sum_{\tau=1}^{T} p_L(\tau) l^{WT} Y^Q(\tau) \psi_Q^{RP} \\ & + \frac{\vartheta}{\vartheta Y^Q(\bar{\tau})} \sum_{\tau=1}^{T} p_L(\tau) l^{WT} Y^Q(\tau) \sum_{i=1}^{2} \frac{\tilde{\psi}_i^Q \kappa_i^Q}{c(\overline{N}_i^Q(0))} e^{\varphi_i^Q \kappa_i^Q \sum\limits_{t=1}^{\tau-1} Y^Q(\tau)} \\ & + \frac{\vartheta}{\vartheta Y^Q(\bar{\tau})} \sum_{\tau=1}^{T} p_L(\tau) l^{WT} Y^Q(\tau) k^Q \left\{ d + \gamma^\tau \right\} \sum_{\mu=1}^{2} \frac{\tilde{\psi}_\mu^K \kappa_\mu^K}{c(\overline{N}_\mu^K(0))} e^{\varphi_\mu^K \kappa_\mu^K \sum\limits_{t=1}^{\tau-1} k^Q Y^Q(t)\{d+\gamma^t\})} . \end{aligned}$$

Three terms make up the total, actual marginal human labour costs:

B.1.1.1 Actual Marginal Human Labour Costs of Water Used for the Reproduction Activities

$$MCL_{\underset{Qrep}{W}}^Q(\bar{\tau}) = p_L(\bar{\tau}) l^{WT} \psi_Q^{RP},$$

B.1.1.2 Actual Marginal Human Labour Costs of Water Used for Extracting the Raw Materials for the Consumption Good Process

$$MCL^{Q}_{W\;N^{Q}_{i=1+2}}(\bar{\tau}) = p_L(\bar{\tau})\, l^{WT} \sum_{i=1}^{2} \frac{\tilde{\psi}^{Q}_{i}\, \kappa^{Q}_{i}}{c(\overline{N}^{Q}_{i}(\bar{\tau}))},$$

B.1.1.3 Actual Marginal Human Labour Costs of Water for Extracting the Raw Materials to Sustain and Develop the Capital Good Stock

$$\begin{aligned}
MCL^{Q}_{W\;N^{K}_{\mu=1+2}}(\bar{\tau}) &= p_L(\bar{\tau})\, l^{WT} k^{Q} \left\{ d + \gamma^{\bar{\tau}} \right\} \sum_{\mu=1}^{2} \frac{\tilde{\psi}^{K}_{\mu}\, \kappa^{K}_{\mu}}{c(\overline{N}^{K}_{\mu}(\bar{\tau}))} \\
&= MCL^{Q}_{W\;N^{K_{dep}}_{\mu=1+2}}(\bar{\tau}) + MCL^{Q}_{W\;N^{K_{dev}}_{\mu=1+2}}(\bar{\tau}) \text{ with} \\
MCL^{Q}_{W\;N^{K_{dep}}_{\mu=1+2}}(\bar{\tau}) &= p_L(\bar{\tau})\, l^{WT} k^{Q} d^{Q} \sum_{\mu=1}^{2} \frac{\tilde{\psi}^{K}_{\mu}\, \kappa^{K}_{\mu}}{c(\overline{N}^{K}_{\mu}(\bar{\tau}))} \\
MCL^{Q}_{W\;N^{K_{dev}}_{\mu=1+2}}(\bar{\tau}) &= p_L(\bar{\tau})\, l^{WT} k^{Q} \gamma^{\bar{\tau}} \sum_{\mu=1}^{2} \frac{\tilde{\psi}^{K}_{\mu}\, \kappa^{K}_{\mu}}{c(\overline{N}^{K}_{\mu}(\bar{\tau}))}.
\end{aligned}$$

The coefficient for water used during reproduction activities is time-invariant, and therefore inter-temporal marginal human labour costs for water production only occur when non-renewable natural resources are used to produce the consumption goods and the required capital goods.

B.1.2 Total Inter-Temporal Marginal Labour Costs for Water Production

$$\begin{aligned}
&ITMCL^{Q}_{W}(\bar{\tau} \to T) \\
&= \frac{\vartheta}{\vartheta Y^{Q}(\bar{\tau})} \sum_{\tau=1}^{T} p_L(\tau) l^{WT}\, Y^{Q}(\tau) \sum_{i=1}^{2} \frac{\tilde{\psi}^{Q}_{i}\, \kappa^{Q}_{i}}{c(\overline{N}^{Q}_{i}(0))}\, e^{\varphi^{Q}_{i} \kappa^{Q}_{i} \sum_{t=1}^{\tau-1} Y^{Q}(\tau)} \\
&+ \frac{\vartheta}{\vartheta Y^{Q}(\bar{\tau})} \sum_{\tau=1}^{T} p_L(\tau) l^{W}\, Y^{Q}(\tau) \sum_{\mu=1}^{2} \frac{\tilde{\psi}^{K}_{\mu}\, \kappa^{K}_{\mu}}{c(\overline{N}^{K}_{\mu}(0))}\, e^{\varphi^{K}_{\mu} \kappa^{K}_{\mu} \sum_{t=1}^{\tau-1} k^{Q} Y^{Q}(t) \{ d + \gamma^{t} \})} \\
&= ITMCL^{Q}_{W\;N^{Q}_{i=1+2}}(\bar{\tau} \to T) + ITMCL^{Q}_{W\;N^{K}_{\mu=1+2}}(\bar{\tau} \to T).
\end{aligned}$$

B.1.2.1 Inter-Temporal Marginal Labour Costs for Water Production Caused by the Non-renewable Resources for the Consumption Process

$$ITMCL^{Q}_{W \atop N^{Q}_{i=1+2}}(\bar{\tau} \rightarrow T) = \sum_{\tau^f=\bar{\tau}+1}^{T} p_L(\tau^f)\, l^W\, Y^Q(\tau^f) \sum_{i=1}^{2} \frac{\tilde{\psi}^Q_i \kappa^Q_i \varphi^Q_i \kappa^Q_i}{c(\overline{N}^Q_i(\tau^f))},$$

B.1.2.2 Inter-Temporal Marginal Labour Costs for Water Production Required to Extract the Raw Material to Sustain and Develop the Capital Stock

$$\begin{aligned}
&ITMCL^{Q}_{W \atop N^{K}_{\mu=1+2}}(\bar{\tau} \rightarrow T)\\
&= k^Q \left\{ d^Q + \gamma^Q_{\bar{\tau}} \right\} \sum_{\tau^f=\bar{\tau}+1}^{T} p_L(\tau^f) l^W Y^Q(\tau^f) k^Q \left\{ d^Q + \gamma^Q_{\tau^f} \right\} \left\{ \sum_{\mu=1}^{2} \frac{\tilde{\psi}^K_\mu \kappa^K_\mu \varphi^K_\mu \kappa^K_\mu}{c(\overline{N}^K_\mu(\tau^f))} \right\}\\
&= ITMCL^{Q}_{W \atop N^{K_{dep}}_{\mu=1+2}}(\bar{\tau} \rightarrow T) + ITMCL^{Q}_{W \atop N^{K_{dev}}_{\mu=1+2}}(\bar{\tau} \rightarrow T),
\end{aligned}$$

with

$$\begin{aligned}
&ITMCL^{Q}_{W \atop N^{K_{dep}}_{\mu=1+2}}(\bar{\tau} \rightarrow T)\\
&\quad = k^Q d^Q \sum_{\tau^f=\bar{\tau}+1}^{T} p_L(\tau^f)\, l^W Y^Q(\tau^f) k^Q \left\{ d^Q + \gamma^Q_{\tau^f} \right\} \left\{ \sum_{\mu=1}^{2} \frac{\tilde{\psi}^K_\mu \kappa^K_\mu \varphi^K_\mu \kappa^K_\mu}{c(\overline{N}^K_\mu(\tau^f))} \right\}\\
&ITMCL^{Q}_{W \atop N^{K_{dev}}_{\mu=1+2}}(\bar{\tau} \rightarrow T)\\
&\quad = k^Q \gamma^Q_{\bar{\tau}} \sum_{\tau^f=\bar{\tau}+1}^{T} p_L(\tau^f)\, l^W Y^Q(\tau^f) k^Q \left\{ d^Q + \gamma^Q_{\tau^f} \right\} \left\{ \sum_{\mu=1}^{2} \frac{\tilde{\psi}^K_\mu \kappa^K_\mu \varphi^K_\mu \kappa^K_\mu}{c(\overline{N}^K_\mu(\tau^f))} \right\}.
\end{aligned}$$

The use of non-renewable natural resources leads not only to inter-temporal marginal human labour costs in the production sector, but also to inter-temporal marginal human labour costs in the water production sector. The decreases in concentrations of these natural resources in the natural environment are combined with an increase in the water amounts, which are required to transport the residuals from the production and reproduction activities to the wastewater sector.

B.2 Marginal Energy Costs for Water Production

The marginal energy costs for water production are easily derived by replacing the human labour input with an energy input coefficient and by using a shadow price for energy. The water amounts determined using consumption good amounts are the same for human labour and energy inputs.

When only the non-profit conditions are considered, the marginal energy costs for the water production process that is needed to produce one unit of the consumption good are equal to

$$MCL_E^Q(\bar{\tau} \mapsto T)$$

$$= \frac{\vartheta}{\vartheta Y^Q(\bar{\tau})} \sum_{\tau=1}^{T} p_E(\tau)\ e^W\ Y^Q(\tau) \left\{ \begin{array}{l} \left\{ \psi_Q^{RP} + \sum_{i=1}^{2} \frac{\tilde{\psi}_i^Q \kappa_i^Q}{c(\bar{N}_i^Q(0))} e^{\varphi_i^Q \kappa_i^Q \sum_{t=1}^{\tau-1} Y^Q(\tau)} \right\} \\ + k^Q \{d + \gamma^\tau\} \sum_{\mu=1}^{2} \frac{\tilde{\psi}_\mu^K \kappa_\mu^K}{c(\bar{N}_\mu^K(0))} e^{\varphi_\mu^K \kappa_\mu^K \sum_{t=1}^{\tau-1} k^Q Y^Q(t) \{d+\gamma^t\})} \end{array} \right\}$$

$$MCL_E^Q(\bar{\tau} \mapsto T)$$

$$= \frac{\vartheta}{\vartheta Y^Q(\bar{\tau})} \sum_{\tau=1}^{T} p_E(\tau)\ e^W\ Y^Q(\tau)\ \psi_Q^{RP}$$

$$+ \frac{\vartheta}{\vartheta Y^Q(\bar{\tau})} \sum_{\tau=1}^{T} p_E(\tau)\ e^W\ Y^Q(\tau) \sum_{i=1}^{2} \frac{\tilde{\psi}_i^Q \kappa_i^Q}{c(\bar{N}_i^Q(0))} e^{\varphi_i^Q \kappa_i^Q \sum_{t=1}^{\tau-1} Y^Q(\tau)} +$$

$$+ \frac{\vartheta}{\vartheta Y^Q(\bar{\tau})} \sum_{\tau=1}^{T} p_E(\tau)\ e^W\ Y^Q(\tau) k^Q\ \{d + \gamma^\tau\} \sum_{\mu=1}^{2} \frac{\tilde{\psi}_\mu^K \kappa_\mu^K}{c(\bar{N}_\mu^K(0))} e^{\varphi_\mu^K \kappa_\mu^K \sum_{t=1}^{\tau-1} k^Q Y^Q(t) \{d+\gamma^t\})},$$

Where the abbreviation for the energy input coefficient for water production is

$$e^W := \left\{ \tilde{\varepsilon}^W \sum_{\omega} \ln \frac{c_\omega^*(0)}{c_\omega(0)} \right\}.$$

B.2.1 Total Actual Marginal Energy Costs for Water Production

$$MCE_W^Q(\bar{\tau}) = p_E(\bar{\tau}) e^W \left\{ \left\{ \psi_Q^{RP} + \sum_{i=1}^{2} \frac{\tilde{\psi}_i^Q \kappa_i^Q}{c(\bar{N}_i^Q(\bar{\tau}))} \right\} + k^Q \left\{ d + \gamma^{\bar{\tau}} \right\} \sum_{\mu=1}^{2} \frac{\tilde{\psi}_\mu^K \kappa_\mu^K}{c(\bar{N}_\mu^K(\bar{\tau}))} \right\}$$

$$= MCE_W^Q(\bar{\tau}) = MCE_{W_{Q_{rep}}}^Q(\bar{\tau}) + MCE_{W_{N_{i=1+2}^Q}}^Q(\bar{\tau}) + MCE_{W_{N_{\mu=1+2}^K}}^Q(\bar{\tau}).$$

B.2.1.1 Actual Marginal Energy Costs of Water Used for Reproduction Activities

$$MCE^{Q}_{W_{Q_{rep}}}(\bar{\tau}) = p_E(\bar{\tau})\, e^W \psi^{RP}_{Q}.$$

B.2.1.2 Actual Marginal Energy Costs of Water for Extracting the Raw Materials Required to Produce the Consumption Good Amounts

$$MCE^{Q}_{W_{N^Q_{i=1+2}}}(\bar{\tau}) = p_E(\bar{\tau})\, e^W \sum_{i=1}^{2} \frac{\tilde{\psi}^Q_i\, \kappa^Q_i}{c(\overline{N}^Q_i(\bar{\tau}))}.$$

B.2.1.3 Actual Marginal Energy Costs of Water for Extracting the Raw Materials to Sustain and to Develop the Capital Stock

$$\begin{aligned}
MCLE^{Q}_{W_{N^K_{\mu=1+2}}}(\bar{\tau}) &= p_E(\bar{\tau})\, e^W k^Q \left\{ d + \gamma^{\bar{\tau}} \right\} \sum_{\mu=1}^{2} \frac{\tilde{\psi}^K_\mu\, \kappa^K_\mu}{c(\overline{N}^K_\mu(\bar{\tau}))} \\
&= MCE^{Q}_{W_{N^{K_{dep}}_{\mu=1+2}}}(\bar{\tau}) + MCE^{Q}_{W_{N^{K_{dev}}_{\mu=1+2}}}(\bar{\tau}) \text{ with}
\end{aligned}$$

$$MCE^{Q}_{W_{N^{K_{dep}}_{\mu=1+2}}}(\bar{\tau}) = p_E(\bar{\tau})\, e^W k^Q d^Q \sum_{\mu=1}^{2} \frac{\tilde{\psi}^K_\mu\, \kappa^K_\mu}{c(\overline{N}^K_\mu(\bar{\tau}))}$$

$$MCE^{Q}_{W_{N^{K_{dev}}_{\mu=1+2}}}(\bar{\tau}) = p_E(\bar{\tau})\, e^W k^Q \gamma^{\bar{\tau}} \sum_{\mu=1}^{2} \frac{\tilde{\psi}^K_\mu\, \kappa^K_\mu}{c(\overline{N}^K_\mu(\bar{\tau}))}.$$

B.2.2 Total Inter-Temporal Energy Costs for Water Production

$$\begin{aligned}
&ITMCE^{Q}_{W}(\bar{\tau} \to T) \\
&= \frac{\vartheta}{\vartheta Y^Q(\bar{\tau})} \sum_{\tau=1}^{T} p_E(\tau)\, e^W\, Y^Q(\tau) \sum_{i=1}^{2} \frac{\tilde{\psi}^Q_i\, \kappa^Q_i}{c(\overline{N}^Q_i(0))}\, e^{\varphi^Q_i \kappa^Q_i \sum_{t=1}^{\tau-1} Y^Q(\tau)} \\
&\quad + \frac{\vartheta}{\vartheta Y^Q(\bar{\tau})} \sum_{\tau=1}^{T} p_E(\tau)\, e^W\, Y^Q(\tau) \sum_{\mu=1}^{2} \frac{\tilde{\psi}^K_\mu\, \kappa^K_\mu}{c(\overline{N}^K_\mu(0))}\, e^{\varphi^K_\mu \kappa^K_\mu \sum_{t=1}^{\tau-1} k^Q Y^Q(t) \{d+\gamma^t\})} \\
&= ITMCE^{Q}_{W_{N^Q_{i=1+2}}}(\bar{\tau} \to T) + ITMCE^{Q}_{W_{N^K_{\mu=1+2}}}(\bar{\tau} \to T).
\end{aligned}$$

B.2.2.1 Inter-Temporal Marginal Energy Costs for Water Production Caused by the Non-renewable Resources for the Consumption Process

$$ITMCE^{Q}_{\substack{W \\ N^{Q}_{i=1+2}}}(\bar{\tau} \rightarrow T) = \sum_{\tau^{f}=\bar{\tau}+1}^{T} p_{E}(\tau^{f})\, e^{W}\, Y^{Q}(\tau^{f}) \sum_{i=1}^{2} \frac{\tilde{\psi}^{Q}_{i} \kappa^{Q}_{i}\, \varphi^{Q}_{i}\, \kappa^{Q}_{i}}{c(\overline{N}^{Q}_{i}(\tau^{f})}.$$

B.2.2.2 Inter-Temporal Marginal Labour for Water Production Caused by the Natural Resources for Sustaining and Developing the Capital Stock

$$\begin{aligned} &ITMCE^{Q}_{\substack{W \\ N^{K}_{\mu=1+2}}}(\bar{\tau} \rightarrow T) \\ &= k^{Q}\left\{d^{Q} + \gamma^{Q}_{\bar{\tau}}\right\} \sum_{\tau^{f}=\bar{\tau}+1}^{T} p_{E}(\tau^{f}) e^{W} Y^{Q}(\tau^{f}) k^{Q}\left\{d^{Q} + \gamma^{Q}_{\tau^{f}}\right\} \left\{\sum_{\mu=1}^{2} \frac{\tilde{\psi}^{K}_{\mu} \kappa^{K}_{\mu} \varphi^{K}_{\mu} \kappa^{K}_{\mu}}{c(\overline{N}^{K}_{\mu}(\tau^{f})}\right\} \\ &= ITMCE^{Q}_{\substack{W \\ N^{K_{dep}}_{\mu=1+2}}}(\bar{\tau} \rightarrow T) + ITMCE^{Q}_{\substack{W \\ N^{K_{dev}}_{\mu=1+2}}}(\bar{\tau} \rightarrow T). \end{aligned}$$

with

$$\begin{aligned} &ITMCE^{Q}_{\substack{W \\ N^{K_{dep}}_{\mu=1+2}}}(\bar{\tau} \rightarrow T) \\ &= k^{Q} d^{Q} \sum_{\tau^{f}=\bar{\tau}+1}^{T} p_{E}(\tau^{f}) e^{W} Y^{Q}(\tau^{f}) k^{Q}\left\{d^{Q} + \gamma^{Q}_{\tau^{f}}\right\} \left\{\sum_{\mu=1}^{2} \frac{\tilde{\psi}^{K}_{\mu} \kappa^{K}_{\mu} \varphi^{K}_{\mu} \kappa^{K}_{\mu}}{c(\overline{N}^{K}_{\mu}(\tau^{f})}\right\} \\ &ITMCE^{Q}_{\substack{W \\ N^{K_{dev}}_{\mu=1+2}}}(\bar{\tau} \rightarrow T) \\ &= k^{Q} \gamma^{Q}_{\bar{\tau}} \sum_{\tau^{f}=\bar{\tau}+1}^{T} p_{E}(\tau^{f}) e^{W} Y^{Q}(\tau^{f}) k^{Q}\left\{d^{Q} + \gamma^{Q}_{\tau^{f}}\right\} \left\{\sum_{\mu=1}^{2} \frac{\tilde{\psi}^{K}_{\mu} \kappa^{K}_{\mu} \varphi^{K}_{\mu} \kappa^{K}_{\mu}}{c(\overline{N}^{K}_{\mu}(\tau^{f})}\right\}. \end{aligned}$$

The use of non-renewable natural resources leads to inter-temporal marginal energy costs in the production sector, and in the water production sector. The decrease in concentrations of these natural resources in the natural environment is combined with an increase in the water amounts, which are required to transport the residuals of the production and reproduction activities to the wastewater sector.

6.3.2.2 Marginal Human Labour and Energy Costs for Wastewater Treatment

C.1 Marginal Human Labour Costs for Wastewater Treatment

When only the non-profit conditions are considered, the total marginal human labour costs for wastewater treatment caused by the reproduction and production activities are equal to

$$MCL^{Q}_{WWT}(\bar{\tau} \mapsto T)$$
$$p_L(\tau) l^{WWT}\, Y^Q(\tau)^*$$
$$= \frac{\vartheta}{\vartheta Y^Q(\bar{\tau})} \sum_{\tau=1}^{T} \ldots^* \left\{ \begin{array}{l} \left\{ \psi^{RP}_Q + \sum_{i=1}^{2} \frac{\tilde{\psi}^Q_i \kappa^Q_i}{c(\overline{N}^Q_i(0))} e^{\varphi^Q_i \kappa^Q_i \sum_{t=1}^{\tau-1} Y^Q(\tau)} \right\} \\ + k^Q \{d + \gamma^\tau\} \sum_{\mu=1}^{2} \frac{\tilde{\psi}^K_\mu \kappa^K_\mu}{c(\overline{N}^K_\mu(0))} e^{\varphi^K_\mu \kappa^K_\mu \sum_{t=1}^{\tau-1} k^Q Y^Q(t) \{d+\gamma^t\})} \end{array} \right\}$$

$$= \frac{\vartheta}{\vartheta Y^Q(\bar{\tau})} \sum_{\tau=1}^{T} p_L(\tau) l^{WWT}\, Y^Q(\tau)\, \psi^{RP}_Q +$$

$$+ \frac{\vartheta}{\vartheta Y^Q(\bar{\tau})} \sum_{\tau=1}^{T} p_L(\tau) l^{WWT}\, Y^Q(\tau) \sum_{i=1}^{2} \frac{\tilde{\psi}^Q_i \kappa^Q_i}{c(\overline{N}^Q_i(0))} e^{\varphi^Q_i \kappa^Q_i \sum_{t=1}^{\tau-1} Y^Q(\tau)}$$

$$+ \frac{\vartheta}{\vartheta Y^Q(\bar{\tau})} \sum_{\tau=1}^{T} p_L(\tau) l^{WWT}\, Y^Q(\tau) k^Q \{d + \gamma^\tau\} \sum_{\mu=1}^{2} \frac{\tilde{\psi}^K_\mu \kappa^K_\mu}{c(\overline{N}^K_\mu(0))} e^{\varphi^K_\mu \kappa^K_\mu \sum_{t=1}^{\tau-1} k^Q Y^Q(t) \{d+\gamma^t\})}.$$

C.1.1 Total Actual Marginal Human Labour Costs for Wastewater Treatment

$$MCL^{Q}_{WWT}(\bar{\tau})$$
$$= p_L(\bar{\tau})\, l^{WWT} \left\{ \left\{ \psi^{RP}_Q + \sum_{i=1}^{2} \frac{\tilde{\psi}^Q_i \kappa^Q_i}{c(\overline{N}^Q_i(\bar{\tau}))} \right\} + k^Q \{d + \gamma^{\bar{\tau}}\} \sum_{\mu=1}^{2} \frac{\tilde{\psi}^K_\mu \kappa^K_\mu}{c(\overline{N}^K_\mu(\bar{\tau}))} \right\}$$
$$= MCL^{Q}_{WWT}(\bar{\tau}) = MCL^{Q}_{\substack{WWT \\ Qrep}}(\bar{\tau}) + MCL^{Q}_{\substack{WWT \\ N^Q_{i=1+2}}}(\bar{\tau}) + MCL^{Q}_{\substack{WWT \\ N^K_{\mu=1+2}}}(\bar{\tau}).$$

C.1.1.1 Actual Marginal Human Labour Costs of Wastewater Treatment Caused by the Reproduction Activities

$$MCL^{Q}_{\substack{WWT \\ Qrep}}(\bar{\tau}) = p_L(\bar{\tau})\, l^{WWT} \psi^{RP}_Q.$$

C.1.1.2 Actual Marginal Human Labour Costs for Wastewater Treatment Caused by the Raw Materials for the Consumption Good Process

$$MCL^{Q}_{\substack{WWT\\N^{Q}_{i=1+2}}}(\bar{\tau}) = p_L(\bar{\tau})\, l^{WWT} \sum_{i=1}^{2} \frac{\tilde{\psi}^{Q}_{i}\, \kappa^{Q}_{i}}{c(\overline{N}^{Q}_{i}(\bar{\tau}))}.$$

C.1.1.3 Actual Marginal Human Labour Costs for Wastewater Treatment Caused by the Raw Materials Needed to Sustain and Develop the Capital Good Stock

$$\begin{aligned}
MCL^{Q}_{\substack{WWT\\N^{K}_{\mu=1+2}}}(\bar{\tau}) &= p_L(\bar{\tau})\, l^{WWT} k^{Q} \left\{ d + \gamma^{\bar{\tau}} \right\} \sum_{\mu=1}^{2} \frac{\tilde{\psi}^{K}_{\mu}\, \kappa^{K}_{\mu}}{c(\overline{N}^{K}_{\mu}(\bar{\tau}))} \\
&= MCL^{Q}_{\substack{W\\N^{K_{dep}}_{\mu=1+2}}}(\bar{\tau}) + MCL^{Q}_{\substack{W\\N^{K_{dev}}_{\mu=1+2}}}(\bar{\tau}) \text{ with} \\
MCL^{Q}_{\substack{WWT\\N^{K_{dep}}_{\mu=1+2}}}(\bar{\tau}) &= p_L(\bar{\tau})\, l^{WWT} k^{Q}\, d^{Q} \sum_{\mu=1}^{2} \frac{\tilde{\psi}^{K}_{\mu}\, \kappa^{K}_{\mu}}{c(\overline{N}^{K}_{\mu}(\bar{\tau}))} \\
MCL^{Q}_{\substack{WWT\\N^{K_{dev}}_{\mu=1+2}}}(\bar{\tau}) &= p_L(\bar{\tau})\, l^{WWT} k^{Q}\, \gamma^{\bar{\tau}} \sum_{\mu=1}^{2} \frac{\tilde{\psi}^{K}_{\mu}\, \kappa^{K}_{\mu}}{c(\overline{N}^{K}_{\mu}(\bar{\tau}))}.
\end{aligned}$$

C.1.2 Total Inter-Temporal Marginal Labour Costs for Wastewater Treatment

$$\begin{aligned}
&ITMCL^{Q}_{WWT}(\bar{\tau} \to T) \\
&= \frac{\vartheta}{\vartheta Y^{Q}(\bar{\tau})} \sum_{\tau=1}^{T} p_L(\tau) l^{WWT}\, Y^{Q}(\tau) \sum_{i=1}^{2} \frac{\tilde{\psi}^{Q}_{i}\, \kappa^{Q}_{i}}{c(\overline{N}^{Q}_{i}(0))}\, e^{\varphi^{Q}_{i} \kappa^{Q}_{i} \sum_{t=1}^{\tau-1} Y^{Q}(\tau)} + \\
&+ \frac{\vartheta}{\vartheta Y^{Q}(\bar{\tau})} \sum_{\tau=1}^{T} p_L(\tau) l^{WWT}\, Y^{Q}(\tau) \sum_{\mu=1}^{2} \frac{\tilde{\psi}^{K}_{\mu}\, \kappa^{K}_{\mu}}{c(\overline{N}^{K}_{\mu}(0))}\, e^{\varphi^{K}_{\mu} \kappa^{K}_{\mu} \sum_{t=1}^{\tau-1} k^{Q} Y^{Q}(t) \{d+\gamma^{t}\})} \\
&= ITMCL^{Q}_{\substack{WWT\\N^{Q}_{i=1+2}}}(\bar{\tau} \to T) + ITMCL^{Q}_{\substack{WWT\\N^{K}_{\mu i=1+2}}}(\bar{\tau} \to T).
\end{aligned}$$

C.1.2.1 Inter-Temporal Marginal Labour Costs for Wastewater Treatment Caused by the Non-renewable Resources for the Consumption Process

$$\underset{N_{i=1+2}^{Q}}{ITMCL_{WWT}^{Q}}\ (\bar{\tau} \rightarrow T) = \sum_{\tau^f=\bar{\tau}+1}^{T} p_L(\tau^f)\, l^{WWT}\, Y^Q(\tau^f) \sum_{i=1}^{2} \frac{\tilde{\psi}_i^Q \kappa_i^Q \varphi_i^Q \kappa_i^Q}{c(\overline{N}_i^Q(\tau^f)}.$$

C.1.2.2 Inter-Temporal Marginal Labour Costs for Wastewater Treatment Caused by the Natural Resources for Sustaining and Developing the Capital Stock

$$\underset{N_{\mu=1+2}^{K}}{ITMCL_{WWT}^{Q}}\ (\bar{\tau} \rightarrow T)$$

$$= k^Q \left\{ d^Q + \gamma_{\bar{\tau}}^Q \right\} \sum_{\tau^f=\bar{\tau}+1}^{T} p_L(\tau^f)\, l^{WWT} Y^Q(\tau^f) k^Q \left\{ d^Q + \gamma_{\tau^f}^Q \right\} \left\{ \sum_{\mu=1}^{2} \frac{\tilde{\psi}_\mu^K \kappa_\mu^K \varphi_\mu^K \kappa_\mu^K}{c(\overline{N}_\mu^K(\tau^f)} \right\}$$

$$= \underset{N_{\mu=1+2}^{K_{dep}}}{ITMCL_{WWT}^{Q}}\ (\bar{\tau} \rightarrow T) + \underset{N_{\mu=1+2}^{K_{dev}}}{ITMCL_{WWT}^{Q}}\ (\bar{\tau} \rightarrow T)\, with$$

$$\underset{N_{\mu=1+2}^{K_{dep}}}{ITMCL_{WWT}^{Q}}\ (\bar{\tau} \rightarrow T)$$

$$= k^Q d^Q \sum_{\tau^f=\bar{\tau}+1}^{T} p_L(\tau^f)\, l^{WWT} Y^Q(\tau^f) k^Q \left\{ d^Q + \gamma_{\tau^f}^Q \right\} \left\{ \sum_{\mu=1}^{2} \frac{\tilde{\psi}_\mu^K \kappa_\mu^K \varphi_\mu^K \kappa_\mu^K}{c(\overline{N}_\mu^K(\tau^f)} \right\}$$

$$\underset{N_{\mu=1+2}^{K_{dev}}}{ITMCL_{WWT}^{Q}}\ (\bar{\tau} \rightarrow T)$$

$$= k^Q \gamma_{\bar{\tau}}^Q \sum_{\tau^f=\bar{\tau}+1}^{T} p_L(\tau^f)\, l^{WWT} Y^Q(\tau^f) k^Q \left\{ d^Q + \gamma_{\tau^f}^Q \right\} \left\{ \sum_{\mu=1}^{2} \frac{\tilde{\psi}_\mu^K \kappa_\mu^K \varphi_\mu^K \kappa_\mu^K}{c(\overline{N}_\mu^K(\tau^f)} \right\}.$$

The use of non-renewable natural resources leads to inter-temporal marginal human labour costs in the production sector and to inter-temporal marginal human labour costs for wastewater treatment, because the decrease of concentrations of these natural resources in the natural environment is combined with an increase in the water amounts.

C.2 Marginal Energy Costs for Wastewater Treatment

When only the non-profit conditions are considered, the marginal energy costs for wastewater treatment are equal to

$$MCE^{Q}_{WWT}(\bar{\tau} \mapsto T) = \frac{\vartheta \sum_{\tau=1}^{T} p_E(\tau)\, e^{WWT}\, Y^Q(\tau) \left\{ \begin{array}{l} \left\{ \psi_Q^{RP} + \sum_{i=1}^{2} \frac{\tilde{\psi}_i^Q \kappa_i^Q}{c(\overline{N}_i^Q(0))} e^{\varphi_i^Q \kappa_i^Q \sum_{t=1}^{\tau-1} Y^Q(\tau)} \right\} \\ + k^Q \{d + \gamma^{\tau}\} \sum_{\mu=1}^{2} \frac{\tilde{\psi}_\mu^K \kappa_\mu^K}{c(\overline{N}_\mu^K(0))} e^{\varphi_\mu^K \kappa_\mu^K \sum_{t=1}^{\tau-1} k^Q Y^Q(t) \{d+\gamma^t\}} \end{array} \right\}}{\vartheta Y^Q(\bar{\tau})}$$

$$= \frac{\vartheta}{\vartheta Y^Q(\bar{\tau})} \sum_{\tau=1}^{T} p_E(\tau)\, e^{WWT}\, Y^Q(\tau)\, \psi_Q^{RP}$$

$$+ \frac{\vartheta}{\vartheta Y^Q(\bar{\tau})} \sum_{\tau=1}^{T} p_L(\tau) l^{WT}\, Y^Q(\tau) \sum_{i=1}^{2} \frac{\tilde{\psi}_i^Q \kappa_i^Q}{c(\overline{N}_i^Q(0))} e^{\varphi_i^Q \kappa_i^Q \sum_{t=1}^{\tau-1} Y^Q(\tau)}$$

$$+$$

$$+ \frac{\vartheta}{\vartheta Y^Q(\bar{\tau})} \sum_{\tau=1}^{T} p_L(\tau)\, e^{WWT}\, Y^Q(\tau) k^Q \{d + \gamma^{\tau}\} \sum_{\mu=1}^{2} \frac{\tilde{\psi}_\mu^K \kappa_\mu^K}{c(\overline{N}_\mu^K(0))} e^{\varphi_\mu^K \kappa_\mu^K \sum_{t=1}^{\tau-1} k^Q Y^Q(t) \{d+\gamma^t\}},$$

Where the energy input coefficient for wastewater treatment is abbreviated as

$$e^{WWT} = \tilde{\varepsilon}^{WWT} \left\{ \ln \frac{c_Q^*(0)}{c_Q(0)} + \ln \frac{c_{\hat{j}}^*(0)}{c_{\hat{j}}(0)} \right\}.$$

C.2.1 Total Actual Marginal Energy Costs for Wastewater Treatment

$$MCE^{Q}_{WWT}(\bar{\tau}) = p_E(\bar{\tau})\, e^{WWT} \left\{ \left\{ \psi_Q^{RP} + \sum_{i=1}^{2} \frac{\tilde{\psi}_i^Q \kappa_i^Q}{c(\overline{N}_i^Q(\bar{\tau}))} \right\} + k^Q \{d + \gamma^{\bar{\tau}}\} \sum_{\mu=1}^{2} \frac{\tilde{\psi}_\mu^K \kappa_\mu^K}{c(\overline{N}_\mu^K(\bar{\tau}))} \right\}$$

$$= MCE^{Q}_{WWT}(\bar{\tau}) = MCE^{Q}_{\substack{WWT \\ Qrep}}(\bar{\tau}) + MCE^{Q}_{\substack{WWT \\ N^Q_{i=1+2}}}(\bar{\tau}) + MCE^{Q}_{\substack{WWT \\ N^K_{\mu=1+2}}}(\bar{\tau})$$

C.2.1.1 Actual Marginal Energy Costs for Treatment of Wastewater Generated by the Reproduction Activities

$$MCE^{Q}_{\substack{WWT \\ Qrep}}(\bar{\tau}) = p_E(\bar{\tau})\, e^{WWT}\, \psi_Q^{RP}$$

C.2.1.2 Actual Marginal Energy Costs for Treatment of Wastewater Generated by Producing the Raw Materials for the Consumption Process

$$MCE^{Q}_{\substack{WWT\\N^{Q}_{i=1+2}}}(\bar{\tau}) = p_E(\bar{\tau})\, e^{WWT} \sum_{i=1}^{2} \frac{\tilde{\psi}^{Q}_{i}\, \kappa^{Q}_{i}}{c(\bar{N}^{Q}_{i}(\bar{\tau}))}$$

C.2.1.3 Actual Marginal Energy Costs for Wastewater Treatment of Wastewater Generated by Producing the Raw Materials for the Capital Good Process

$$\begin{aligned} MCLE^{Q}_{\substack{WWT\\N^{K}_{\mu=1+2}}}(\bar{\tau}) &= p_E(\bar{\tau})\, e^{WWT} k^{Q} \left\{ d + \gamma^{\bar{\tau}} \right\} \sum_{\mu=1}^{2} \frac{\tilde{\psi}^{K}_{\mu}\, \kappa^{K}_{\mu}}{c(\overline{N}^{K}_{\mu}(\bar{\tau}))} \\ &= MCE^{Q}_{\substack{W\\N^{K_{dep}}_{\mu=1+2}}}(\bar{\tau}) + MCE^{Q}_{\substack{W\\N^{K_{dev}}_{\mu=1+2}}}(\bar{\tau})\ with \end{aligned}$$

$$\begin{aligned} MCE^{Q}_{\substack{WWT\\N^{K_{dep}}_{\mu=1+2}}}(\bar{\tau}) &= p_E(\bar{\tau})\, e^{WWT} k^{Q}\, d^{Q} \sum_{\mu=1}^{2} \frac{\tilde{\psi}^{K}_{\mu}\, \kappa^{K}_{\mu}}{c(\overline{N}^{K}_{\mu}(\bar{\tau}))} \\ MCE^{Q}_{\substack{WWT\\N^{K_{dev}}_{\mu=1+2}}}(\bar{\tau}) &= p_E(\bar{\tau})\, e^{WWT} k^{Q}\, \gamma^{\bar{\tau}} \sum_{\mu=1}^{2} \frac{\tilde{\psi}^{K}_{\mu}\, \kappa^{K}_{\mu}}{c(\overline{N}^{K}_{\mu}(\bar{\tau}))}. \end{aligned}$$

C.2.2 Total Inter-Temporal Energy Costs for Wastewater Treatment

$$\begin{aligned} & ITMCE^{Q}_{WWT}(\bar{\tau} \to T) \\ &= \frac{\vartheta}{\vartheta Y^{Q}(\bar{\tau})} \sum_{\tau=1}^{T} p_E(\tau)\, e^{WWT}\, Y^{Q}(\tau) \sum_{i=1}^{2} \frac{\tilde{\psi}^{Q}_{i}\, \kappa^{Q}_{i}}{c(\overline{N}^{Q}_{i}(0))}\, e^{\varphi^{Q}_{i} \kappa^{Q}_{i} \sum_{t=1}^{\tau-1} Y^{Q}(\tau)} + \\ &+ \frac{\vartheta}{\vartheta Y^{Q}(\bar{\tau})} \sum_{\tau=1}^{T} p_E(\tau)\, e^{WWT}\, Y^{Q}(\tau) \sum_{\mu=1}^{2} \frac{\tilde{\psi}^{K}_{\mu}\, \kappa^{K}_{\mu}}{c(\overline{N}^{K}_{\mu}(0))}\, e^{\varphi^{K}_{\mu} \kappa^{K}_{\mu} \sum_{t=1}^{\tau-1} k^{Q} Y^{Q}(t) \{ d + \gamma^{t} \})} \\ &= ITMCE^{Q}_{\substack{WWT\\N^{Q}_{i=1+2}}}(\bar{\tau} \to T) + ITMCE^{Q}_{\substack{WWT\\N^{K}_{\mu=1+2}}}(\bar{\tau} \to T). \end{aligned}$$

C.2.2.1 Inter-Temporal Marginal Energy Costs for Wastewater Treatment Caused by the Non-renewable Resources for the Consumption Process

$$ITMCE^{Q}_{N^{Q}_{i=1+2}}(\bar{\tau} \rightarrow T) = \sum_{\tau^f=\bar{\tau}+1}^{T} p_E(\tau^f)\, e^{WT}\, Y^Q(\tau^f) \sum_{i=1}^{2} \frac{\tilde{\psi}^Q_i \kappa^Q_i \varphi^Q_i \kappa^Q_i}{c(\overline{N}^Q_i(\tau^f))}.$$

C.2.2.2 Inter-Temporal Marginal Energy Costs for Treatment of Wastewater Generated by Producing Raw Materials for the Capital Good Process

$$\begin{aligned}
& ITMCE^{Q}_{\substack{WWT \\ N^{K}_{\mu=1+2}}}(\bar{\tau} \rightarrow T) \\
&= k^Q \left\{ d^Q + \gamma^Q_{\bar{\tau}} \right\} \sum_{\tau^f=\bar{\tau}+1}^{T} p_E(\tau^f) e^{WWT} Y^Q(\tau^f) k^Q \left\{ d^Q + \gamma^Q_{\tau^f} \right\} \left\{ \sum_{\mu=1}^{2} \frac{\tilde{\psi}^K_\mu \kappa^K_\mu \varphi^K_\mu \kappa^K_\mu}{c(\overline{N}^K_\mu(\tau^f))} \right\} \\
&= ITMCE^{Q}_{\substack{WWT \\ N^{K_{dep}}_{\mu=1+2}}}(\bar{\tau} \rightarrow T) + ITMCE^{Q}_{\substack{WWT \\ N^{K_{dev}}_{\mu=1+2}}}(\bar{\tau} \rightarrow T).
\end{aligned}$$

with

$$\begin{aligned}
& ITMCE^{Q}_{\substack{WWT \\ N^{K_{dep}}_{\mu=1+2}}}(\bar{\tau} \rightarrow T) \\
&= k^Q d^Q \sum_{\tau^f=\bar{\tau}+1}^{T} p_E(\tau^f)\, e^{WWT} Y^Q(\tau^f) k^Q \left\{ d^Q + \gamma^Q_{\tau^f} \right\} \left\{ \sum_{\mu=1}^{2} \frac{\tilde{\psi}^K_\mu \kappa^K_\mu \varphi^K_\mu \kappa^K_\mu}{c(\overline{N}^K_\mu(\tau^f))} \right\} \\
& ITMCE^{Q}_{\substack{WWT \\ N^{K_{dev}}_{\mu=1+2}}}(\bar{\tau} \rightarrow T) \\
&= k^Q \gamma^Q_{\bar{\tau}} \sum_{\tau^f=\bar{\tau}+1}^{T} p_E(\tau^f)\, e^{WWT} Y^Q(\tau^f) k^Q \left\{ d^Q + \gamma^Q_{\tau^f} \right\} \left\{ \sum_{\mu=1}^{2} \frac{\tilde{\psi}^K_\mu \kappa^K_\mu \varphi^K_\mu \kappa^K_\mu}{c(\overline{N}^K_\mu(\tau^f))} \right\}.
\end{aligned}$$

The use of non-renewable natural resources leads to inter-temporal marginal energy costs in the production sector and to inter-temporal marginal energy costs for wastewater treatment, because the decrease of concentrations of these natural resources in the natural environment is combined with an increase in the water amounts.

6.4 Conclusions

The marginal cost components for water production and wastewater treatment can be summarized in the following matrix.

process $\hat{g}$	actual m. lab. costs	intertemp. lab costs $(\bar{\tau} \to T)$	actual m. energy costs	intertemporal m. energy costs $(\bar{\tau} \to T)$
${}^{Q}_{Q}$	$MCL^{Q}_{Q}(\bar{\tau})$	$ITCL^{Q}_{K}$	$MCE^{Q}_{Q}(\bar{\tau})$	
${}^{Q}_{N_{i=1+2}}$	$MCL^{Q}_{N_{i=1+2}}(\bar{\tau})$	$ITCL^{Q}_{N_{i=1+2}}$	$MCE^{Q}_{N_{i=1+2}}(\bar{\tau})$	$ITCE^{Q}_{N_{i=1+2}}$
${}^{Q}_{K}$	$MCL^{Q}_{K}(\bar{\tau})$		$MCE^{Q}_{K}(\bar{\tau})$	
${}^{Q}_{\substack{K_{dep}\\N_{\mu=1+2}}}$	$MCL^{Q}_{\substack{K_{dep}\\N_{\mu=1+2}}}(\bar{\tau})$	$ITCL^{Q}_{\substack{K_{dep}\\N_{\mu=1+2}}}$	$MCE^{Q}_{\substack{K_{dep}\\N_{\mu=1+2}}}(\bar{\tau})$	$ITCE^{Q}_{\substack{K_{dep}\\N_{\mu=1+2}}}$
${}^{Q}_{\substack{K_{dev}\\N_{\mu=1+2}}}$	$MCL^{Q}_{\substack{K_{dev}\\N_{\mu=1+2}}}(\bar{\tau})$	$ITCL^{Q}_{\substack{K_{dev}\\N_{\mu=1+2}}}$	$MCE^{Q}_{\substack{K_{dev}\\N_{\mu=1+2}}}(\bar{\tau})$	$ITCE^{Q}_{\substack{K_{dev}\\N_{\mu=1+2}}}$
${}^{Q}_{\substack{W\\Q_{rep}}}$	$MCL^{Q}_{\substack{W\\Q}}(\bar{\tau})$	$ITCL^{Q}_{\substack{W\\Q}}$	$MCE^{Q}_{\substack{W\\Q}}(\bar{\tau})$	$ITCL^{Q}_{\substack{W\\Q}}$
${}^{Q}_{\substack{W\\N^{Q}_{i=1+2}}}$	$MCL^{Q}_{\substack{W\\N^{Q}_{i=1+2}}}(\bar{\tau})$	$ITCL^{Q}_{\substack{W\\N^{Q}_{i=1+2}}}$	$MCE^{Q}_{\substack{W\\N^{Q}_{i=1+2}}}(\bar{\tau})$	$ITCE^{Q}_{\substack{W\\N^{Q}_{i=1+2}}}$
${}^{Q}_{\substack{W\\N^{K_{dep}}_{\mu=1+2}}}$	$MCL^{Q}_{\substack{W\\N^{K_{dep}}_{\mu=1+2}}}(\bar{\tau})$	$ITCL^{Q}_{\substack{W\\N^{K_{dep}}_{\mu=1+2}}}$	$MCE^{Q}_{\substack{W\\N^{K_{dep}}_{\mu=1+2}}}(\bar{\tau})$	$ITCE^{Q}_{\substack{W\\N^{K_{dep}}_{\mu=1+2}}}$
${}^{Q}_{\substack{W\\N^{K_{dev}}_{\mu=1+2}}}$	$MCL^{Q}_{\substack{W\\N^{K_{dev}}_{\mu=1+2}}}(\bar{\tau})$	$ITCL^{Q}_{\substack{W\\N^{K_{dev}}_{\mu=1+2}}}$	$MCE^{Q}_{\substack{W\\N^{K_{dev}}_{\mu=1+2}}}(\bar{\tau})$	$ITCL^{Q}_{\substack{W\\N^{K_{dev}}_{\mu=1+2}}}$
${}^{Q}_{\substack{WWT\\Q}}$	$MCL^{Q}_{\substack{WWT\\Q}}(\bar{\tau})$	$ITCL^{Q}_{\substack{WWT\\Q}}$	$MCE^{Q}_{\substack{WWT\\Q}}(\bar{\tau})$	$ITCL^{Q}_{\substack{WWT\\Q}}$
${}^{Q}_{\substack{WWT\\N^{Q}_{i=1+2}}}$	$MCL^{Q}_{\substack{WWT\\N^{Q}_{i=1+2}}}(\bar{\tau})$	$ITCL^{Q}_{\substack{WWT\\N^{Q}_{i=1+2}}}$	$MCE^{Q}_{\substack{WWT\\N^{Q}_{i=1+2}}}(\bar{\tau})$	$ITCE^{Q}_{\substack{WWT\\N^{Q}_{i=1+2}}}$
${}^{Q}_{\substack{WWT\\N^{K_{dep}}_{\mu=1+2}}}$	$MCL^{Q}_{\substack{WWT\\N^{K_{dep}}_{\mu=1+2}}}(\bar{\tau})$	$ITCL^{Q}_{\substack{WWT\\N^{K_{dep}}_{\mu=1+2}}}$	$MCE^{Q}_{\substack{WWT\\N^{K_{dep}}_{\mu=1+2}}}(\bar{\tau})$	$ITCE^{Q}_{\substack{WWT\\N^{K_{dep}}_{\mu=1+2}}}$
${}^{Q}_{\substack{WWT\\N^{K_{dev}}_{\mu=1+2}}}$	$MCL^{Q}_{\substack{WWT\\N^{K_{dev}}_{\mu=1+2}}}(\bar{\tau})$	$ITCL^{Q}_{\substack{WWT\\N^{K_{dev}}_{\mu=1+2}}}$	$MCE^{Q}_{\substack{WWT\\N^{K_{dev}}_{\mu=1+2}}}(\bar{\tau})$	$ITCL^{Q}_{\substack{WWT\\N^{K_{dev}}_{\mu=1+2}}}$
MC^{TOT}	$\sum_{g} MCL^{TOT}_{g}(\bar{\tau})$	$\sum_{g} ITCL^{TOT}_{g}$	$\sum_{g} MCE^{TOT}_{g}(\bar{\tau})$	$\sum_{g} ITCE^{TOT}_{g}$

There are no capital stocks for the water and wastewater sector in the water use model yet, but the information is already quite complex. Therefore, we will use a less disaggregated form for presenting the marginal costs on the process level for each of the relevant sectors in the dynamic water infrastructure model in Part II.

In addition to the explanations already provided in the context of the model constraints and optimal conditions, some general conclusions and remarks are worth highlighting.

1. Although our model only has two desired goods, the use of non-renewable natural resources, water, and the introduction of technical changes and environmental protection measures results in a large number of marginal cost components. We must distinguish between the actual and inter-temporal, direct and indirect marginal costs for human labour and energy inputs on the production levels. The production levels of the raw material extraction processes, the water production and wastewater treatment processes, where the extracted raw materials are converted to the desired consumption and capital good, are determined by the development path of the consumption good amounts.
2. The capital stock contributes to reducing the human labour input for the production of the consumption good, and therefore, the technical progress is an integral part of the water use model. The knowledge embedded in the capital stock could also influence other input coefficients such as the coefficient of the energy, water, and raw material inputs. Energy and water saving strategies are particularly important subjects, which will be analysed in the context of the water infrastructure model in Part II.
3. The marginal costs of using and developing the capital stock are composed of four terms: the actual and inter-temporal marginal costs caused by the depreciation of the capital stock and the actual and inter-temporal marginal costs of the acquired new investment into the capital stock. The capital stock development is determined by the time preference for future consumption good amounts given by the characteristics of the inter-temporal welfare function.
4. In the case of renewable resources, no inter-temporal marginal costs for the basic input factors exist because the concentrations for extracting the raw materials for production would not be a declining function and the inter-temporal marginal costs in this matrix would be zero. In the case of non-renewable resources, the past and actual activities will influence future activities. When these inter-temporal marginal costs are ignored, the future economic development cannot be considered sustainable.
5. Technical progress expressed in the form of a decrease in human labour input as a function of capital stock reduces the marginal costs for production. This reduction in the marginal costs of the labour and energy input for the production process of the consumption good compensates for the other inter-temporal effects that work against a sustainable development of the economic system.
6. The marginal costs of the capital good input for producing one unit of the consumption good process with the expressions show that in case of comparable a low capital good depreciation rate and a low capital good input coefficient , the

indirect actual and inter-temporal marginal costs of the capital good would also be lower, where the influences of producing the required raw materials and water and the effect on wastewater treatment have also to be taken into account.

7. Often, the water and wastewater amounts required to sustain and to develop the capital stock for producing one unit of the consumption good are not known. The high water and energy amounts required for extracting the raw materials are also often ignored. Since the concentrations of the raw materials for production and energy generation are declining functions, the required water amounts and the amount of unavoidable by-products are increasing. If we add the marginal costs of the raw materials and water amounts needed for producing the capital good, then we learn that the actual and future marginal costs for wastewater treatment normally increase for achieving the exogenously environmental water quality standards for discharged domestic and industrial wastewater pollution components.
8. Certain measures can be implemented to reduce the inter-temporal marginal costs. New production technologies could be developed to substitute non-renewable resources with renewable resources. Other measures could be water and energy saving strategies, the introduction of recycling processes, and the change of the treatment and disposal concept of unavoidable by-products of production and consumption.
9. Our dynamic model can be extended so that certain coefficients, e.g. for water and energy uses, that we assumed to be time invariant are allowed to change over time. This might be done in the form of a function that builds up sector-specific capital stocks.

Part II
The Water Infrastructure Model

Chapter 7
Case Studies Guiding the Integration of Water Infrastructure

Abstract The integration of water infrastructure into the natural sciences-consistent dynamic models is guided by two case studies. The first case study deals with the contamination problems of the Leuna aquifer. The problems were caused by past activities and can only be solved using external technological intervention because the aquifer has an extraordinarily low self-purification capability for the specific pollutants. The groundwater treatment technology that was chosen for the site must operate for a minimum of 20 years. Since the polluter-pay-principle cannot be applied, the estimated dynamic prime costs for solving this problem must be understood as political prices, or social costs, which have to be paid by future generations. The second case study focuses on the revision and extension of the water and wastewater system for the city of Adana in Turkey. This project emphasizes how past activities determine the future development of the water infrastructure for urban centres. The case study shows in detail which essential components must be integrated into the water infrastructure model and which aggregation level is acceptable for closing the gap between theory and practical application.

Our objective in Part II is to extend the water use model with essential water infrastructure components within the framework of the basic natural sciences consistent dynamic model that has been described thus far. Since water uses are already an integral subject in the basic model, few modifications are needed to close the gap between theoretical concepts and concrete applications. This is best achieved if sufficient background information and relevant practical experience are available. We will use two case studies that meet these criteria through our involvement with them over the last decade. The first case study is based on M. Schirmer's scientific investigation of groundwater pollution in the Leuna aquifer. The second case study is from H. Niemes' contributions to the studies, design, and implementation of the rehabilitation and extension of the water supply and wastewater system of the urban centre of Adana, Turkey. It is not our objective to model the measures already implemented, but to show that our dynamic models are useful instruments for deepening the understanding of the complexity of solving water problems.

Consider the whole cycle of water uses, starting with extraction, treatment, distribution, consumption, discharge of used water, and ending with wastewater treatment

H. Niemes, M. Schirmer, *Entropy, Water and Resources*,
DOI 10.1007/978-3-7908-2416-2_7, © Springer-Verlag Berlin Heidelberg 2010

and discharge back to the natural environment. In the context of this entire cycle, the selected case studies, and more generally groundwater pollution, address only one sub-system of the whole water infrastructure system. Groundwater and other water pollution problems are concerned with the interface between the ecological and embedded economic systems. These interrelations will be analysed within the framework of our natural sciences consistent dynamic models. Since the rehabilitation and construction of an urban water supply and wastewater system include all essential system elements of the water sector, we will use the Adana case study to extend our water use model by urban water infrastructure components.

7.1 The MTBE Contamination of the Leuna Aquifer[1]

7.1.1 Characteristics of the MTBE Contamination Problem

Leuna is an old industrial site in Eastern Germany that was a centre of chemical production for approximately 100 years. Figure 7.1 shows an overview of the site. Methyl tertiary-butyl ether (MTBE) is a chemical compound with the molecular formula $C_5H_{12}O$. It replaced tetra-ethyl lead to increase the octane rating of gasoline and help prevent engine knocking. It has been used since 1981 and was produced on a large scale in the former refinery starting in 1984. The refinery was closed in 1996 and a new one was built in the new plant area. During the production period, large quantities of gasoline containing MTBE were introduced into the subsurface by spills during the filling processes or from leaking underground storage tanks that

Fig. 7.1 Leuna site in plan view (photograph used with permission of Gerhard Lammer, WFL GmbH, Rottendorf)

---

[1]We are grateful to Mrs. Dr. B. Harpke from the "Landesanstalt für Altlastenfreistellung des Landes Sachsen-Anhalt" (LAF) (State Agency for Redemption of Liability Saxony-Anhalt) for supporting us with information about the investment and operation costs for the rehabilitation of the Leuna aquifer and comments on the MTBE case study.

contained blended gasoline. As a result, the subsurface at the former refinery is contaminated with mineral oil free phases. From these free oil phases, different gasoline components, particularly BTEX (benzene, toluene, methylbenzene and xylems) aromatics and MTBE are released into the groundwater. A further significant pollutant at the site is ammonia, which came from a former ammonia production site located upstream of the old refinery plant.

The dissolved pollutants are transported by groundwater flow. Downstream of the oil spill are different objects that could potentially be impacted by the groundwater pollutants. The groundwater is flowing towards the Saale River, which is located approximately 2,000 m downstream of the pollution source. The city of Leuna is located to the northeast of the industrial site. MTBE was already detected in low concentrations in the drinking water wells located 1,500 m downstream of the source. To prevent MTBE contaminations in the drinking water, a protective remediation well was installed.

The hydro-geological and geochemical structures of this site were investigated in detail. The main aquifer thickness is 2–4 m and the groundwater table is located approximately 3–4 m below the ground surface. The aquifer is relatively heterogeneous and is composed of fine to coarse sand and gravel. The hydraulic conductivity, K, was calculated from hydraulic tests, grain size analyses, column experiments, and in situ tracer experiments. The mean K value is 4×10^{-4} m/s (middle to coarse sand and gravel), but it varies in different aquifer structures and ranges between 6×10^{-2} m/s in coarse gravel and 9×10^{-5} m/s in middle to fine sand areas. The groundwater flow velocity was estimated using water level data, pumping, and tracer tests and it varies between 0.3 and 1.0 m/day. The main flow direction is southwest to northeast. Due to the presence of hydraulic barriers, it changes within the investigation site to a west to east direction. The groundwater temperature varies between 7°C and 14°C (mean 11°C) depending on the location of the sampling wells.

The economic rationale behind the joint use of an aquifer is based on economies of scale for the essential stock components, which have the characteristics of local public goods because no rivalry and exclusion between users exist. In reality, however, a limited number of beneficiaries have access to a public good because of natural, technical, economical, administrative, or other constraints. Under these circumstances, a groundwater aquifer has local public or local club good if jointly used for the water supply services of a community.

In economic theory, the handling of public goods is independent of its specific type. Samuelson (1954) and Musgrave (1956) derived optimal conditions for public goods and expenditure on a governmental level. The sum of marginal benefits for each of the consumers must be equal to the total marginal benefit of the offered public good. Tiebout (1956:419), however, showed that the optimal amounts of local public goods in different locations, which are in competition to attract people to one location, can only be achieved under less realistic and more extreme assumptions.

The amount of water required at the different consumption points has private good characteristics, because the water that is consumed by one household or an industry is not available for other users. Water infrastructure components normally have specific natural or technical capacity limits. Therefore, the individual consumption causes increasing negative externalities or social costs to the other consumers when the total water flow reaches such limits.

In the context of the joint use of local public goods, we must distinguish clearly between stock and flow components. We must also consider the economy of scale effect on public goods that can be partly compensated by flow related externalities and social costs caused by the increasing number of beneficiaries. Consequently, modelling the joint use of local public goods addresses different disciplines in economics, such as the theories of public goods, externalities, and capital goods.

After the general economic pattern of local public goods has been described, some specific patterns of the MTBE contaminated groundwater aquifer at Leuna must be added. The decisions made after the unification of West and East Germany clarified the responsibility for these rehabilitation measures. It was decided that the economic reactivation of East Germany, and at this site in particular, which was and is now again one of the main chemical industry centres, should not be hindered by social costs normally paid in accordance with the polluter-pay-principle. The publicly owned agency "Landesanstalt für Altlastenfreistellung des Landes Sachsen-Anhalt" (LAF) (State Agency for Redemption of Liability Saxony-Anhalt) was established by law on 25.10.1999. It was declared responsible for solving these contamination problems over the next decades with the financial resources of the public community, but almost free of charge for the private companies settling down at this site and at other sites within Saxony-Anhalt.

As illustrated in Fig. 7.2, the sanitation measures already implemented and financed via the LAF divide the aquifer into different sub-systems. The wall, which is 400 m long and 15 m deep, separates the aquifer into local "bad" and "good" sections. The "bad" section is upstream of the wall and the "good" section is downstream of the wall. The pollution stock (a) in Fig. 7.2 is the highly MTBE-contaminated surface sediment of the upstream aquifer and it is permanently polluting the aquifer (b). The MTBE concentrations in the groundwater are up to 125 mg/L, and average about 20 mg/L. These levels are significantly above acceptable environmental standards for this and other pollution parameters. The combination of the large amount of MTBE stock in the surface soil along with the limited dilution and degradation capacity as well as the slow groundwater velocity means that this serious pollution problem will not be solved naturally over the next few 100 years. There is, therefore, a need to remediate the contaminated groundwater extracted up-stream of the wall (c) in a treatment plant (d) and to infiltrate the treated groundwater (e) into the downstream section of the aquifer (f).

The treatment technology that was chosen after many alternatives were tested is air stripping, which brings the MTBE concentration in water down to the level of 200 μg/L. The drinking water standard is 5 μg/L and this is the concentration that must be achieved at the drinking water wells of the water production plant G located 1.5 km downstream. Reduction of the concentration from 200 to 5 μg/L will be achieved through dilution and the self-purification capability of the downstream aquifer (for more details see Martienssen et al. 2006 and Schirmer and Martienssen

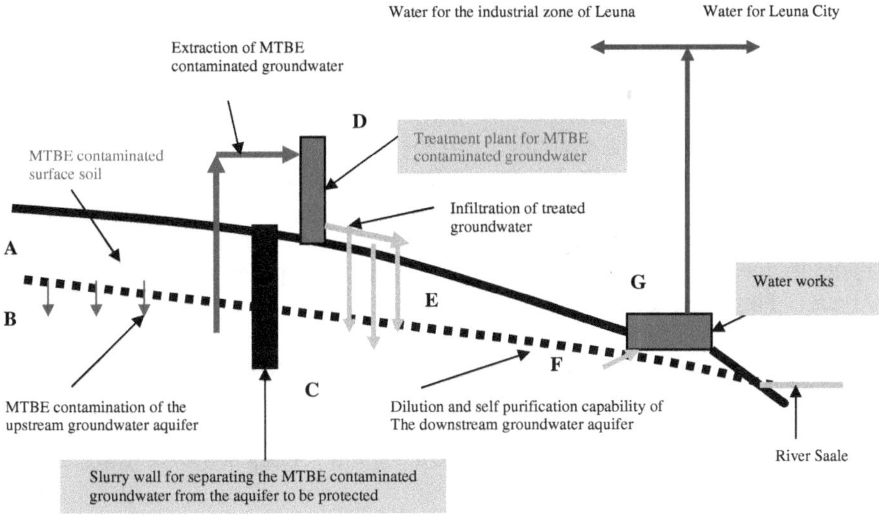

Fig. 7.2 Water infrastructure implemented to solve the contamination problem of the Leuna aquifer

2007). Other technologies with higher treatment efficiency, such as activated carbon, are available. Therefore, the chosen technology considered the most efficient from a technical and economic point of view because the remaining risks can be reduced by additional measures. These additional measures include eliminating local pollution sources in the downstream aquifer and removing the MTBE-contaminated soil over the next 2 decades.

There are no known MTBE problems in the new drinking water works, which have a capacity of 220 m^3 per hour or 5,280 m^3 per day and serves 7,500 inhabitants in the city of Leuna as well as the industrial complex of Leuna. The chosen raw water quality standard and sanitation strategy therefore seems to be acceptable.

7.1.2 Technical Solutions to Reduce MTBE Contamination

After a long period of investigation, the many MTBE treatment options were narrowed down to two. The two concepts that were chosen were an external treatment that uses so-called stripping technologies and an internal biological treatment, where microbiological processes are accelerated by oxygen addition. The second technology required further investigation because it is not state-of-the-art. The so-called oxy-wall application had thus far only been performed within the pollution source zones. As a result, effective downstream results could not be seen. If plume remediation were done, this technology might be more successful, but it would still require a long reaction space.Air stripping was chosen as the most common on-site groundwater treatment technologies. With air stripping, volatile organic contaminants in the dissolved phase are transferred into the vapour phase, by blowing air through the contaminated water. Contaminated water is trickled from the top of a large tower while air is introduced from the bottom. Air stripping is effective with petroleum hydrocarbons but will be less efficient for MTBE because the moderate volatility of MTBE in the aqueous phase. The Henry's Law constant, H, of about 0.04 for MTBE is much lower, for example, than the 0.222 for benzene (Schirmer 1999). The MTBE's lower volatility requires either a higher air-to-water ratio or heated air for accelerating the air stripping removal efficiency. The air is later cleaned by passing it through an activated carbon filter. This technique can meet the regulatory limits for water that is to be re-injected into the aquifer, commonly less than 100 μg/L. However, if MTBE must be completely eliminated, then air stripping is highly expensive.

An alternative to on-site treatment technologies is enhanced natural attenuation (ENA) approach. ENA, by means of direct gas injection or conditioning treatment facilities, is highly effective and can be adapted to either low or high contaminant concentrations. However, using ENA requires large reaction zones in the aquifer.

One of the simplest ENA methods for MTBE remediation is adding oxygen to the groundwater by air sparging. The main advantages of sparging technologies are the simple technical equipment and low costs. Air sparging, as described by Javanmardian and Glasser (1997) and by Leeson et al. (1999), can be used for the elimination of MTBE concentrations up to 40 mg/L. One disadvantage of air sparging is a gas clogging effect. Nitrogen bubbles that remain after the oxygen is consumed reduce the water-filled porosity and the permeability of the aquifer in the intended treatment zone. This reduced permeability may result in a partial bypass of contaminated groundwater around the treatment zone. Therefore, recent treatments inject pure oxygen instead of air. Oxygen sparging has been applied successfully by different investigators and at several different sites (Salanitro et al. 2000, Wilson et al. 2002 and Smith et al. 2005).

The operation of oxygen injection systems can be continuous or intermittent. Continuous oxygen injection commonly provides higher dissolved oxygen concentrations than intermittent injection. On the other hand, increased gas saturation in the aquifer can significantly reduce the hydraulic conductivity. This gas clogging is not only the result of excess oxygen supply but also results from increased accumulation of dissolved nitrogen and other permanent gases in the gas bubbles (Geistlinger et al. 2005). Therefore, periodic oxygen gas injection has been preferred in most applications. Trapped gas bubbles remaining in the aquifer provide the groundwater with oxygen for extended periods between the gas injection cycles. Optimized intervals between subsequent gas injections allow the remaining nitrogen and other permanent gases to re-dissolve and prevent significant gas clogging.

Optimal conditions depend on the concentrations of pollutants, oxygen demand of the different site reactions, and the geological and hydro geological setting. For oxygen sparging, injection wells are installed into the aquifer at different locations and depths depending on the geological structure of the aquifer and the measured MTBE concentration.

7.1.3 The Target Group of the Rehabilitation Measures

The town of Leuna is mainly served by the new Daspig drinking water works, which extract raw water from the Saale River bed infiltration and groundwater from the Leuna aquifer. Although there are no known, direct, negative influences from the MTBE pollution stock on the raw water quality, the town of Leuna and the industrial site can be considered as target groups for rehabilitating this local public good, for which a potential risk still remains. As a result, the rehabilitation of the Leuna aquifer is not a question of economic optimization, but a political decision to reduce future risks caused by economic activities in the past. This means that the emission standard price approach can be applied instead of optimizing a welfare function under given constraints for the target group. It is sufficient to estimate the dynamic prime costs of the rehabilitation measures, which reflects the social costs caused by polluting the Leuna aquifer. The time it takes to rehabilitate the aquifer is determined by the natural aquifer conditions, the technology chosen to eliminate MTBE pollution, and also political decisions. The dynamic prime costs can also be interpreted as a shadow price of the willingness to return the aquifer over time to its original natural state.

In this context, it could be said that the costs for the rehabilitation should be covered directly by the beneficiary groups, for example, in the form of additional costs on the price for producing drinking water. A strong argument supporting this position could also be that many citizens of Leuna previously earned income at the chemical factories at the Leuna industrial site. Today, however, fewer Leuna citizens are employed at the new chemical factories. Furthermore, the "benefits" from ignoring the social costs in the past are not only gained by the population of Leuna city but by the society as a whole. Since the polluter-pay-principle can not be applied in this special case, it is a matter of justice that these social costs are not only being paid by the population of Leuna, but also by the government from the tax income. This means that the social costs are not directly paid by individuals, but indirectly paid as part of the taxes transferred to the government, which subsidize these rehabilitation measures.

7.1.4 Estimation of the MTBE Contamination Amounts

The MTBE has accumulated and is now considered a serious pollution stock because the self-purification capacity of the aquifer is limited. The actual amount of MTBE that was discharged to the Leuna aquifer is not known. We can therefore only estimate the amounts indirectly. The Daspig water works have a maximum capacity of 5,280 m^3/day. However, this water is extracted from wells on the left and right riverbanks of the Saale River. Therefore, we cannot use data from the water works for such estimating MTBE masses. We must first distinguish between water coming from the left and right banks of the river. A wall that is 400 m long and 15 m deep divides the aquifer. If we assume a 200 m length of wall with an aquifer thickness of 2–4 m and a groundwater velocity between 0.4 and 0.8 m per day, we get groundwater flow volumes in the range of 160 and 640 m^3 per day.

The wells constructed for collecting the polluted groundwater were constructed to have a capacity of about 12–15 m^3 per hour or 288–360 m^3 per day. This range fits with the estimation mentioned previously. Due to the lack of precise groundwater flow volumes, a pilot plant has been in operation for 15 months to collect data for optimizing the final groundwater treatment plant (for more details see Harpke et al. 2006).

If we use the estimated average MTBE concentration of 20 mg per litre (20 g per m^3) then the daily MTBE pollution amount transported by the aquifer and extracted by groundwater treatment would be in the range of 5.76–7.3 kg per day (2.1–2.6 tonnes per year). The treatment facility reduces the average MTBE concentration from 20 mg/L to 200 μg/L. The self-purification and dilution capability then reduces the concentration further to 5 μg/L. It is estimated that the treatment facility will operate for 20–30 years. Using these data, the total MTBE amounts are estimated to be between 42 and 78 tonnes. MTBE production took place over a period of 10 years before environmental protection measures were put in place. Therefore, it is estimated that 4.2–7.8 tonnes per year were discharged into the Leuna aquifer. Although this is a rough estimate, it is a realistic value when considering past production practises.

Even though only a rough estimate of MTBE pollution is possible, the change in concentration from 20 mg/L to 200 μg/L after groundwater treatment and 5 μg/L for the drinking water standard are very ambitious targets from a thermodynamic point of view. Equations (2.17) and (2.18) in Part I, where exergy is expressed as a function of raw material concentrations, show that high exergy is needed to remove a component that is present at a low concentration. If we focus on only two components, MTBE pollution (M) and water (W), we get the following system of equations

$$\frac{d(Ex)}{dt} = NRT_0 \left[\ln \frac{c_M}{c_M^0} \frac{dc_M}{dt} + \ln \frac{c_W}{c_W^0} \frac{dc_W}{dt} \right]. \tag{7.1}$$

$$Ex(t) = NRT_0 \left[c_M \ln \frac{c_M}{c_M^0} - \left(c_M - c_M^0 \right) + c_W \ln \frac{c_W}{c_W^0} - \left(c_W - c_W^0 \right) \right]. \tag{7.2}$$

The concentrations of the two components c_M and c_W depend on each other through the relation $c_M + c_W = 1 \Leftrightarrow c_W = 1 - c_M$. Therefore, only the concentration of one component, such as MTBE, is required. We receive

$$\frac{d(Ex)}{dt} = NRT_0 \left[\ln \frac{c_M}{c_M^0} + \ln \frac{1 - c_M}{1 - c_M^0} \right] \frac{dc_M}{dt}. \tag{7.3}$$

$$Ex(t) = NRT_0 \left[c_M \ln \frac{c_M}{c_M^0} + (1 - c_M) \ln \frac{1 - c_M}{1 - c_M^0} \right]. \tag{7.4}$$

The average MTBE concentration of 20 mg/L (or g/m^3) can be used as the value for the steady state condition, or thermodynamic equilibrium, of a complex open system with three subsystems: the aquifer; its surface soil; and the MTBE pollution stock. If the concentrations of MTBE are very low, the second term of Eq. (7.4) converges to zero and can be neglected.

Reducing the MTBE concentration from $c_M^0 = 20\text{mg/L}$ to the emission standard of $c_M^T = 200\ \mu\text{g/L}$ requires free energy amounts for returning this irreversible process. In the case of low concentrations, the approximation in Eq. (3.18)

$$\left.\frac{\Delta \mathrm{Ex}^i}{\Delta \mathrm{c}^i}\right|_{\mathrm{c}^i(t)}^{\mathrm{c}^i(t+1)\wedge \mathrm{c}^i_i(t+1)=1} \approx -\mathrm{RT_0N_i^i}\left\{\ln \mathrm{c}^i(t) - 1\right\} \leq -\mathrm{RT_0N_i^i}\left\{\ln \mathrm{c}^i(t)\right\}$$

Also confirms the marginal increase of free energy required to separate water pollution components from water. Therefore, the energy costs are the main cost components of the dynamic prime costs for water treatment.

7.1.5 Estimation of Costs for Solving the MTBE Problem

The social costs of the groundwater contamination are influenced by political decisions. In the case of the Leuna aquifer, there were many of these decisions. One example was the choice of the 20–30 year time horizon of the cleanup. Another decision was to use the self-purification capacity of the aquifer. Finally, the technological and financial opportunities of reducing the MTBE pollution were also factors in the decision making process. When calculating the so-called dynamic price costs, which reflect the shadow prices of the social costs caused in the past and the future prices of the groundwater rehabilitation measures, we should keep in mind that the costs for treating the groundwater and infiltrating it back into the downstream aquifer are political prices.

The dynamic prime costs are defined as follows

$$\mathrm{DPC} = \frac{\mathrm{NPVC}}{\mathrm{NPVW}} = \frac{\left\{\sum_{\tau=1}^{T} \frac{\{I(\tau)+O(\tau)\}}{(1+\delta)^{\tau-1}}\right\} - \frac{\mathrm{RVI(T)}}{(1+\delta)^{T}}}{\sum_{\tau=1}^{T} \frac{W(\tau)}{(1+\delta)^{\tau-1}}} \tag{7.5}$$

Where $W(\tau)$ is the amount of groundwater to be treated, T the time horizon, $I(\tau)$ the investment cost, RVI(T) the residual value of the investment costs at the end of the planning horizon, $O(\tau)$ the operation costs, and $\delta \geq 0$ the exogenously given discount rate.

The sum of the discounted investment and the operation costs minus the discounted residual value of the investment costs are called the net present values of the costs. The sum of the discounted groundwater amounts is called the net present value of the groundwater amounts.

7.1.5.1 The Groundwater Amounts and the Time Horizon for Groundwater Treatment

Based on the experience with the pilot plant that was constructed to optimize the groundwater treatment technology the groundwater amounts to be treated by the stripping plant range from 12 to 15 m^3/h. The variation in the amount of extracted groundwater is caused by meteorological conditions. The average groundwater amount of 14 m^3/h (336 m^3/d or 122.640 m^3/a) fit within the range of 160–640 m^3/d, which was estimated in Sect. (7.1.4). This average groundwater amount will be assumed to remain constant during the time period of groundwater treatment.

It is estimated that the groundwater treatment plant will operate over a time horizon of at least 20, and up to 30 years. When calculating net present values or dynamic prime costs, the values of the costs and groundwater amounts must be discounted, that means that the values at later time intervals are less relevant. In order to conservative, a time horizon of 30 years will be assumed.

7.1.5.2 The Investment Costs

The main financial investment for the groundwater treatment measures was done in 2004 and included the following items:

1.	The 400 m slurry separation wall	980,000€
2.	the water storage during construction	60,000€
3.	the system for draining groundwater to the treatment plant	490,000€
4.	the groundwater treatment plant	195,000€
5.	the pilot plant for optimizing the treatment process	64,000€
6.	supplementary measures	250,000€
7.	Total	2,039,000€

These investment measures were based on hydro-geological investigations, which cost approximately 400,000€ from 2001 to 2004. These preparatory investment costs can be aggregated and added to the first three items of the investment costs, to obtain a total investment cost up until 2004 of 1,530,000€. For the other three investment costs, which total 509,000€ (the water treatment plant and its pilot phase, and the supplementary measures), we assume that they went into operation in 2005. The investment costs for 2004 and 2005 are 2,439,000€.

For the elimination of the remaining contaminant, additional investment costs of 4–5 million € are expected. We will assume that these additional costs are equally distributed over 6 years, which means that approximately 643,000€ will be invested over the time interval from 2009 to 2015.

There are complementary measures such as the excavation, treatment, and disposal of the contaminated soils that have been conducted since 1991 and will be continued in the future. These complementary measures are estimated to cost 40 million €, but are not considered part of the aquifer restoration. Therefore, these costs are not included in the calculation of the dynamic prime costs. In this context, it should be mentioned that the companies that move onto the industrial site at Leuna must pay for a portion of these complementary measures.

7.1.5.3 Residual Values of the Investment Costs at the End of the Planning Horizon

The additional investment costs of 4.5 million € between 2009 and 2015 are assumed as costs at 2009 prices and then discounted when calculating the dynamic prime costs. The 250,000€ for supplementary measures and the future investment costs are mainly for eliminating pollution sources. These investments can be considered as modifying the natural conditions, which will remain forever. As a result, the residual values of these investments at the end of the planning horizon are equal to the investment costs. The separation wall and the drainage system are typical civic works and can be expected to last at least 60 years. Therefore, their residual value will be assumed to have half of the value of the previous investment costs at the end of the planning horizon, 735,000€. The residual value of the groundwater treatment plant, however, will be set to be zero. With these assumptions, the total residual value of all investment costs sum to a value of 5,485,000€.

7.1.5.4 Fixed and Variable Operation Costs

The fixed operation costs for the maintenance and operation of the groundwater treatment facilities are approximately 120,000€/a. Estimates of the variable operation costs for energy, chemicals, etc. range from 0.6 to 1.2€/m^3 of treated groundwater. Based on an average of 0.9€/m^3, the annual variable operations costs amount to 110,376€/a (0.9€/m^3 * 122,640 m^3/a). For the year 2004, however, only half the operations costs will be assumed because the facilities were in full operation at the beginning of 2005.

7.1.5.5 Net Present Values and Dynamic Prime Costs for Groundwater Treatment

Figure 7.3 is a table outlining the dynamic prime costs for the treatment of the Leuna aquifer. Based on a discount rate of 5% (equal to a discount factor of 1.05), Fig. 7.3 shows the net present value of all investment and operations costs result in a value of 7,809,963€ for the entire planning period. When dividing this net present value by the net present value of the groundwater amounts (1,979,541 m^3), we obtain a dynamic prime cost (DPC) of 3.95€/m^3. This value is relatively high and it is also very sensitive to the additional investment over the next 7 years. If these additional investment costs can not be related directly to the aquifer under consideration, the dynamic prime costs are much lower. They cannot be lower than 2.93€/m^3 because this would create the situation where no additional investment costs would be required. If the total investment costs are not included in the calculation, the dynamic prime costs for only the operation costs would have a value of

Year	Investment Costs Original	Investment Costs Additional	Operation Costs Fixed	Operation Costs Variable	Total Costs	Groundwater Amount	Discount Factor 1,05	Discounted Costs	Discounted Groundwater Amounts
	€/a	€/a	€/a	€/a	€/a	m³/a		€/a	m³/a
2004	1.930.000		60.000	55.188	2.045.188	122.640	1,00	2.045.188	122.640
2005	509.000		120.000	110.376	739.376	122.640	1,05	704.168	116.800
2006			120.000	110.376	230.376	122.640	1,10	208.958	111.238
2007			120.000	110.376	230.376	122.640	1,16	199.007	105.941
2008			120.000	110.376	230.376	122.640	1,22	189.531	100.896
2009		643.000	120.000	110.376	873.376	122.640	1,28	684.313	96.092
2010		643.000	120.000	110.376	873.376	122.640	1,34	651.727	91.516
2011		643.000	120.000	110.376	873.376	122.640	1,41	620.692	87.158
2012		643.000	120.000	110.376	873.376	122.640	1,48	591.135	83.008
2013		643.000	120.000	110.376	873.376	122.640	1,55	562.986	79.055
2014		643.000	120.000	110.376	873.376	122.640	1,63	536.177	75.290
2015		643.000	120.000	110.376	873.376	122.640	1,71	510.645	71.705
2016			120.000	110.376	230.376	122.640	1,80	128.282	68.291
2017			120.000	110.376	230.376	122.640	1,89	122.173	65.039
2018			120.000	110.376	230.376	122.640	1,98	116.356	61.942
2019			120.000	110.376	230.376	122.640	2,08	110.815	58.992
2020			120.000	110.376	230.376	122.640	2,18	105.538	56.183
2021			120.000	110.376	230.376	122.640	2,29	100.512	53.507
2022			120.000	110.376	230.376	122.640	2,41	95.726	50.959
2023			120.000	110.376	230.376	122.640	2,53	91.168	48.533
2024			120.000	110.376	230.376	122.640	2,65	86.826	46.222
2025			120.000	110.376	230.376	122.640	2,79	82.692	44.021
2026			120.000	110.376	230.376	122.640	2,93	78.754	41.924
2027			120.000	110.376	230.376	122.640	3,07	75.004	39.928
2028			120.000	110.376	230.376	122.640	3,23	71.432	38.027
2029			120.000	110.376	230.376	122.640	3,39	68.031	36.216
2030			120.000	110.376	230.376	122.640	3,56	64.791	34.491
2031			120.000	110.376	230.376	122.640	3,73	61.706	32.849
2032			120.000	110.376	230.376	122.640	3,92	58.767	31.285
2033			120.000	110.376	230.376	122.640	4,12	55.969	29.795
2034	-985.000	-4.500.000	0	0	-5.485.000	0	4,32	-1.269.105	0

Net Present Values of the Costs	€	7.809.963	
Net Present Values of the Groundwater Amounts	m³		1.979.541
Dynamic Prime Cost including the additional Investment	€/m³	3.95	
Dynamic Prime Cost excluding the additional Investment	€/m³	2.93	
Dynamic Prime Cost excluding the Investment	€/m³	1.82	

Fig. 7.3 Calculation of the dynamic prime costs for water treatment of the Leuna aquifer

1.82€/m^3. Because the dynamic prime costs are so sensitive to the investment costs, it might have been possible to implement these additional investments over a longer time scale. We can therefore say that, depending on the implementation strategy, the dynamic prime costs of the aquifer rehabilitation measures are in the range of 3.5€/m^3 +/– 0.5€/m^3.

Although dynamic prime costs have the characteristics of shadow or political prices, the calculation of the dynamic prime costs confirms that eliminating the groundwater contamination that was caused in the past is quite expensive, when comparing the dynamic prime costs with the actual water tariff of 1.2€/m^3 charged by the downstream water works. It would not be correct, however, to argue that the water production is highly subsidized by public investment for protecting the Leuna aquifer because this serious environmental problem was created by past activities. This is a public investment, which balances the faults of the past with future interests of the society.

7.2 Water Infrastructure to Serve Adana in Turkey[2,3]

7.2.1 Urbanization and Water Infrastructure of Mega-Cities

Urban areas are a focus of increasing conflict with regard to water use and water protection. Half of the world's population, and approximately 73% of Europeans, live in cities. Currently, approximately 82% of the total population growth in the world occurs in the cities of the developing countries (UN 2005). Such a rapid growth of urban areas will lead to a growing number of mega-cities, which often spawn huge peripheral urban slums and deplete the environmental resources in the surrounding areas.

The urban economic literature focuses on the explanation of local concentration processes. They mainly refer to economies of scale or positive externalities as well as discussing the advantages of the joint use of local public infrastructure. An often analysed example to explain urbanization processes is urban traffic infrastructure (see e.g. Newberry 1990, Atkinson and Stiglitz 1980, Jacobs 1985, Fujita 1996, Mills and Hamilton 1994). One crucial restriction for satisfactory living conditions in urban agglomerations is access to safe drinking water. This includes wastewater collection, treatment, and acceptable discharge into the natural environment. The availability of the drinking water and wastewater infrastructure is important for the development of urban agglomerations.

In contrast to its importance for quality of life in mega-cities, the problems associated with supplying of an appropriate water infrastructure are seldom analysed in the urban economic literature (see e.g. Gläeser 1998, Benndorf 2003, Graw and Maggio 2001, Green 2003, Krebs 2003). Our aim is to illustrate that the availability and implementation of water infrastructure could be essential to the development

[2] The background information has been extracted from the paper: Niemes, Jöst, and Schirmer (2007): Bi-directional Relations between Urbanization and Water Infrastructure. The urban centre of Adana in Turkey as case study, presented at the Symposium of the European Society of Ecological Economics, 2007, Leipzig/Germany.

[3] We are grateful to Malte Faber, Robert Holländer, and Martin Quaas for helpful comments.

of urban areas. We would also like to show the complexity of planning and implementing solutions for water supply and wastewater systems in rapidly growing urban centres.

7.2.2 Private and Local Public Welfare Properties of Water

We will use an example to help us analyse the characteristics of water infrastructure from an economic point of view. Let us assume two farmers are living on adjacent pieces of land, served by two separate private wells for getting fresh water for production (farming) and consumption purposes (e.g. drinking, washing, cleaning, transporting waste, and other purposes). Let us further assume that they pump the fresh water from the same groundwater aquifer. This aquifer can be seen as a local public good assuming neither of the farmers is in the position to exclude his neighbour from the use of the aquifer. If the amount of water pumped out of this aquifer is not restricted by natural conditions or other technical constraints, the two farmers should be willing to cooperate. The farmers might recognize that the time, energy, and money spent on operating two separate wells and transporting the water to the houses and stables can be reduced by having one joint well equipped with a motorized pump and some water pipes to the farmers houses and animal stables. If the local water resource capacity seems to be unlimited when compared to the water required, they might even invite their neighbours to share further investments for increasing the pumping and water pipe capacities.

From an economic point of view, the water infrastructure is a local public or a club good (for more details see Nauges 2003). In contrast with a pure public good, the non-rivalry in consumption depends on the number of users. There are a maximum number of users (club members), which could consume the good simultaneously without a decrease in utility. Furthermore, some essential components of the water supply system are characterised by economies of scale. Because of its public good character and the economies of scale technology, people benefit from a joint investment in local infrastructure measures. Hence, similar to the transport infrastructure, the water infrastructure could be seen as an advantage of agglomerations and could therefore contribute to explaining the existence of urban areas.

However, the benefits of joint use for a water infrastructure do not rise monotonically with an increasing number of users. For a given stock of water pipelines, there are a maximum number of users that could simultaneously use the infrastructure without a decline in utility. If an existing water pipeline system comes to its hydraulic limits, congestion effects occur such that additional users cause negative externalities, namely by hydraulic losses.

There is a clear analogy to the use of a local traffic system. Each driver carries not only his private costs, but also the traffic flow related externalities, or social costs, caused by other drivers (for more details see Newberry 1990, Tiebout 1956). Water use and its infrastructure have some unique characteristics. As an example, in 2001, only 2% of the 129L of daily domestic per capita water consumption in Germany was for drinking and cooking purposes. If the water quality is poor, which is the

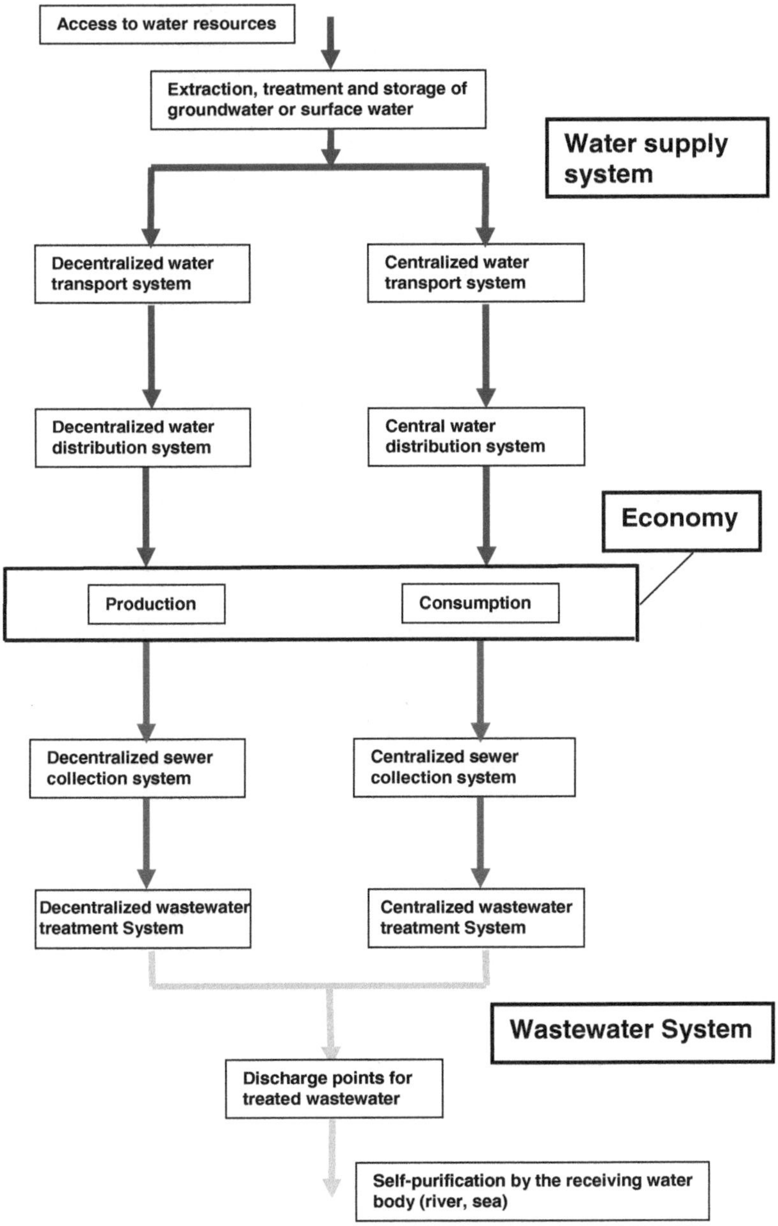

Fig. 7.4 Major components of the urban water and wastewater infrastructure

case in many mega-cities, this minor portion can even be substituted by good quality bottled water. Most of the domestic water consumed is for hygienic purposes, washing, cleaning, and flushing of residuals. The consumed water ultimately becomes wastewater. In developed countries and more densely populated urban centres, we have a wastewater generation rate of approximately 80% of the domestic water consumption. In addition to the water inflow system, a wastewater collection system is required to avoid the hygienic and health risks of uncontrolled wastewater discharge. Compared to traffic systems, which serve bidirectional flows, a separate wastewater collection system is required in urban centres.

Figure 7.4 illustrates the major components of the urban water and wastewater infrastructure. As it is shown in Fig. 7.4, certain essential components of the water supply and wastewater system, such as the water extraction and treatment facilities, storage tanks, water and wastewater pumping and lifting stations, and finally the wastewater treatment plant, carry some economies of scale and externalities.

7.2.3 Implementation Concept for Adana's Water Infrastructure

The urban centre of Adana is located in the Mediterranean Region of Turkey, as illustrated in Fig. 7.5. It is chosen as a case study because H. Niemes has been involved in the study, design, and implementation of a water infrastructure over a period of 12 years. In addition, the Adana water supply and wastewater systems were significantly restructured.

Some general background information about Adana is quoted below (see http://en.wikipedia.org/wik/Adana, //www.allaboutturkey.com/mediterranean).

- Adana is the fourth largest city in Turkey and the capital of the province of Adana. Being one of the larger cities of Asia Minor, and located about 30.5 km from the Mediterranean sea, Adana derives its importance from its location as the gateway

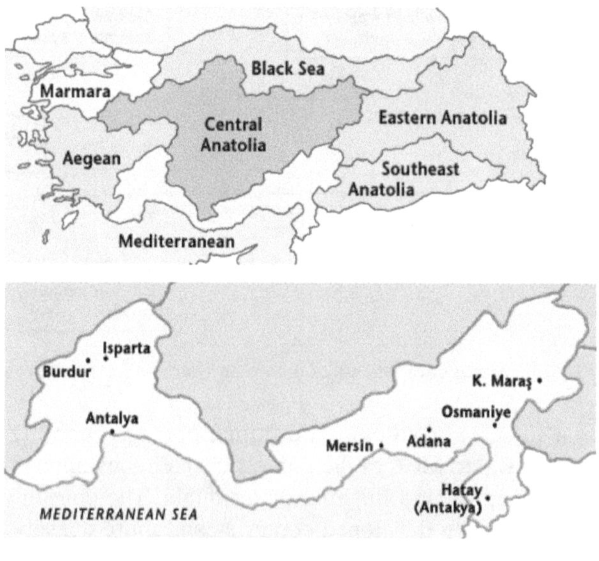

Fig. 7.5 Location of Adana in the Mediterranean region of Turkey

to the Clinician plain, that great flat stretch of fertile land, possibly the most productive in this part of the world on the east side of the Taurus Mountains.
- Adana is the marketing and distribution centre for an agricultural region, where cotton, wheat, barley, grapes, citrus fruits, olives, and tobacco are produced. The main industries in the city are textile manufacturing, tanning, and the processing of wool and various foods.

The population figures for Adana were used as the design criteria for implementing an improved water supply and wastewater system for different districts and target years. Based on these population figures, Adana has a growth rate of 2.5% per year, whereas the average population in Turkey increased by only 1.4% per year over the last decade. One reason for the rapid urbanization is the accelerated migration rate in some big cities in Turkey. The migration rate increased to levels between 4 and 9% for several years during the period of political instabilities of the 1980s. During this time, the separation movements of the Kurdish population in East Anatolia and the strategy to settle this conflict forced people to move from their villages into cities controlled by the Turkish government. Another reason for the rapid population growth in Turkish cities is the unemployment problem in the south eastern part of the country.

The first priority in the cities was to construct multi story houses financed from the mass housing programme, which left the cities responsible for arranging the water infrastructure needed to handle growth. The water infrastructure in Adana would not have been able to handle the rapid addition of 500,000 users without the technical and financial assistance of the Turkish Government and international financing agencies. In 1993, Germany and Turkey, decided to include Adana in their technical and financial assistance programme.

The major results of the whole planning and implementation process are summarized in the following sections. Our main intention in providing this summary is to illustrate the complexity of this subject and how it relates to the future urbanization and water infrastructure in Adana as well as other rapidly growing urban centres.

7.2.3.1 Planning and Implementation Concept for the Wastewater System

The German Financing Agency "Kreditanstalt für Wiederausbau" (KfW) began its involvement in Adana by financing the very detailed Adana Wastewater Treatment Plant Feasibility Study in 1993–1994. It included, but was not limited to, the following project components (GKW Consult 1993/1994):

- Environmental investigations and the application of water quality simulation models (Niemes 1995), for the determination of potential sites and technical alternatives for the construction of wastewater treatment facilities.
- Hydraulic simulation and design of the wastewater and storm water collection systems.
- Carrying out an industrial cadastre for the industrial and commercial sector, installation of a laboratory for water quality analysis, and training of the staff for environmental monitoring.

- Determination of design criteria and pre-design of technical alternatives for the storm water and wastewater collection system and the wastewater treatment facilities.
- Socio-economic surveys and the economic and financial evaluation of potential technical alternatives.
- Design of the least-cost solutions.
- Institutional analysis.
- Proposals for the future institutional, technical, and financial set-up for the Public Water and Sewerage Work Company of Adana (ASKI).

The major measures proposed in this study, can be summarized as follows

1. The optimal solution for wastewater collection and treatment, from the environmental point of view, would be to construct a single wastewater treatment plant located in the southern irrigation site of the Adana-Tarsus region. This is labelled as Site for Adana West in Fig. 7.6, which is presented at the end of Chap. 7. The main advantage for this wastewater treatment location would be that treated

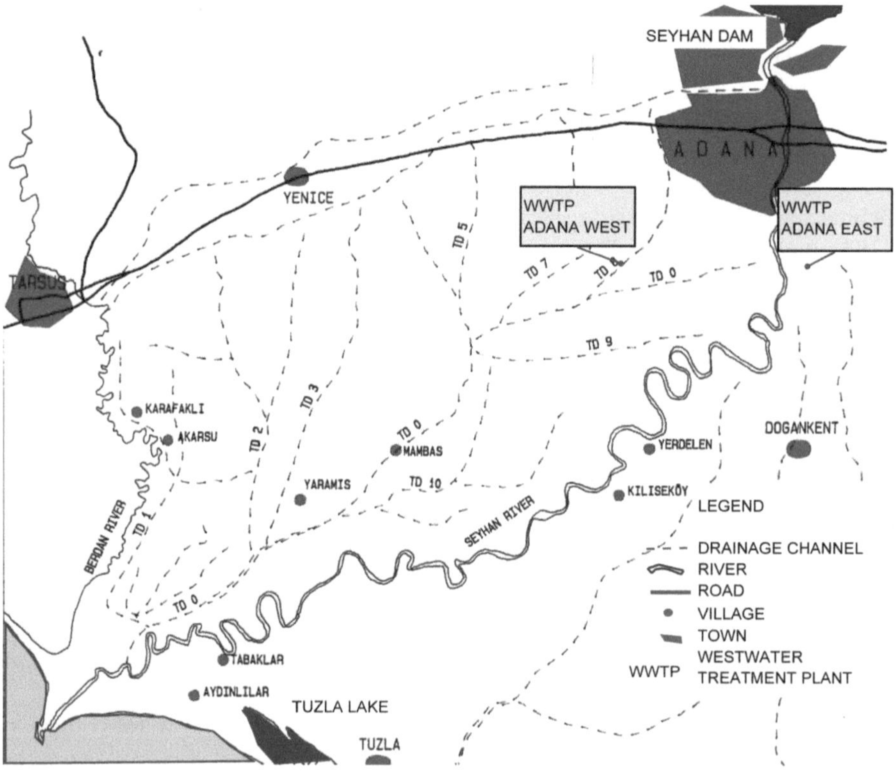

Fig. 7.6 Southern project area of Adana and the selected sites for constructing wastewater treatment plants

wastewater would not be discharged into the Seyhan River, but into the parallel major drainage channel of the large irrigation scheme, which is 120,000 ha.
2. From the technical point of view, the most serious problems facing the wastewater and storm water collection systems were that both had low efficiency, low collection rates, extensive leakage, and insufficient hydraulic capacity. Optimal technical solutions were viewed as difficult. Even secondary solutions would require at least one and possibly several years for implementation.
3. The most crucial discovery of this study phase, however, was that the proposed solutions for the wastewater collection and treatment facilities were neither financially nor administratively feasible. This was not only because of the high investment cost estimated at more than € 125 million, but also because of the weak financial situation of the client due to system water losses of more than 60% at the time.

The dimension of the large irrigation area can be illustrated by comparing the water amount of 120 m^3 /second in the peak month of September with that of the whole city, roughly estimated to be 4 m^3/s. The artificial lakes and dams are therefore mainly constructed for irrigation purposes and electricity production. Serving the city with water is only a minor by-product of these facilities. Seyhan Lake, located close to the city, was built approximately 50 years ago, when the large wetland area south of Adana was transformed into irrigation land.

After extensive discussions and negotiations, the Water and Sewage Company of Adana (ASKI), KfW, and the European Investment Bank (EIB) came to the conclusion that a detailed Water Supply Feasibility and a Wastewater Amendment Study were required. The most important conclusions in the Wastewater Amendment Study (GKW Consult 1996/1997) are

1. The environmental objectives could also be reached step by step by constructing two wastewater treatment plants, one for Adana West and one for Adana East. The Adana West plant would have a designed capacity of 1.2 million population equivalent (P.E.) for the target year 2010 (and 1.8 P.E. for the target year 2025). The Adana East plant would have a designed capacity of 0.7 million P.E. for the target year 2015 (and 0.924 million P.E. for the target year 2025). These two plants would replace one central plant and eliminate the need to transfer water from Adana East to West by crossing the Seyhan River.
2. The investment costs for the first stage could be reduced by more than € 50 million based on the following modifications:
    (a) Prioritize the improvement of the eastern wastewater collection system and the development of the wastewater treatment facilities for Adana East. This can be done because the wastewater collection rate for Adana East is comparably low.
    (b) Construct a by-pass and new sewer lines (Fig. 7.7) to solve the serious hydraulic constraints of the western sewer and storm water system. This

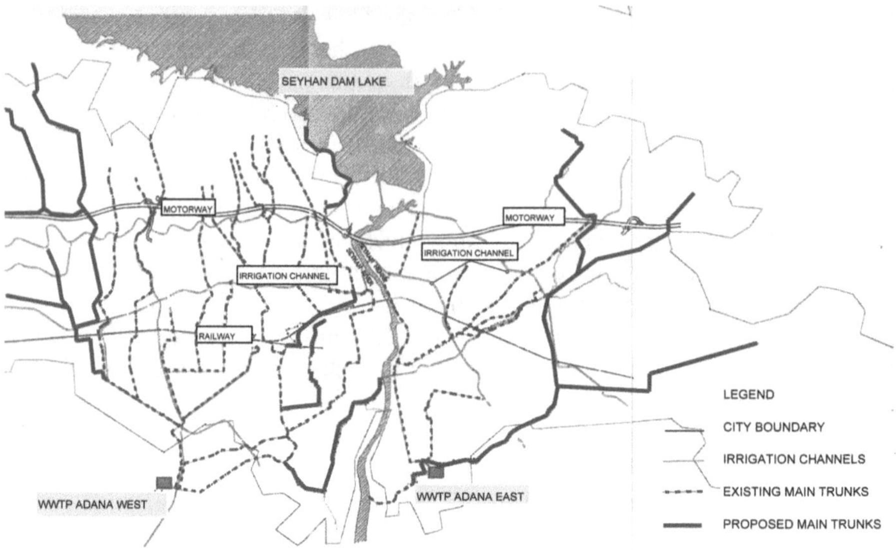

Fig. 7.7 Main sewer trunks and sites of the Adana wastewater treatment plants West and East

would help avoid overloading the combined system in the southern parts of the older city.

(c) Do not connect the western industrial zone to the treatment plant in Adana West, but rather install separate wastewater collection, treatment, and discharge facilities for the larger factories. The main advantages of this approach would be: making treatment of the specific industrial wastewater more efficient; saving the transfer lines to the western central wastewater treatment plant; reducing the capacity of the central plant by roughly 20%; allocating the costs for investment and operation of the industrial wastewater facilities according to the polluter-pay-principle.

These modifications are outlined in more detail below.

1. 177.3 km of additional storm water channels need to be constructed in stages: 44.6 km in the first stage, serving an area of 6,780 ha (55% of the 2005 development target of 12,328 ha); 42.3 km in the second stage, serving an area of 9,368 ha (65% of the 2015 development target of 14,413 ha); and 90.4 km in the third stage, serving an area of 13,554 ha (85% of the 2025 development target of 15,946 ha).
2. The storm water system of Adana East must be extended by 17.5 km in the first stage, by 5.1 km in the second stage, and 23.9 km in the third stage, for a total of 46.5 km. For the storm water system of Adana West, the implementation schedule for the 110.8 km extension will be: 27.1 km (2005), 17.2 km (2015) and 66.5 km (2025).

3. For improving the wastewater collection system, 95.6 km new sewer lines will be implemented in stages: 77.1 km for the target year 2005, 13.5 km for the year 2015, and 5.0 km for the year 2025. The wastewater collection rate of less than 65% in 1995 will increase to a level of 90% in 2025.
4. The three largest textile factories located on the main road between Adana West and Tarsus, have meanwhile constructed one joint wastewater treatment plant and they obtain their water from private groundwater wells. Some industries located in the centre have been shut down or moved into the organized industrial zones, where factories are obliged to arrange their water infrastructure in accordance with the polluter-pay-principle.

The essential revision to the previous wastewater collection and treatment system is the two-city concept. This revision divides the system into two independent ones, one for Adana West and one for Adana East.

7.2.3.2 Planning and Implementation of the Water Supply System

The Water and Sewage Work Company (ASKI) was obtaining water from 148 deep wells at a rate of 4.2 m^3/s while the industrial sector demand was mainly provided by their own wells. Within the scope of the Water Supply Feasibility Study in 1995–1997 (Lahmeyer International et al. 1996/1997), it was found that the groundwater wells were exposed to organic and chemical contamination from leaking sewerage and septic tanks, industrial wastes, fertilizers, and various agricultural compounds. Furthermore, it was learned that the existing water supply network was not structurally sound. The study area was divided into pressure zones because the elevations increase from 14 m above sea level in the south to 160 m in the north.

Pumping water directly into the municipal network from wells served by the 80 m deep aquifer leads to hydraulically unstable conditions. Therefore, it was recommended that the use of the groundwater aquifer be replaced gradually by a long-distance surface water supply system. If necessary, rehabilitated wells with a high yield can be integrated at a later time. The most suitable source of water was found to be the Catalan Dam constructed upstream of the Seyhan River, and located about 15 km away from the city. If the contamination problems of the groundwater resources can be solved in ten to 20 years, the option would remain to remobilize the groundwater resources later at a larger scale.

The previous water transfer concept of constructing only one transmission line from the Catalan Dam to Adana West, and to supply Adana East using transfer lines from the western area by crossing the Seyhan River was modified during the final design stage in 1998–1999 (GKW Consult 1999–2004). Further investigations confirmed that the "two city" concept, which was chosen for the wastewater system, would also be the more flexible and stable solution for the water supply system.

The main components of the new Adana Water Supply System (Fig. 7.8) are: the water intake structure, the tunnels and transmission lines for the raw water, the water treatment plant, transmission lines for the treated water, water reservoirs, pump stations, and the primary and secondary water distribution system. The main

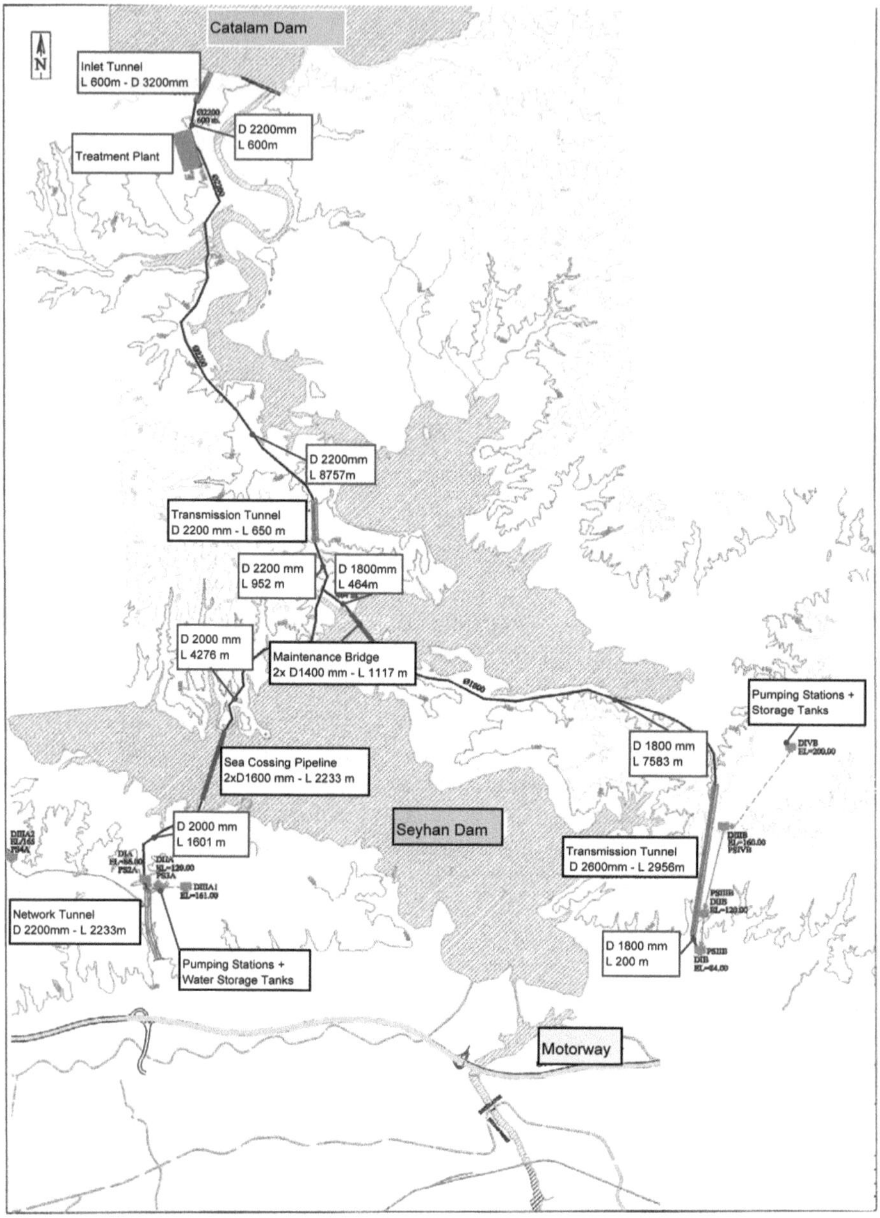

Fig. 7.8 General layout, with some major components of the "Two City" water supply strategy serving Adana West and East

water distribution network is now divided into eleven pressure zones with the following elevations: below 40 m; between 60–100; 100–140; 140–180; 180–220; and 220–270 m. The supply zones with higher elevations are located around Seyhan Lake and require pumping facilities. More than 75% of the population are therefore living in the below 40 and 60–100 m elevation zones. Those supply zones can be served by gravity (see Fig. 7.8), an aspect that is less important for the future development.

The major works implemented between 2000 and 2003 are:

1. Construction and commissioning of works on a turnkey potable water treatment plant with 250,000 m^3/day capacity in the first stage. This is to be extended to 1,000,000 m^3/day capacity in 2025 in three subsequent stages.
2. Construction of a 25 km transmission line that includes pipes with diameters of 1,600, 1,800, 2,000, and 2,200 mm.
3. Installation of 1,117 m of twin pipes with a diameter of 1,400 mm crossing Seyhan Lake to supply the eastern region.
4. Installation of 2,233 m of twin pipes with a diameter of 1,600 mm crossing Seyhan Lake to supply the western region.
5. Construction of two 2,200 mm diameter transmission tunnels, 2,662 and 1,482 m in length.
6. Construction of 9 reservoirs with a total capacity of 57,000 m^3.
7. Construction of pump stations and booster pump stations at 7 different locations.
8. Construction of primary distribution networks in pressure zones I, II, III and IV, with a total length of 55 km.

Instead of developing Adana to the north, the new water infrastructure supports that (see Fig. 7.9) the city should be developed in the western and eastern directions. The long-term development strategy for Adana should be to develop the supply

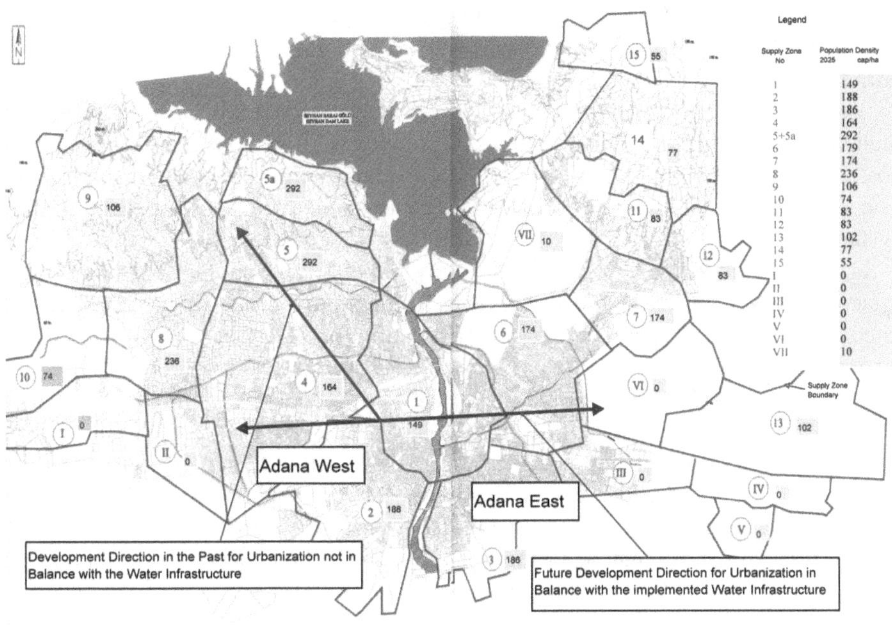

Fig. 7.9 Water supply zones with projected population densities in the year 2025, with past and future town development directions

districts in Adana East, where the population density is comparably low, even zero in some areas. All these measures contribute to balance the economies of scale (or positive externalities and social benefits) of the water supply and wastewater system with the water-flow related negative externalities (or social costs). This dynamic balancing process of externalities might be a more helpful approach in the context of optimizing local public goods, especially water infrastructure.

7.2.4 Dynamic Prime Costs of Adana's Water Infrastructure

The rehabilitation of the Leuna aquifer involved the installation of technical facilities to work with the existing system. The water infrastructure of Adana and the other urban centres, however, already contain all the essential components. Groundwater contamination and water treatment are only a sub-component of the urban water infrastructure.

Groundwater or raw water is first extracted from natural lakes, rivers, or artificial lakes. It is treated to drinking water quality and then pumped or transported into the urban water distribution system. The urban system consists of main pipes, pumping stations and storage tanks as well as secondary and tertiary networks with domestic connections and in-house facilities. The used water ultimately becomes wastewater, which must be collected and transported to wastewater treatment plants. The treated wastewater will then be discharged into the drainage channels of the irrigation ditches, into the Seyhan river delta and finally into the Mediterranean ocean. The water and sludge residuals from wastewater treatment must also be disposed into the natural environment. The stabilized sludge from the Adana wastewater and sludge treatment plants could be used for soil improvement and agriculture purposes,

The calculation of dynamic prime costs for the urban water infrastructure should include all relevant system components. The method of calculating dynamic prime costs, already shown in Sect. 7.1.5 for only one sub-component, remains the same. The calculation will simply be extended using more details about the essential system elements.

For the Adana case, a special tariff study for the water supply and wastewater system (GKW Consult 1997) was prepared. The study looked at using the calculated dynamic prime costs as the basic information for a cost recovery tariff system, where social aspects such as household income, are taken into account. To repeat the study's calculations here would go too far. We will outline the major results that can be used to extend our dynamic model.

Based on 1997 prices and a discount rate of 5%, the dynamic prime costs for the rehabilitation and extension of the water supply systems are 0.25€/m^3. The cost is the sum of two components: the dynamic prime costs for the investments (0.09€/m^3) and the dynamic prime costs for the maintenance and operation of the new system (0.16€/m^3). The dynamic prime costs for investments include the costs for the rehabilitation and improvement measures of the existing system, the new transmission lines, the water treatment plant, pumping stations, new storage tanks, the primary network, land acquisition, and consulting services. The dynamic prime costs for maintenance and operation of the new system include costs for maintenance, energy, chemicals, personnel, and administration.

The dynamic prime costs for wastewater collection and treatment are estimated at 0.21€/m^3, with 0.12€/m^3 for investments and 0.09€/m^3 for maintenance and operation. The investment costs include costs for the wastewater treatment plants, main sewer lines, vehicles, inventory, and replacement of existing sewers.

An updated tariff study is not available, but the actual tariffs are roughly three times higher than the dynamic prime costs described previously. This shows that tariffs are very sensitive to increases in essential input prices, such as the prices for chemicals, personnel, or especially energy. Technical and administrative water losses could be reduced from roughly 50% to less than 30% in Adana. This decrease in water loss would reduce energy costs, which are permanently increasing because energy production is based on non-renewable resources.

7.3 Conclusions for Constructing the Water Infrastructure Model

As illustrated for the urban centre of Adana in Turkey, returning the water infrastructure to a level where economies of scale are not overcompensated by water flow related externalities is a very complicated and time consuming process. Instead of trying to find an optimal solution for the water infrastructure, the strategy developed and implemented in the city of Adana was to restructure the water supply and wastewater collection system into clearly defined service zones. The restructuring separates the overwhelming externalities into problem zones for which individual priorities can be set.

In more detail, the solution for Adana is to implement a "two city" concept. This "two city" concept avoids negative feedbacks between the rehabilitation and restructuring of the existing systems, mainly located in Adana West, and the new urban development zone in Adana East. Furthermore, this concept allows time for rehabilitating and restructuring the existing systems. It also allows time to develop the new settlement areas with a modern water infrastructure that has low flow-related externalities combined with economies of scale. Over the next 2 decades, it will be possible to return the water infrastructure of Adana West to the economies of scale level. It might be even possible to use the groundwater aquifer again at a larger scale, when the contamination problems will have been solved and reasonable protection zones around important wells are established.

The Adana case study shows that there are clear interrelations between the urbanization process and water infrastructure. It also highlights some important ideas. First, water treatment should be considered in addition to water extraction and distribution. Second, the reduction of water losses over time should be an integral part of the dynamic model. Finally, because of the interrelation of water and exergy losses, energy efficiency should also be taken into account.

Chapter 8
Specifications for Constructing the Water Infrastructure Model

Abstract The model structure is extended by introducing capital stocks with innovation properties for the water production and distribution sector and for the wastewater collection and treatment sectors. Besides water saving strategies and wastewater collection rate improvements, the increase in the energy efficiency will also be connected to the development of these capital stocks. The development of the capital stocks for the different sectors depends on the development path for producing the quantities of the desired consumption good. These are mainly determined by the optimality conditions under the model constraints. In particular, for the water infrastructure model, there is a need to reduce the number of flow and stock variables. This is achieved using a substitution strategy in advance. Otherwise, determining model constraints and the derivation of optimality conditions becomes a complicated procedure.

We will now show that the dynamic water use model has the flexibility to be extended in different directions and can be used to address problems facing urban water infrastructures in rapidly growing cities. From the environmental perspective, the model can be used to study the long-term effects of water loss reduction; water saving strategies; or the relationship between water and wastewater treatment.

The MTBE contamination case study addresses one specific component. The rehabilitation and extension measures of the water supply and wastewater system of Adana, however, include all the essential components of urban water infrastructure. The elimination of sand and other residuals by the new water treatment plant of Adana and the reduction of the MTBE contamination of Leuna aquifer are identical problems, being solved by different technologies. Wastewater treatment is similar for both cases because the used water becomes wastewater. It must be treated to achieve certain water quality standards that are exogenously determined by political decisions and supported by different environmental instruments.

Even though water uses and wastewater treatment aspects have already been included, the dynamic water use model still needs a more detailed structure for the water supply and wastewater sector. To integrate water treatment aspects, for example, we have to release the assumption that the quality of the extracted water is high enough that it does not require treatment. As confirmed by the

H. Niemes, M. Schirmer, *Entropy, Water and Resources*,
DOI 10.1007/978-3-7908-2416-2_8, © Springer-Verlag Berlin Heidelberg 2010

Adana case, groundwater resources can be replaced by surface water, in particular if the groundwater is contaminated and requires the construction of large-scale water treatment facilities. For the integration of additional components, our dynamic model will be extended accordingly. However, the details, or aggregation, of the urban water infrastructure depends on the kind of problems being analysed.

8.1 Structure of the Water Infrastructure Model

The structure of the water use model is determined by the assumptions outlined in Chaps. 4 and 5. The input factors for the water use model are human labour and free energy for water production and wastewater. No water infrastructure stock variables accumulate over time. From the case studies, we know that a large number of different types of stock variables exist in reality. These variables can be integrated into the model in order to reflect urban water infrastructure problems. From the tariff studies, however, we learned that the detailed information needed for dimensioning, designing, and constructing the urban water infrastructure are ultimately aggregated to total investments and operation costs for the civil work and mechanical and electrical equipment components of the water supply and wastewater systems.

The aggregation level, however, should not go too far. The minimum requirements for integrating the urban infrastructure successfully are:

1. to distinguish between the water supply and wastewater sector;
2. to introduce a minimum of two additional production processes for producing the capital goods, specifically one for civil work and the other for mechanical & electrical equipment;
3. to add a limited number of extraction processes of raw materials that are required for the production of capital goods;
4. to keep the water and wastewater treatment processes that already exist in our basic model separate from the other investment components of the urban water infrastructure because the operation costs are not only a function of the water and wastewater amounts but they also depend on the water quality levels that the water and wastewater treatment must achieve.

Within the framework of these requirements, the assumptions of the water use model will be adopted, or modified as follows:

Assumption 8.1 (for the time structure) The entire planning horizon will be divided into time intervals $\tau = 1,...,T$. Each interval will contain discrete time periods that can be considered very short to allow for distinguishing between them. The consumption and capital goods produced in the activity interval τ sustain the requirements for the next time intervals.

Assumption 8.2 (for the number of desired goods) The basic model has desired good 1, called the consumption good, with amounts $Q(\tau) = Q_1(\tau)$ and good 2, called the capital good, with amounts $K^Q(\tau){=}Q_2(\tau)$ in the activity interval τ. Two additional capital goods, the civil work capital good with amounts $K^{CW}(\tau) = Q_3(\tau)$ and the mechanical & electrical equipment capital good with amounts $K^{ME}(\tau) = Q_4(\tau)$ are required for building up the urban water and wastewater infrastructure.

Assumption 8.3 (for the number of processes that require capital goods) The capital good $K^Q(\tau)$ is only required to build up the capital stock for the production process ^{Q}, where the extracted raw materials are converted to the desired consumption good. The capital goods $K^{CW}(\tau) \wedge K^{ME}$ are used to build up the capital stock, required as input for four processes: the water distribution and water treatment processes ^{WD}, ^{WT} and the wastewater collection and treatment processes ^{WWC}, ^{WWT}.

Assumption 8.4 (for the number and amounts of required raw materials) The raw materials $N^Q_{i=1,2}(\tau)$ and $N^K_{\mu i=1,2}(\tau)$ are needed to produce the consumption and capital good; and the raw materials $N^{CW}_{\nu=1,2}(\tau)$ and $N^{ME}_{o=1,2}(\tau)$ to build up the urban infrastructure, namely the civil work

$$K^{CW}(\tau) = K^{CW^{WD}}(\tau) + K^{CW^{WT}}(\tau) + K^{CW^{WWC}}(\tau) + K^{CW^{WWT}}(\tau), \quad (8.1)$$

And the mechanical & electrical capital good amounts

$$K^{ME}(\tau) = K^{ME^{WD}}(\tau) + K^{ME^{WT}}(\tau) + K^{ME^{WWC}}(\tau) + K^{ME^{WWT}}(\tau). \quad (8.2)$$

Assumption 8.5 (for the number and concentrations of unavoidable by-products) The number of unavoidable by-products could be arbitrarily chosen. The lowest number could be one unique by-product or a mixture of a finite number of by-products, j being generated by the extraction processes of the different raw materials. A unit mixture of unavoidable by-products j will be generated by the different extraction processes, with average concentrations that are related to concentrations of the corresponding desired raw materials

$$\begin{aligned} \bar{c}_j^{\overset{Q}{i=1,2}}(\tau) &= 1 - c^Q_{i=1,2}(\tau), \; \bar{c}_j^{\overset{K}{\mu=1,2}}(\tau) = 1 - c^K_{\mu=1,2}(\tau) \\ \bar{c}_j^{\overset{CW}{\nu=1,2}}(\tau) &= 1 - c^{CE}_{\nu=1,2}(\tau), \; \bar{c}_j^{\overset{ME}{\nu=1,2}}(\tau) = 1 - c^{ME}_{o=1,2}(\tau). \end{aligned} \quad (8.3)$$

Assumption 8.6 (for water and wastewater treatment) The water extracted from the ecological system will be treated to a water concentration standard $\bar{c}^W$ that is acceptable for the production and reproduction activities. The concentration of unavoidable by-products that remain after water and wastewater treatment will not be taken into consideration.

Assumption 8.7 (for the use of water for production and reproduction activities) The produced water will be used to transport unavoidable by-products from the production and the reproduction sector to the wastewater treatment sector. No water will be used to transport the by-products from the extraction processes that produce the civil work and mechanical & electrical capital goods.

Assumption 8.8 (for the concentrations of the materials appropriated from the ecological system and used for extracting the required raw materials) The concentrations of the raw materials in the natural environment are assumed to be declining functions of activities in the past within the total planning period of the following type

$$\begin{aligned} c(\bar{N}_i^Q(\tau)) &= c(\bar{N}_i^Q(0))\, e^{-\varphi_i^Q \bar{N}_i^Q(\tau)} >> \underline{c}_i^Q > 0 \\ \text{with } \bar{N}_i^Q(\tau) &= \sum_{t=1}^{\tau-1} N_i^Q(t) \\ c(\bar{N}_\mu^K(\tau)) &= c(\bar{N}_\mu^K(0))\, e^{-\varphi_\mu^K \bar{N}_\mu^K(\tau)} >> \underline{c}_\mu^K > 0 \\ \text{with } \bar{N}_\mu^K(\tau) &= \sum_{t=1}^{\tau-1} N_\mu^K(t). \end{aligned} \tag{8.4}$$

$$\begin{aligned} c(\bar{N}_\nu^{CW}(\tau)) &= c(\bar{N}_\nu^{CW}(0))\, e^{-\varphi_\nu^{CW} \bar{N}_\nu^{CW}(\tau)} >> \underline{c}_\nu^{CW} > 0 \\ \text{with } \bar{N}_\nu^{CW}(\tau) &= \sum_{t=1}^{\tau-1} N_\nu^{CW}(t) \\ c(\bar{N}_o^{ME}(\tau)) &= c(\bar{N}_o^{ME}(0))\, e^{-\varphi_o^K \bar{N}_o^{ME}(\tau)} >> \underline{c}_o^{ME} > 0 \\ \text{with } \bar{N}_o^{ME}(\tau) &= \sum_{t=1}^{\tau-1} N_o^{ME}(t). \end{aligned} \tag{8.5}$$

Assumption 8.9 (for the human labour inputs for producing one unit of desired output) The human labour input coefficient for producing the consumption good, which requires the capital good, is a declining function that is closely connected to the newly acquired investment into the capital stock. This action permanently modifies the interface between the economic and ecological system. The human labour input coefficient starts with an initial value determined by the initial capital stock $\bar{K}^Q(0)$ and is a declining function of the following type

$$l^Q(\bar{K}^Q(0), \bar{Y}_K^Q(\tau)) = l^Q(\bar{K}(0))\, e^{-\varphi_l \bar{Y}_K^Q(\tau)} > 0 \text{ with } \bar{Y}_K^Q(\tau) := \sum_{t=1}^{\tau-1} Y_K^Q(t). \tag{8.6}$$

The human labour input coefficient for the other processes of the economic system can also change. We therefore assume

Assumption 8.10 (for the total amount of human labour inputs for the production and reproduction activities) The total amount of human labour input for production activities $L^P(\tau)$ can be assumed to remain constant or, alternatively, to increase according to the following function with a population increase rate $\gamma > 0$

$$L^P(\tau) = L(0) , \tau \in \{1,...,T\} \vee L^P(\tau) = L(0)(1+\gamma)^{\tau}. \quad (8.7)$$

Assumption 8.11 (for building up the capital stock) The capital stocks should have the characteristics of permanent technological modernization and evolution.

Assumption 8.12 (the restriction for building up the capital stocks) The capital stocks will be maintained and evolved to such a degree that the properties of the existing capital stock and the newly invested capital good used for building up the capital stocks are marginally different from each other.

Assumption 8.13 (for increasing the production efficiency using the capital stock of the consumption good process) The permanently modernized capital stock for the consumption good process leads to a reduction in the human labour input coefficient for the economic transformation processes that use capital goods as input factors.

Assumption 8.14 (for increasing the efficiency using the capital stocks for the water infrastructure) The permanently modernized civil work capital stocks for the water infrastructure lead to a reduction in water losses from the water distribution system and the wastewater collection system. The permanently modernized mechanical and electrical equipment capital leads to an increase of the energy efficiencies for the water infrastructure.

The relations between these capital stocks and the water losses and energy efficiency will be introduced in the context of process coefficients for the urban water infrastructure. For the energy content of the raw material that is used to generate free energy, we assume

Assumption 8.15 (for producing free energy for the economic activities) One unit of raw material used for energy production should have time-invariant, embedded specific energy content $\sigma^E(N_e^E)$. This leads us to the simple linear relation between the energy and raw materials amounts

$$\begin{aligned} E_e^E(\tau) &= \sigma^E(N_e^E) N_e^E(\tau) = \frac{\sigma^E(N_e^E) N^E(\tau)}{c(\bar{N}^E(\tau))} \\ \textit{with } c(\bar{N}^E(\tau)) &= \frac{N_e^E(\tau)}{N^E(\tau)}. \end{aligned} \quad (8.8)$$

$c(\bar{N}^E(\tau))$, is the average concentration of the material $\bar{N}^E(\tau)$ that is extracted from the natural environment for producing free energy in the energy sector E during production interval τ. $c(\bar{N}^E(\tau))$, could also be a declining function of the following type with a constant coefficient $\varphi_e \geq 0$ in relation to the total amount of material already extracted during previous periods

$$\begin{aligned} c(\bar{N}^E(\tau)) &= c(\bar{N}^E(0))\, e^{-\varphi_e \bar{N}^E(\tau)} >> \underline{c}^E > 0 \\ \textit{with } \bar{N}^E(\tau) &= \sum_{t=1}^{\tau-1} N^E(t). \end{aligned} \tag{8.9}$$

We also have

Assumption 8.16 (for the quality of the produced free energy) The produced free energy should be of the highest quality attainable. For example, electricity should be the dominant energy form to transfer low exergy inputs to the high exergy outputs of the economic transformation process.

8.2 Process Coefficients of the Water Infrastructure Sectors

The input and output coefficients of the processes for generating free energy, extracting raw materials, and producing the capital good used to convert the raw materials to the consumption good in the basic dynamic model remain unchanged. The assumptions made for the integration of the urban water infrastructure lead to an extension of our basic model by additional transformation processes. The assumptions also lead to some revisions of the water production and wastewater treatment processes that already exist.

As expressed in Assumptions 8.2 and 8.3, two additional capital good processes, the civil work and the mechanical & electrical equipment capital good processes, will be introduced for building up the capital stock of the urban water infrastructure. In these processes, two raw materials are converted into either the civil work or the mechanical& electrical equipment capital goods.

8.2.1 Coefficients of the CW and M&E Production Processes

For each of the additional capital good processes, two separate raw materials are required. The set of indices $g \in \left\{ {}^{E}, {}^{W}, {}^{Q}_{1}; {}^{Q}_{2}, {}^{K}_{1}; {}^{K}_{2}, {}^{Q}, {}^{K}, {}^{RP}, {}^{WWT} \right\}$ used to specify the different processes in the water use model must be extended to the set of indices $\tilde{g} \in \left\{ {}^{E}, {}^{CW}_{1}, {}^{CW}_{2}, {}^{ME}_{1}, {}^{ME}_{2}, {}^{CW}, {}^{ME}, {}^{W}, {}^{Q}_{1}; {}^{Q}_{2}, {}^{K}_{1}; {}^{K}_{2}, {}^{Q}, {}^{K}, {}^{RP}, {}^{WWT} \right\}$ for the urban water infrastructure model. The extraction processes for the raw materials that are converted by the civil work and mechanical & electrical equipment process into the capital goods that are required for building up the urban infrastructure are represented by the indices: ${}^{CW}_{\nu}, \nu = 1,2 \wedge {}^{ME}_{o}, o = 1,2$.

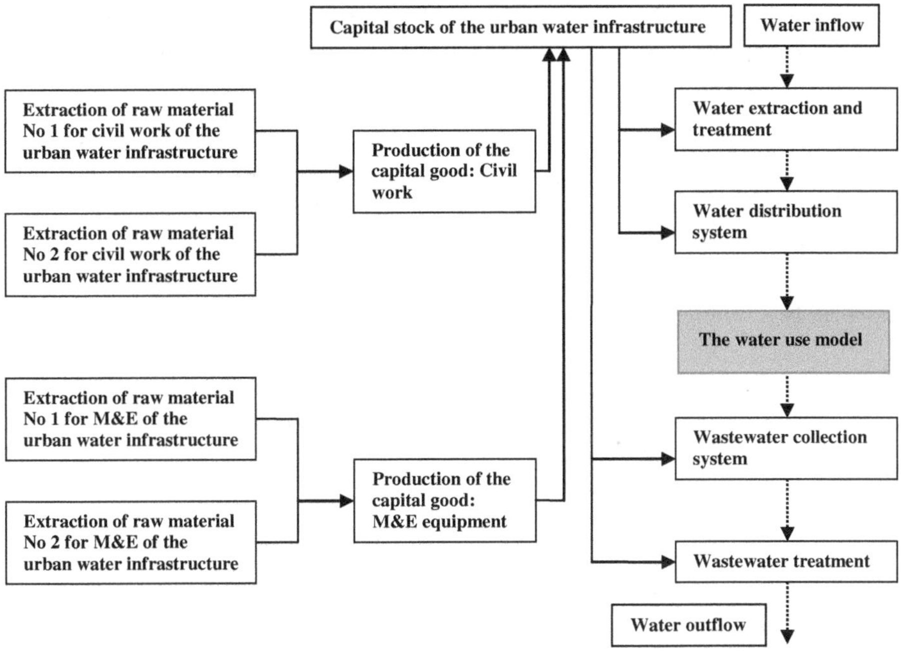

Fig. 8.1 Extension of the water use model by processes of the urban water infrastructure

When comparing the extraction processes for producing the capital and consumption goods with these extraction processes, Assumption 8.7 means that no specific water inputs are required to transport the residuals into the natural environment. This assumption is similar to the assumption made for energy production, where we argued that the internal energy consumption for production is included in the gross energy and only the available net free energy will be considered.

8.2.1.1 The Energy Coefficients and Amounts of the Processes for the Civil Work and Mechanical and Electrical Equipment Capital Goods

Equation (3.22), derived for the energy input coefficient for the extraction processes in accordance with the thermodynamic laws, could also be used to specify the energy inputs of the four additional extraction processes as follows

$$
\begin{aligned}
&e_\nu(c\,(\bar{N}_\nu^{CW}(\tau))) \\
&= \frac{\frac{ex_\nu^{app}}{c_\nu\,(\bar{N}_\nu^{CW}(0))}\, e^{\varphi_\nu^{CW}\,\bar{N}_\nu^{CW}(\tau)} - \varepsilon\,\ln c_\nu\,(\bar{N}_\nu^{CW}(0)) + \varepsilon\,\varphi_\nu^{CW}\,\bar{N}_\nu^{CW}(\tau)}{\eta_{\,\nu}(c\,(\bar{N}_\nu^{CW}(\tau)))}, \nu = 1,2 \\
&e_o(c\,(\bar{N}_o^{ME}(\tau))) \\
&= \frac{\frac{ex_o^{app}}{c_o\,(\bar{N}_o^{ME}(0))}\, e^{\varphi_o^{ME}\,\bar{N}_o^{ME}(\tau)} - \varepsilon\,\ln c_o\,(\bar{N}_o^{ME}(0)) + \varepsilon\,\varphi_o^{ME}\,\bar{N}_o^{ME}(\tau)}{\eta_{\,o}(c_{\,o}(\bar{N}_o^{ME}(\tau)))}, o = 1,2.
\end{aligned}
\tag{8.10}
$$

In the case where the availability of the natural resources for the production of the civil work and mechanical & electrical equipment goods are not critical constraints, we have assumed

Assumption 8.17 (for the raw materials for the production of the civil work and mechanical & electrical equipment capital good) The natural resources for the production of the civil work and mechanical & electrical equipment capital goods are considered renewable.

Therefore, the coefficients have the values $\varphi_\nu^{CW} = 0 \wedge \varphi_o^{ME} = 0$, Eq. (8.10) can then be simplified using the following factors

$$
\begin{aligned}
e_\nu := \ & e_\nu(c(\bar{N}_\nu^{CW}(\tau))) = \frac{f_\nu^{CW}}{\eta_\nu^{CW}(\tau)\, c_\nu\, (\bar{N}_\nu^{CW}(\tau))}, \ \nu = 1,2 \\
e_o := \ & e_o(c(\bar{N}_o^{ME}(\tau))) = \frac{f_o^{ME}}{\eta_\nu^{ME}(\tau) c_o\, (\bar{N}_o^{ME}(\tau))}, \ o = 1,2.
\end{aligned} \tag{8.11}
$$

As a result, no inter-temporal negative effects will be caused by the use of these natural resources.

Instead of using energy efficiency coefficients, it is more practical to split the energy input into the component doing useful work (exergy) and the component of lost exergy (called anergy), and to replace Eq. (8.10) by the functions

$$
\begin{aligned}
e_\nu := \ & ex_\nu + an_V = \frac{\tilde{f}_\nu^{CW}}{c_\nu\, (\bar{N}_\nu^{CW}(\tau))} + an_V, \ \nu = 1,2 \\
e_o := \ & ex_o + an_o = \frac{\tilde{f}_o^{ME}}{c_o\, (\bar{N}_o^{ME}(\tau))} + an_o, \ o = 1,2.
\end{aligned} \tag{8.12}
$$

These functions show that efforts to increase the energy efficiency can synonymously be interpreted as reductions of anergy. This alternative interpretation will be applied for those processes, where energy efficiencies are influenced by capital stock innovations.

When taking into consideration

Assumption 8.18 (for the input and output characteristics of the civil work and mechanical & electrical equipment capital good processes) The processes ^{CW} and ^{ME} have input and output characteristics similar to the capital good process^K. This means that coefficients for the civil work and mechanical and electrical equipment processes for the human labour, free energy, and raw material inputs are time-invariant.

The energy input amounts of these extraction processes $_\nu^{CW} \wedge {}_o^{ME}$, $\nu = 1,2 \wedge o = 1,2$ depend on the activity levels of the civil work $Y^{CW}(\tau)$ and mechanical and electrical processes $Y^{ME}(\tau)$. The corresponding input coefficients $\kappa_\nu^{CW} \wedge \kappa_o^{ME}$, $\nu = 1,2 \wedge o = 1,2$ of the raw materials that are to be converted to the civil work and mechanical & electrical equipment are given by

$$
\begin{aligned}
E_{\nu}^{CW}(\tau) &= e_{\nu}\kappa_{\nu}^{CW} Y^{CW}(\tau) = \kappa_{\nu}^{CW} Y^{CW}(\tau)\left\{\frac{\tilde{f}_{\nu}^{CW}}{c_{\nu}\,(\bar{N}_{\nu}^{CW}(\tau))} + an_{V}\right\} \\
E_{o}^{ME}(\tau) &= \kappa_{o}^{ME} Y^{ME}(\tau)\left\{\frac{\tilde{f}_{o}^{CW}}{c_{o}\,(\bar{N}_{o}^{CW}(\tau))} + an_{V}\right\}, \; o = 1,2.
\end{aligned}
\tag{8.13}
$$

Finally, the energy amounts inputs for these combination processes are determined by the simple relations

$$
\begin{aligned}
E^{CW}(\tau) &= e^{CW} Y^{CW}(\tau) = \left\{ex^{CW} + an^{CW}\right\} Y^{CW}(\tau) \\
E^{ME}(\tau) &= e^{ME} Y^{ME}(\tau) = \left\{ex^{CW} + an^{CW}\right\} Y^{ME}(\tau).
\end{aligned}
\tag{8.14}
$$

8.2.1.2 The Human Labour Coefficients and Amounts of the Processes for the Civil Work and Mechanical and Electrical Equipment Capital Goods

The expressions for the human labour input and water input coefficients for the extraction processes $_{i}^{Q} \wedge {}_{\mu}^{K}$ are

$$
l_i^{ext}(c_i(\bar{N}_i^{Q}(\tau)) = \frac{l_i^{app}}{c_i(\bar{N}_i^{Q}(0))}\, e^{\varphi_i^{Q}\bar{N}_i^{Q}(\tau)};\; \psi_i^{ext}(c_i(\bar{N}_i^{Q}(\tau)) = \frac{\psi_i^{app}}{c_i(\bar{N}_i^{Q}(0))}\, e^{\varphi_i^{Q}\bar{N}_i^{Q}(\tau)};
$$

$$
l_\mu^{ext}(c_\mu(\bar{N}_\mu^{K}(\tau)) = \frac{l_\mu^{app}}{c_\mu(\bar{N}_\mu^{K}(0))}\, e^{\varphi_\mu^{K}\bar{N}_\mu^{K}(\tau)};\; \psi_\mu^{ext}(c_\mu(\bar{N}_\mu^{K}(\tau)) = \frac{\psi_\mu^{app}}{c_\mu(\bar{N}_\mu^{K}(0))}\, e^{\varphi_\mu^{K}\bar{N}_\mu^{K}(\tau)},
$$

Similarly, the following expressions could also be introduced for the additional extraction processes $_{\nu}^{CW} \wedge {}_{o}^{ME}$

$$
\begin{aligned}
l_\nu^{ext}(c(\bar{N}_\nu^{CW}(\tau)) &= \frac{l_\nu^{app}}{c_\nu(\bar{N}_\nu^{CW}(0))}\, e^{\varphi_\nu^{CW}\bar{N}_\nu^{CW}(\tau)} \\
l_\nu^{ext}(c(\bar{N}_\nu^{CW}(\tau)) &= \frac{l_\nu^{app}}{c_\nu(\bar{N}_\nu^{CW}(0))}\, e^{\varphi_\nu^{CW}\bar{N}_\nu^{CW}(\tau)} \\
\psi_\nu^{ext}(c_\nu(\bar{N}_\nu^{CW}(\tau)) &= \frac{\psi_\nu^{app}}{c_\nu(\bar{N}_\nu^{CW}(0))}\, e^{\varphi_\nu^{CW}\bar{N}_\nu^{CW}(\tau)} \\
l_o^{ext}(c(\bar{N}_o^{ME}(\tau)) &= \frac{l_o^{app}}{c_o(\bar{N}_o^{ME}(0))}\, e^{\varphi_o^{ME}\bar{N}_o^{ME}(\tau)} \\
\psi_o^{ext}(c_o(\bar{N}_o^{ME}(\tau)) &= \frac{\psi_o^{app}}{c_o(\bar{N}_o^{ME}(0))}\, e^{\varphi_o^{ME}\bar{N}_o^{ME}(\tau)}.
\end{aligned}
\tag{8.15}
$$

Assumption 8.16 simplifies the free energy input coefficients. It also simplifies the human labour input coefficient for extracting the raw materials that will be converted to the civil work and mechanical & electrical equipment capital goods. In addition, Assumption 4.5 means that no water is used for the raw materials that are required for producing the civil work, and that the mechanical & electrical equipment components for the urban water infrastructure are taken into account. Equation (8.14) for the human labour input coefficients of the extraction processes $_{\nu}^{CW} \wedge _{o}^{ME}$ can be simplified to

$$l_{\nu}^{ext}(c_{\nu}(\bar{N}_{\nu}^{CW}(\tau)) = \frac{l_{\nu}}{c_{\nu}(\bar{N}_{\nu}^{CW}(0))} \wedge l_{o}^{ext}(c_{o}(\bar{N}_{o}^{ME}(\tau)) \ . \tag{8.16}$$

The human labour inputs for the extraction processes for producing the civil work and mechanical & electrical equipment goods, however, depend reciprocally on the concentration of the required raw materials as follows

$$L_{\nu}^{CW}(\tau) = \frac{l_{\nu}\,\kappa_{\nu}^{CW} Y^{CW}(\tau)}{c_{\nu}(\bar{N}_{\nu}^{CW}(0))}, \nu = 1,2 \wedge L_{o}^{ME}(\tau) = \frac{l_{o}\kappa_{o}^{ME} Y^{ME}(\tau)}{c_{o}(\bar{N}_{o}^{ME}(0))}, o = 1,2. \tag{8.17}$$

The human labour inputs for converting these raw materials to the civil work and mechanical & electrical *(ME)* equipment goods are simple linear functions

$$L^{CW}(\tau) = l^{CW} Y^{CW}(\tau) \wedge L^{ME}(\tau) = l^{ME} Y^{ME}(\tau). \tag{8.18}$$

8.2.2 Coefficients for the Water Infrastructure Processes

The additional capital goods *CW* and *ME* build up the urban water infrastructure. The Adana case study has shown that the urban water infrastructure is composed of many system components. Feasibility studies, final design, implementation, or tariff studies are based on the detailed information about the investment and operation costs of these components. Nevertheless, this detailed information is aggregated when calculating the so-called dynamic prime costs for the water supply and wastewater system as a whole, or explicitly, for some essential components of the system. This information is the basis for political decisions concerning the tariff level and its structure as well as for cost recovery strategies that take into account the social aspects of the different user groups.

It is always difficult to define an appropriate aggregation level for practical and theoretical work. The chosen aggregation level for integrating the urban water infrastructure into the model will be very high when compared with the engineering details available for calculating the dynamic prime costs of the water and wastewater system.

The water that is consumed ultimately becomes wastewater. It is therefore reasonable to divide the urban water infrastructure into a water supply system and a wastewater system. The function and dimensions of the water treatment plant are determined by both the amount of water to be treated and also by the water quality that is to be achieved in accordance with the given water quality standards. This aspect distinguishes the water treatment plant from the other elements of the water supply system. The dimensions and function of the other elements (surface water intake structure; groundwater extraction wells; long distance pipelines; pumping stations; storage tanks; primary, secondary, or tertiary distribution pipes and facilities) depend only on the volume of water to be handled.

The guidelines for the maximum aggregation level of the water infrastructure model are

1. To separate the system element(s), whose dimensions are determined by the required water quantity and quality; and
2. To aggregate the other system elements, whose dimensions depend only on the water quantity

These guidelines can be used for both the water and the wastewater systems because water treatment is similar to wastewater treatment and water distribution is similar to wastewater collection from an engineering point of view. Both treatment processes, for example, reduce and eliminate residuals, which influence either the quality of the drinking water or the emission standard of the discharged wastewater. Therefore, the water supply sector can be represented by the water treatment process WT and the water distribution process WD and the wastewater sector can be represented by the wastewater collection process WWC and the wastewater treatment process WWT.

Determining the input coefficients of these four processes for the energy, human labour, civil work, and mechanical & electrical equipment inputs can be based on how the input coefficient for the production process of the consumption good and the wastewater treatment process were approximated in the water use model. The energy input coefficient for reducing the unavoidable by-products of the production and reproduction activities can be approximated using Eq. (3.37), which was derived for two pollution components

$$e_{Q,\hat{j}}^{WWT}(\tau) = e_{Q}^{WWT}(\tau) + e_{\hat{j}}^{WWT}(\tau) = \frac{\tilde{\varepsilon}^{WWT}}{\eta^{WWT}(\tau)} \left\{ \ln \frac{c_Q^*(\tau)}{c_Q(\tau)} + \ln \frac{c_{\hat{j}}^*(\tau)}{c_{\hat{j}}(\tau)} \right\} .$$

In the case where there are more than two pollution components, the approximation must be extended accordingly.

We will later analyse how to improve the energy efficiency of the wastewater treatment process, so the alternate version of this relation can be used

$$\begin{aligned} e_{Q,\hat{j}}^{WWT}(\tau) &= ex_{Q\hat{j}}^{WWT}(\tau) + an_{Q\hat{j}}^{WWT}(\tau) \\ &= \tilde{\varepsilon}^{WWT} \left\{ \ln \frac{c_Q^*(\tau)}{c_Q(\tau)} + \ln \frac{c_{\hat{j}}^*(\tau)}{c_{\hat{j}}(\tau)} \right\} + an_{Q\hat{j}}^{WWT}(\tau) . \end{aligned} \tag{8.19}$$

The energy coefficients for water treatment, where a certain number of residuals have to be shifted to exogenously given water quality standards, are equal to

$$\begin{aligned} e^{WT}(\tau) &= \sum_{\omega} e_{\omega}^{WT}(\tau) = ex^{WT}(\tau) + an^{WT}(\tau) \\ &= \tilde{\varepsilon}^{WT} \left\{ \sum_{\omega} \ln \frac{c_{\omega}^{*}(\tau)}{c_{\omega}(\tau)} \right\} + an^{WT}(\tau) . \end{aligned} \tag{8.20}$$

The energy coefficient could also change over time, but this aspect will only be mentioned. We therefore assume

Assumption 8.19 (for the wastewater and water treatment shifting factors) The shifting factors for the concentrations of the wastewater and treated water are exogenously given by specific target concentrations.

To obtain energy input amounts for the urban water processes, the water and wastewater amounts have to be determined more precisely if water and wastewater losses exist. The water and wastewater amounts for the water use model

$$W^{nW} = W^{nRP} + W^{nP} = \psi_Q^{RP} Y_Q^{RP} + \sum_{i=1}^{2} \frac{\tilde{\psi}_i^Q}{c_i^Q} \kappa_i^Q Y^Q + \sum_{\mu=1}^{2} \frac{\tilde{\psi}_{\mu}^K}{c_{\mu}^K} \kappa_{\mu}^K Y^K = W^{nWW}$$

$$W^{nW} = W^{nWW} \approx Y^Q \left\{ \psi_Q^{RP} + \sum_{i=1}^{2} \frac{\tilde{\psi}_i^Q}{c_i^Q} \kappa_i^Q \right\} + \sum_{\mu=1}^{2} \frac{\tilde{\psi}_{\mu}^K}{c_{\mu}^K} \kappa_{\mu}^K Y^K$$

Can be interpreted and marked as net amounts. This can be done because water and wastewater losses are not considered in the basic model. A wastewater generation rate of 100% is also used in the basic model.

When replacing this assumption of the water use for the wastewater generation rate by

Assumption 8.20 (for the wastewater generation rate of the production and reproduction activities) The wastewater generation rate $\rho^{WW}(\tau)$ is in the range $0 \leq \rho^{WW}(\tau) < 1$. The wastewater amounts, which enter into the wastewater collection system, would be equal to

$$W^{WWC}(\tau) = Y^Q(\tau) \left\{ \rho^{WW}(\tau) \psi_Q^{RP} + \sum_{i=1}^{2} \frac{\tilde{\psi}_i^Q}{c_i^Q(\tau)} \kappa_i^Q \right\} + \sum_{\mu=1}^{2} \frac{\tilde{\psi}_{\mu}^K}{c_{\mu}^K(\tau)} \kappa_{\mu}^K Y^K(\tau). \tag{8.21}$$

When multiplying the collected wastewater amounts by the energy input coefficient of the wastewater collection process, we obtain the following expression for the energy inputs for the wastewater collection system

$$E^{WWC}(\tau) = \left[ex^{WWC}(\tau) + an^{WWC}(\tau)\right] \left\{ \begin{array}{l} Y^{Q}(\tau)\left\{\rho^{WW}(\tau)\psi_{Q}^{RP} + \sum\limits_{i=1}^{2} \frac{\tilde{\psi}_i^{Q}}{c_i^{Q}(\tau)} \kappa_i^{Q}\right\} \\ + \sum\limits_{\mu=1}^{2} \frac{\tilde{\psi}_{\mu}^{K}}{c_{\mu}^{K}(\tau)} \kappa_{\mu}^{K} Y^{K}(\tau) \end{array} \right\}. \tag{8.22}$$

There is likely to be leakage from the wastewater collection system, which means less wastewater will enter the treatment system. We therefore assume

Assumption 8.21 (for the leakage rate of the wastewater collection system) The leakage rate of the wastewater collection system is within the range $0 \leq r^{WWC}(\tau) < 1$.

The wastewater amounts are equal to

$$W^{WWT}(\tau) = \left\{1 - r^{WWC}(\tau)\right\} \left\{ \begin{array}{l} Y^{Q}(\tau)\left\{\rho^{WW}(\tau)\psi_{Q}^{RP} + \sum\limits_{i=1}^{2} \frac{\tilde{\psi}_i^{Q}}{c_i^{Q}(\tau)} \kappa_i^{Q}\right\} \\ + \sum\limits_{\mu=1}^{2} \frac{\tilde{\psi}_{\mu}^{K}}{c_{\mu}^{K}(\tau)} \kappa_{\mu}^{K} Y^{K}(\tau) \end{array} \right\}, \tag{8.23}$$

And the required energy amounts for wastewater treatment are given by

$$\begin{aligned} E^{WWT}(\tau) &= \left[ex^{WWT}(\tau) + an^{WWT}(\tau)\right] W^{WWT}(\tau) \\ &= \left[\tilde{\varepsilon}^{WWT}\left\{\ln\frac{c_Q^*(\tau)}{c_Q(\tau)} + \ln\frac{c_j^*(\tau)}{c_j(\tau)}\right\} + an^{WWT}(\tau)\right]^* \\ &\quad ...^*\left[1 - r^{WWC}(\tau)\right] \left\{ \begin{array}{l} Y^{Q}(\tau)\left\{\rho^{WW}(\tau)\psi_{Q}^{RP} + \sum\limits_{i=1}^{2} \frac{\tilde{\psi}_i^{Q}}{c_i^{Q}(\tau)} \kappa_i^{Q}\right\} \\ + \sum\limits_{\mu=1}^{2} \frac{\tilde{\psi}_{\mu}^{K}}{c_{\mu}^{K}(\tau)} \kappa_{\mu}^{K} Y^{K}(\tau) \end{array} \right\}. \end{aligned} \tag{8.24}$$

On the water supply side, the losses from the water supply system are taken into account by

Assumption 8.22 (for the water loss rate of the water distribution system) The water loss rate from the water supply system is in the range $0 \leq r^{WD}(\tau) < 1$.

For the internal water losses of the water treatment process, however, we assume

Assumption 8.23 (for the water losses of the water and wastewater treatment systems) The internal water losses for water and wastewater treatment, which could be in the ranges $0 \leq r^{WT}(\tau) < 1$ and $0 \leq r^{WWT}(\tau) < 1$, are minor amounts.

These assumptions mean that a surplus of water, namely the amount of water lost during the distribution process, has to be treated and distributed by the water supply system. The water amounts, for which the water supply system is dimensioned, are equal to

$$W^{WT}(\tau) = W^{WD}(\tau) = \left[1 + r^{WD}(\tau)\right] \left\{ \begin{array}{l} Y^{Q}(\tau) \left\{ \psi_{Q}^{RP} + \sum\limits_{i=1}^{2} \frac{\tilde{\psi}_{i}^{Q}}{c_{i}^{Q}(\tau)} \kappa_{i}^{Q} \right\} \\ + \sum\limits_{\mu=1}^{2} \frac{\tilde{\psi}_{\mu}^{K}}{c_{\mu}^{K}(\tau)} \kappa_{\mu}^{K} Y^{K}(\tau) \end{array} \right\}. \tag{8.25}$$

The energy amounts for the water supply distribution and water treatment processes are calculated in the same way that the energy amounts were calculated for the wastewater system. The water amounts are multiplied by the corresponding energy input coefficients.

$$\begin{aligned} E^{WD}(\tau) &= e^{WD}(\tau)\, W^{WD}(\tau) \\ &= \left[ex^{WD}(\tau) + an^{WD}(\tau)\right] \left[1 + r^{WD}(\tau)\right] W^{WD}(\tau) \\ &= \left[ex^{WD}(\tau) + an^{WD}(\tau)\right] \left[1 + r^{WD}(\tau)\right] \left\{ \begin{array}{l} Y^{Q}(\tau) \left\{ \psi_{Q}^{RP} + \sum\limits_{i=1}^{2} \frac{\tilde{\psi}_{i}^{Q}}{c_{i}^{Q}(\tau)} \kappa_{i}^{Q} \right\} \\ + \sum\limits_{\mu=1}^{2} \frac{\tilde{\psi}_{\mu}^{K}}{c_{\mu}^{K}(\tau)} \kappa_{\mu}^{K} Y^{K}(\tau) \end{array} \right\}. \end{aligned} \tag{8.26}$$

$$\begin{aligned} E^{WT}(\tau) &= e^{WT}(\tau)\, W^{WT}(\tau) \\ &= \left[ex^{WT}(\tau) + an^{WT}(\tau)\right] \left[1 + r^{WD}(\tau)\right] W^{WT}(\tau) \\ &= \left[\tilde{\varepsilon}^{WD} \sum_{\omega} \ln \frac{c_{\omega}^{*}(\tau)}{c_{\omega}(\tau)} + an^{WT}(\tau)\right] \left[1 + r^{WD}(\tau)\right] \left\{ \begin{array}{l} Y^{Q}(\tau) \left\{ \psi_{Q}^{RP} + \sum\limits_{i=1}^{2} \frac{\tilde{\psi}_{i}^{Q}}{c_{i}^{Q}(\tau)} \kappa_{i}^{Q} \right\} \\ + \sum\limits_{\mu=1}^{2} \frac{\tilde{\psi}_{\mu}^{K}}{c_{\mu}^{K}(\tau)} \kappa_{\mu}^{K} Y^{K}(\tau) \end{array} \right\}. \end{aligned} \tag{8.27}$$

If we compare the treated and distributed water amounts with the collected and treated wastewater amounts, we obtain the following relation

$$\begin{aligned} W^{WT}(\tau) &= W^{WD}(\tau) \\ &= \left[1 + r^{WD}(\tau)\right] \left\{ Y^{Q}(\tau) \left\{ \psi_{Q}^{RP} + \sum_{i=1}^{2} \frac{\tilde{\psi}_{i}^{Q}}{c_{i}^{Q}(\tau)} \kappa_{i}^{Q} \right\} + \sum_{\mu=1}^{2} \frac{\tilde{\psi}_{\mu}^{K}}{c_{\mu}^{K}(\tau)} \kappa_{\mu}^{K} Y^{K}(\tau) \right\} \end{aligned}$$

$$
\begin{aligned}
&> W^{WWC}(\tau) \\
&= Y^Q \left\{ \rho^{WW}(\tau)\psi_Q^{RP} + \sum_{i=1}^{2} \frac{\tilde{\psi}_i^Q}{c_i^Q(\tau)} \kappa_i^Q \right\} + \sum_{\mu=1}^{2} \frac{\tilde{\psi}_\mu^K}{c_\mu^K(\tau)} \kappa_\mu^K Y^K \\
&> W^{WWT}(\tau) \\
&= \left[1 - r^{WWC}(\tau)\right] \left\{ Y^Q(\tau) \left\{ \rho^{WW}(\tau)\psi_Q^{RP} + \sum_{i=1}^{2} \frac{\tilde{\psi}_i^Q}{c_i^Q(\tau)} \kappa_i^Q \right\} + \sum_{\mu=1}^{2} \frac{\tilde{\psi}_\mu^K}{c_\mu^K(\tau)} \kappa_\mu^K Y^K(\tau) \right\} .
\end{aligned}
\tag{8.28}
$$

If the wastewater generation rate and the efficiencies of the water distribution and wastewater collection system are low, then the discrepancies between the produced water and the treated wastewater amounts are high. This is equivalent to the situations where there are large water distribution and wastewater collection losses in the water supply and wastewater systems. Losses from the water and wastewater system can create environmental and hygienic risks. For example, the pollution of the groundwater resources necessitated a switch to a surface water supply system in the Adana case study. Water losses also lead to a surplus of capital good stocks for the water supply system and to a reduced capital stock for the wastewater system. This means that there is allocation inefficiency for the water infrastructure capital stocks.

The capital goods invested in the water supply and wastewater systems can be considered an investment in the natural environment. This investment modifies the natural conditions of the local public good in such a way that rapidly growing urban centres can be served more efficiently by sustainable water supply and wastewater services. The urban water infrastructure, as part of the surrounding natural environment, has the characteristics of a local public good. The water used for production and reproduction activities has the characteristics of a private good. These goods can be differentiated using the energy and human labour input factors. The energy input amounts correlate with the amounts and quality (pollutant concentrations) of the water and wastewater amounts being treated, distributed, and collected. The energy input costs have the properties of typical variable costs. Other input factors not considered in our model such as chemicals used in water, wastewater, or sludge treatment would also belong to this group of variable components and costs. The maintenance and human labour inputs depend on the capacity and size of the urban water and wastewater systems. As a result, these inputs correlate with the water and generated wastewater amounts, but not with the water quality as measured in the form of concentrations.

We could assume that the human labour inputs of the water and wastewater processes are decreasing functions of the accumulated civil work and mechanical & electrical equipment capital stocks for the urban water infrastructure. However, we prefer to focus our attention on capital stock innovations for reducing the water and wastewater losses as well as increasing the energy efficiencies for the urban water infrastructure. We therefore have

Assumption 8.24 (for the human labour and capital good input coefficients and the depreciation rates of the water infrastructure processes) The human labour, capital good input coefficients, and depreciation rates of the water infrastructure processes, which are assumed identical for the different processes using civil work and mechanical & electrical equipment capital goods, remain constant over time

$$\begin{aligned}
&k_{CW}^{WT}, k_{ME}^{WT}, k_{CW}^{WD}, k_{ME}^{WD}, k_{CW}^{WWC}, k_{ME}^{WWC}, k_{CW}^{WWT}, k_{ME}^{WWT} \\
&\left\{ d^{CW} = d_{CW}^{WT} = d_{CW}^{WD} = d_{CW}^{WWC} = d_{CW}^{WWT} \right\} \\
&\wedge \left\{ d^{ME} = d_{ME}^{WT} = d_{ME}^{WD} = d_{ME}^{WWC} = d_{ME}^{WWT} \right\} \\
&l^{WT}, l^{WD}, l^{WWC}, l^{WWT}.
\end{aligned} \tag{8.29}$$

The input coefficients k_{CW}^{WWT} and k_{ME}^{WWT} of the wastewater treatment process and the input coefficients k_{CW}^{WWC} and k_{ME}^{WWC} of the civil work and mechanical & electrical capital goods for the wastewater system characterize the structure of the capital stock for the wastewater system. The input coefficients k_{CW}^{WT} and k_{ME}^{WT} of the water treatment process and the input coefficients k_{CW}^{WD} and k_{ME}^{WD} of the water distribution process characterize the structure of the water supply system.

Assumption 8.13 states that the capital stock innovations on the water infrastructure should lead to a reduction in the water and exergy losses. We restrict ourselves in this context with

Assumption 8.25 (for the water losses of the water distribution and wastewater collection processes) The accumulated civil work capital stocks for the water and wastewater system carry the innovation. The innovation reduces the water losses measured as water loss rates $0 \leq r_{W \vee WW}^{g^*} < 1$, $g^* \in \{WD, WWC\}$ for the water distribution and wastewater collection system by the relations

$$r_W^{WD}(\tau) = r_W^{WD}(0)\, e^{\varphi_W^{WD} Y^Q(\tau)} \; with \; \varphi_W^{WD} > \frac{1}{Y^Q(\tau)}. \tag{8.30}$$

$$r_{WW}^{WWC}(\tau) = r_{WW}^{WWC}(0)\, e^{\varphi_{WW}^{WWC} Y^Q(\tau)} \; with \; \varphi_{WW}^{WWC} > \frac{1}{Y^Q(\tau)}. \tag{8.31}$$

Assumption 8.26 (for the energy efficiency of the water distribution and wastewater collection processes) The accumulated capital stocks for the water and wastewater system carry the innovation, to reduce the exergy losses measured as anergy rates $0 \leq r_{an}^{g^*} < 1$, $g^* \in \{WD, WWC\}$, by the relations

$$r_{an}^{WT}(\tau) = r_{an}^{WT}(0)\, e^{\varphi_{an}^{WT} Y^Q(\tau)} \; with \; \varphi_{an}^{WD} > \frac{1}{Y^Q(\tau)}, \tag{8.32}$$

$$r_{an}^{WD}(\tau) = r_{an}^{WD}(0)\, e^{\varphi_{an}^{WD} Y^Q(\tau)} \; with \; \varphi_{an}^{WD} > \frac{1}{Y^Q(\tau)}, \tag{8.33}$$

$$r_{an}^{WWC}(\tau) \;=\; r_{an}^{WWC}(0)\, e^{\varphi_{an}^{WWC}\, Y^Q(\tau)} \text{ with } \varphi_{an}^{WWC} \;>\; \frac{1}{Y^Q(\tau)}, \tag{8.34}$$

$$r_{an}^{WWT}(\tau) \;=\; r_{an}^{WWT}(0)\, e^{\varphi_{an}^{WWT}\, Y^Q(\tau)} \; with \; \varphi_{an}^{WWT} \;>\; \frac{1}{Y^Q(\tau)}. \tag{8.35}$$

These relations mean that the innovation effects are considered an implicit function of the achieved consumption, or welfare level. Similar to the innovation effect of the capital stock used for producing one unit of the consumption good, we could assume the innovation effect for the water infrastructure would be an explicit function of the civil work and mechanical & electrical equipment capital stocks. However, an explicit function of this kind leads to quite complicated expressions when deriving the optimal conditions of the model. Therefore, we accept the risk that some interesting information might be lost by this simplification.

The capital stock innovation effects on the water, wastewater, human labour, and energy input amounts lead to expressions that will be presented as an integral part of the model constraints in Chap. 9.

8.3 Reduction of Variables and Dynamics of the Capital Stocks

Before we formulate the constraints of the water infrastructure model, it is reasonable to reduce the number of variables by taking into account the dynamics of the capital stocks.

While the water use model had only the consumption and capital good amounts as variables, the water infrastructure model has additional variables in the system of constraints. The method used to first derive the optimality conditions for the capital and consumption good variables and to then connect them with each other could also be applied to the extended model. This method, however, requires too many calculation steps to get the final matrix where marginal cost components will be summarized. Although some information will not be available explicitly, we prefer to transfer the system of constraints from the status

$$H = F\left\{\begin{array}{l} Y^Q;\, Y^K(Y^Q); \\ Y_{CW}^{WT}\left[W^{WT}(Y^Q, Y^K(Y^Q)\right];\; Y_{CW}^{WD}\left[W^{WD}(Y^Q, Y^K(Y^Q\right]; \\ Y_{CW}^{WWC}\left[W^{WWC}(Y^Q,Y^K(Y^Q)\right];Y_{CW}^{WWT}\left[W^{WT}(Y^Q,Y^K(Y^Q)\right]; \\ Y_{ME}^{WT}\left[(W^{WT}(Y^Q,Y^K(Y^Q)\right];\, Y_{ME}^{WD}\left[W^{WD}(Y^Q,Y^K(Y^Q)\right]; \\ Y_{ME}^{WWC}\left[W^{WWC}(Y^Q, Y^K(Y^Q)\right];\, Y_{ME}^{WWT}\left[W^{WT}(Y^Q, Y^K(Y^Q)\right] \end{array}\right\} \text{To the status}$$

$$H = F\left\{\begin{array}{l} Y^Q; \\ Y_{CW}^{WT}\left[W^{WT}(Y^Q)\right];\, Y_{CW}^{WD}\left[W^{WD}(Y^Q)\right]; \\ Y_{CW}^{WWC}\left[W^{WWC}(Y^Q)\right];Y_{CW}^{WWT}\left[W^{WT}(Y^Q)\right]; \\ Y_{ME}^{WT}\left[(W^{WT}(Y^Q)\right];\, Y_{ME}^{WD}\left[W^{WD}(Y^Q)\right]; \\ Y_{ME}^{WWC}\left[W^{WWC}(Y^Q)\right];\, Y_{ME}^{WWT}\left[W^{WT}(Y^Q)\right] \end{array}\right\} \text{And finally to } H = F\left\{Y^Q\right\}.$$

Non-profit conditions mean that the capital stock is always used completely. The capital stock equations show that, under non-profit conditions, the two variables on the production side are closely connected by relations derived from the capital stock development between two time intervals that follow each other. These relations are

$$
\begin{aligned}
K^Q(0) + \sum_{t=1}^{\tau} Y^K(t) &= d^Q k^Q \sum_{t=1}^{\tau} Y^Q(t) + k^Q Y^Q(\tau+1) \\
- K^Q(0) + \sum_{t=1}^{\tau-1} Y^K(t) &= d^Q k^Q \sum_{t=1}^{\tau-1} Y^Q(t) + k^Q Y^Q(\tau) \\
= Y^K(\tau) &= d^Q k^Q Y^Q(\tau) + k^Q \left\{ Y^Q(\tau+1) - Y^Q(\tau) \right\} \\
= Y^K(\tau) &= k^Q Y^Q(\tau) \left\{ d^Q + \gamma_\tau^Q \right\} \; \textit{with} \; \gamma_\tau^Q := \frac{Y^Q(\tau+1) - Y^Q(\tau)}{Y^Q(\tau)} .
\end{aligned}
\tag{8.36}
$$

Equation (8.36) shows that the development path of the consumption good amounts can substitute for the capital good flow. Starting with the initial capital stock, the increase of the capital stock by new investments can also be based on the development path of the consumption good amounts as follows

$$
\sum_{t=1}^{\tau-1} Y^K(\tau) = k^Q \sum_{t=1}^{\tau-1} Y^Q(t) \left\{ d^Q + \gamma_t^Q \right\} \; \textit{with} \; \gamma_t^Q := \frac{Y^Q(t+1) - Y^Q(t)}{Y^Q(t)} .
\tag{8.37}
$$

When applying the same transformation for the capital stocks of the water infrastructure, we obtain

$$
\begin{aligned}
K_{CW}^{g'''}(0) + \sum_{t=1}^{\tau} Y_{CW}^{g'''}(t) &= d^{CW} \sum_{t=1}^{\tau} k_{CW}^{g'''} W^{g'''}(t) + k_{CW}^{g'''} W^{g'''}(\tau+1) \\
- K_{CW}^{g'''}(0) + \sum_{t=1}^{\tau-1} Y_{CW}^{g'''}(t) &= d^{CW} \sum_{t=1}^{\tau-1} k_{CW}^{g'''} W^{g'''}(t) + k_{CW}^{WT} W^{g'''}(\tau) \\
= Y_{CW}^{g'''}(\tau) &= d^{CW} k_{CW}^{g'''} W^{g'''}(\tau) + k_{CW}^{g'''} \left\{ W^{g'''}(\tau+1) - W^{g'''}(\tau) \right\} \\
= Y_{CW}^{g'''}(\tau) &= k_{CW}^{g'''} W^{g'''}(\tau) \left\{ d^{CW} + \gamma_\tau^{g'''} \right\} \\
\text{with } \gamma_\tau^{g'''} &:= \frac{W^{g'''}(\tau+1) - W^{g'''}(\tau)}{W^{g'''}(\tau)} \wedge g''' \in \left\{ WT, WD, WWC, WWT \right\} .
\end{aligned}
\tag{8.38}
$$

$$
\begin{aligned}
&\sum_{t=1}^{\tau-1} Y_{CW}^{g'''}(t) = k_{CW}^{g'''} \sum_{t=1}^{\tau-1} W^{g'''}(t) \left\{ d^{CW} + \gamma_t^{g'''} \right\} \; \textit{with} \\
&\gamma_t^{g'''} := \frac{W^{g'''}(t+1) - W^{g'''}(t)}{W^{g'''}(t)} \wedge g''' \in \left\{ WT, WD, WWC, WWT \right\} .
\end{aligned}
\tag{8.39}
$$

$$\begin{aligned}
K_{ME}^{g'''}(0) + \sum_{t=1}^{\tau} Y_{ME}^{g'''}(t) &= d^{ME} \sum_{t=1}^{\tau} k_{ME}^{g'''} W^{g'''}(t) + k_{ME}^{g'''} W^{g'''}(\tau+1) \\
- K_{ME}^{g'''}(0) + \sum_{t=1}^{\tau-1} Y_{ME}^{g'''}(t) &= d^{ME} \sum_{t=1}^{\tau-1} k_{ME}^{g'''} W^{g'''}(t) + k_{ME}^{WT} W^{g'''}(\tau) \\
= Y_{ME}^{g'''}(\tau) &= d^{ME} k_{ME}^{g'''} W^{g'''}(\tau) + k_{ME}^{g'''} \left\{ W^{g'''}(\tau+1) - W^{g'''}(\tau) \right\} \\
= Y_{ME}^{g'''}(\tau) &= k_{ME}^{g'''} W^{g'''}(\tau) \left\{ d^{ME} + \gamma_{\tau}^{g'''} \right\} \\
\text{with } \gamma_{\tau}^{g'''} &:= \frac{W^{g'''}(\tau+1) - W^{g'''}(\tau)}{W^{g'''}(\tau)} \wedge g''' \in \{WT, WD, WWC, WWT\}.
\end{aligned} \tag{8.40}$$

$$\begin{aligned}
\sum_{t=1}^{\tau-1} Y_{ME}^{g'''}(t) &= \sum_{t=1}^{\tau-1} k_{ME}^{g'''} W^{g'''}(t) \left\{ d^{ME} + \gamma_t^{g'''} \right\} \text{ with} \\
\gamma_t^{g'''} &:= \frac{W^{g'''}(t+1) - W^{g'''}(t)}{W^{g'''}(t)} \wedge g''' \in \{WT, WD, WWC, WWT\}.
\end{aligned} \tag{8.41}$$

The total civil work and mechanical & electrical equipment amounts produced and invested for maintaining and developing the capital stock of the water infrastructure are equal to the sums

$$Y^{CW}(\tau) = \sum_{g'''} Y_{CW}^{g'''}(\tau) = \sum_{g'''} k_{CW}^{g'''} W^{g'''}(\tau) \left\{ d^{CW} + \gamma^{g'''} \right\}. \tag{8.42}$$

$$Y^{ME}(\tau) = \sum_{g'''} Y_{ME}^{g'''}(\tau) = \sum_{g'''} k_{ME}^{g'''} W^{g'''}(\tau) \left\{ d^{ME} + \gamma_{\tau}^{g'''} \right\}. \tag{8.43}$$

The activity levels for producing the civil work and electrical & mechanical equipment capital goods can be substituted by the water amounts. Since the capital good stocks and flows are determined by the consumption good amounts, the system of constraints can be transformed to have only the consumption good amounts as a variable.

Chapter 9
Constraints of the Water Infrastructure Model

Abstract When formulating the model constraints, special attention must be given to the aggregation of processes to sectors. The aggregation level, however, should not go too far because the model should reflect practical requirements from an engineering point of view. The capital stocks for the water infrastructure will be divided into those for civil work and those for mechanical and electrical equipment facilities. Because water losses in the water distribution system, wastewater generation rates, and leakages change over time, the water and wastewater amounts differ sector by sector within the actual and future time intervals. These structural changes depend on the past and future activities.

As previously discussed in the context of the constraints of the water use model, the number of variables within the different activity intervals can be reduced by substitution under

Assumption 9.1 (for positive quantities and no storage of the natural resources and goods produced) The natural resources appropriated from the natural environment and the goods produced should always have positive quantities, which will not be stored and therefore fully utilized within the activity periods and intervals.

Under this condition, the activity levels of the different processes for extracting the raw materials and for generating the free energy and water are determined by the activity level of the production process for the consumption good. Since wastewater will also not be stored, the activity level of the wastewater treatment process is also determined by the activity levels of the production processes of the consumption good. Water requirements for production, however, are based on the activity levels of the extraction processes.

9.1 Constraints for the Consumption Good Amounts

The constraints of the extended model marked as EM1 for the consumption Y^Q that is produced within the activity interval τ and then available one activity interval later, $\tau + 1$, is identical with constraints BM1 from the basic model

$$H_Q(\tau) := Y^Q(\tau) - Q(\tau + 1) \geq 0 \qquad \text{(EM1)}$$

H. Niemes, M. Schirmer, *Entropy, Water and Resources*,
DOI 10.1007/978-3-7908-2416-2_9, © Springer-Verlag Berlin Heidelberg 2010

9.2 Constraints for Extracting Raw Materials

In addition to the constraints

$$H_{N_{ii}^{Q}}(\tau) = N_{ii}^{Q}(\tau) - \kappa_i^Q Y^Q(\tau) \geq 0 \wedge i = 1,2 \tag{EM2}$$

$$H_{N_{\mu\mu}^{K}}(\tau) = N_{\mu\mu}^{K}(\tau) - \kappa_\mu^K Y^K(\tau) \geq 0 \wedge \mu = 1,2 \tag{EM3}$$

$$H_{N_{i=1,2}^{Q}}(\tau) = N_i^Q(\tau) - Y^Q(\tau)\,\frac{\kappa_i^Q e^{\varphi_i^Q \kappa_i^Q \sum_{t=1}^{\tau-1} Y^Q(t)}}{c(\overline{N}_i^Q(0))} \geq 0 \tag{EM4}$$

$$H_{N_{\mu=1,2}^{K}}(\tau) = N_\mu^K(\tau) - Y^K(\tau)\,\frac{\kappa_\mu^K e^{\varphi_\mu^K \kappa_\mu^K \sum_{t=1}^{\tau-1} Y^K(t)}}{c(\overline{N}_\mu^K(0))} \geq 0 \tag{EM5}$$

we obtain the following constraints of the raw materials, for which no stock constraints have to be formulated because they are assumed to be renewable resources

$$\begin{aligned} &H_{N_{\nu\nu}^{CW}}(\tau) = N_{\nu\nu}^{CW}(\tau) - \kappa_\nu^{CW} Y^{CW}(\tau) \geq 0 \\ &\mathit{with}\ \nu = 1,2 \wedge Y^{CW}(\tau) = Y_{CW}^{WT}(\tau) + Y_{CW}^{WD}(\tau) + Y_{CW}^{WWC}(\tau) + Y_{CW}^{WWT}(\tau) \end{aligned} \tag{EM6}$$

$$\begin{aligned} &H_{N_{oo}^{ME}}(\tau) = N_{oo}^{ME}(\tau) - \kappa_\mu^{ME} Y^{ME}(\tau) \geq 0 \\ &\mathit{with}\ o = 1,2 \wedge Y^{ME}(\tau) = Y_{ME}^{WT}(\tau) + Y_{ME}^{WD}(\tau) + Y_{ME}^{WWC}(\tau) + Y_{ME}^{WWT}(\tau) \end{aligned} \tag{EM7}$$

$$H_{N_\nu^{CW}}(\tau) = N_\nu^{CW}(\tau) - \frac{\kappa_\nu^{CW} Y^{CW}(\tau)}{c(\overline{N}_\nu^{CW}(0))} \geq 0\ \mathit{with}\ \nu = 1,2 \tag{EM8}$$

$$H_{N_o^{ME}}(\tau) = N_o^{ME}(\tau) - \frac{\kappa_\mu^{ME} Y^{ME}(\tau)}{c(\overline{N}_o^{ME}(0))} \geq 0\ \mathit{with}\ o = 1,2 \tag{EM9}$$

9.3 Constraints for the Water and Wastewater Amounts

Since the wastewater generation rate ρ^{WW} for the production and reproduction sector is in the range $0 < \rho^{WW} \leq 1$ and the water loss rates of the water distribution and the wastewater collection system are also in the ranges $0 < r^{WD}(\tau) \leq 1 \wedge 0 < r^{WWC}(\tau) \leq 1$, the activity levels of the water and wastewater processes differ from each other. As a result, we obtain four different constraints

$$
\begin{aligned}
H_{WT}(\tau) = {} & W^{WT}(\tau) \\
& - \left[1 + r_W^{WD}(0) e^{\varphi_r^{WD} Y^Q(\tau)}\right] \left\{ \begin{array}{l} Y^Q(\tau) \left\{ \psi_Q^{RP} + \sum\limits_{i=1}^{2} \dfrac{\tilde{\psi}_i^Q \kappa_i^Q e^{\varphi_i^Q \kappa_i^Q \sum_{t=1}^{\tau-1} Y^Q(t)}}{c(\overline{N}_i^Q(0))} \right\} \\ + Y^K(\tau) \sum\limits_{\mu=1}^{2} \dfrac{\tilde{\psi}_\mu^K \kappa_\mu^K e^{\varphi_\mu^K \kappa_\mu^K \sum_{t=1}^{\tau-1} Y^K(t)}}{c(\overline{N}_\mu^K(0))} \end{array} \right\} \geq 0.
\end{aligned}
\tag{EM10}
$$

$$
\begin{aligned}
H_{WD}(\tau) = {} & W^{WD}(\tau) \\
& - \left[1 + r_W^{WD}(0) e^{\varphi_r^{WD} Y^Q(\tau)}\right] \left\{ \begin{array}{l} Y^Q(\tau) \left\{ \psi_Q^{RP} + \sum\limits_{i=1}^{2} \dfrac{\tilde{\psi}_i^Q \kappa_i^Q e^{\varphi_i^Q \kappa_i^Q \sum_{t=1}^{\tau-1} Y^Q(t)}}{c(\overline{N}_i^Q(0))} \right\} \\ + Y^K(\tau) \sum\limits_{\mu=1}^{2} \dfrac{\tilde{\psi}_\mu^K \kappa_\mu^K e^{\varphi_\mu^K \kappa_\mu^K \sum_{t=1}^{\tau-1} Y^K(t)}}{c(\overline{N}_\mu^K(0))} \end{array} \right\} \geq 0.
\end{aligned}
\tag{EM11}
$$

$$
\begin{aligned}
H_{WWC}(\tau) = {} & W^{WWC}(\tau) \\
& - \left\{ \begin{array}{l} Y^Q(\tau) \left\{ \rho^{WW}(\tau) \psi_Q^{RP} + \sum\limits_{i=1}^{2} \dfrac{\tilde{\psi}_i^Q \kappa_i^Q e^{\varphi_i^Q \kappa_i^Q \sum_{t=1}^{\tau-1} Y^Q(t)}}{c(\overline{N}_i^Q(0))} \right\} \\ + Y^K(\tau) \sum\limits_{\mu=1}^{2} \dfrac{\tilde{\psi}_\mu^K \kappa_\mu^K e^{\varphi_\mu^K \kappa_\mu^K \sum_{t=1}^{\tau-1} Y^K(t)}}{c(\overline{N}_\mu^K(0))} \end{array} \right\} \geq 0.
\end{aligned}
\tag{EM12}
$$

$$
\begin{aligned}
H_{WWT}(\tau) = {} & W^{WWT}(\tau) \\
& - \left[1 - r_{WW}^{WWC}(0)\, e^{\varphi_r^{WWC} Y^Q(\tau)}\right] \left\{ \begin{array}{l} Y^Q(\tau) \left\{ \begin{array}{l} \rho^{WW}(\tau) \psi_Q^{RP} \\ + \sum\limits_{i=1}^{2} \dfrac{\tilde{\psi}_i^Q \kappa_i^Q e^{\varphi_i^Q \kappa_i^Q \sum_{t=1}^{\tau-1} Y^Q(t)}}{c(\overline{N}_i^Q(0))} \end{array} \right\} \\ + Y^K(\tau) \sum\limits_{\mu=1}^{2} \dfrac{\tilde{\psi}_\mu^K \kappa_\mu^K e^{\varphi_\mu^K \kappa_\mu^K \sum_{t=1}^{\tau-1} Y^K(t)}}{c(\overline{N}_\mu^K(0))} \end{array} \right\} \geq 0\ .
\end{aligned}
\tag{EM13}
$$

The dynamics of the water infrastructure are not only characterized by the capital stock for producing the consumption good, but also by developing the civil work and mechanical & electrical equipment capital stocks allocated to the different water and wastewater processes.

9.4 Constraints for Free Energy

The raw material constraint (BM10) for generating energy in the water use model

$$H_{N^E}(\tau) = N^E(\tau) - \frac{E^Q(\tau) + E^K(\tau) + \sum_{i=1}^{2} E_i^Q(\tau) + \sum_{\mu=1}^{2} E_\mu^K(\tau) + E^W(\tau) + E^{WWT}(\tau)}{c^E(0)\,\eta^E \sigma^E} \geq 0,$$

Must be modified and extended as follows

$$H_{N^E}(\tau) = N^E(\tau) - \frac{1}{c^E(0)\,\eta^E \sigma^E} \left\{ \begin{array}{l} E^Q(\tau) + E^K(\tau) + \sum_{i=1}^{2} E_i^Q(\tau) + \sum_{\mu=1}^{2} E_\mu^K(\tau) \\ + E^{CW}(\tau) + E^{ME}(\tau) + \sum_{\nu=1}^{2} E_\nu^{CW}(\tau) + \sum_{o=1}^{2} E_o^{ME}(\tau) \\ + E^{WT}(\tau) + E^{WD}(\tau) + E^{WWC}(\tau) + E^{WWT}(\tau) \end{array} \right\} \geq 0 \tag{EM14}$$

With the detailed expressions

$$E^Q(\tau) = \frac{ex^Q}{\eta^Q} Y^Q(\tau) \wedge E^K(\tau) = \frac{ex^K}{\eta^K} Y^K(\tau),$$

$$E_{i=1+2}^Q(\tau) = Y^Q(\tau) \sum_{i=1}^{2} \frac{\kappa_i^Q}{\eta_i^Q} \left\{ \begin{array}{l} \frac{ex_i^{app}}{c(\overline{N}_i^Q(0))} e^{\varphi_i^Q \kappa_i^Q \sum_{t=1}^{\tau-1} Y^Q(t)} + \\ \varepsilon \left(-\ln c(\overline{N}_i^Q(0)) + \varphi_i^Q \kappa_i^Q \sum_{t=1}^{\tau-1} Y^Q(t) \right) \end{array} \right\},$$

$$E_{\mu=1+2}^K(\tau) = Y^K(\tau) \sum_{\mu=1}^{2} \frac{\kappa_\mu^K}{\eta_\mu^K} \left\{ \begin{array}{l} \frac{ex_\mu^{app}}{c(\overline{N}_\mu^K(0))} e^{\varphi_\mu^K \kappa_\mu^K \sum_{t=1}^{\tau-1} Y^K(t)} + \\ \varepsilon \left(-\ln c(\overline{N}_\mu^K(0)) + \varphi_\mu^K \kappa_\mu^K \sum_{t=1}^{\tau-1} Y^K(t) \right) \end{array} \right\},$$

$$E^{CW}(\tau) = \frac{ex^{CW}}{\eta^{CW}} Y^{CW}(\tau) \wedge E^{ME}(\tau) = \frac{ex^{ME}}{\eta^{ME}} Y^{ME}(\tau),$$

$$E_{\nu=1+2}^{CW}(\tau) = Y^{CW}(\tau) \sum_{\nu=1}^{2} \frac{e_\nu \kappa_\nu^{CW}}{c(\overline{N}_\nu^{CW}(0))} \wedge E_{o=1+2}^{ME}(\tau) = Y^{ME}(\tau) \sum_{o=1}^{2} \frac{e_o \kappa_o^{ME}}{c(\overline{N}_o^{ME}(0))},$$

$$E^{WT}(\tau) = \tilde{\varepsilon}^{WD} \sum_{\omega} \ln \frac{c_\omega^*(0)}{c_\omega(0)} \left[1 + r_{an}^{WT}(0)\, e^{\varphi_{an}^{WT} Y^Q(\tau)} \right] \left[1 + r_W^{WD}(0) e^{\varphi_r^{WD} Y^Q(\tau)} \right] *$$

$$\ldots * \left\{ Y^Q(\tau) \left\{ \psi_Q^{RP} + \sum_{i=1}^{2} \frac{\tilde{\psi}_i^Q}{c_i^Q(\tau)} \kappa_i^Q \right\} + \sum_{\mu=1}^{2} \frac{\tilde{\psi}_\mu^K}{c_\mu^K(\tau)} \kappa_\mu^K Y^K(\tau) \right\},$$

$$E^{WD}(\tau) = ex^{WD}(0)\left[1 + r_{an}^{WD}(0)\, e^{\varphi_{an}^{WD} Y^Q(\tau)}\right]\left[1 + r^{WD}(0) e^{\varphi_r^{WD} Y^Q(\tau)}\right] *$$

$$\ldots * \left\{ Y^Q(\tau)\left\{\psi_Q^{RP} + \sum_{i=1}^{2} \frac{\tilde{\psi}_i^Q \kappa_i^Q e^{\varphi_i^Q \kappa_i^Q \sum_{t=1}^{\tau-1} Y^Q(t)}}{c(\overline{N}_i^Q(0))}\right\} + \sum_{\mu=1}^{2} \frac{\tilde{\psi}_\mu^K}{c_\mu^K(\tau)} \kappa_\mu^K Y^K(\tau)\right\},$$

$$E^{WWC}(\tau) = ex^{WWC}(0)\left[1 + r_{an}^{WWC}(0)\, e^{\varphi_{an}^{WWC} Y^Q(\tau)})\right] *$$

$$\ldots * \left\{ Y^Q(\tau)\left\{\rho^{WW}(\tau)\psi_Q^{RP} + \sum_{i=1}^{2} \frac{\tilde{\psi}_i^Q \kappa_i^Q e^{\varphi_i^Q \kappa_i^Q \sum_{t=1}^{\tau-1} Y^Q(t)}}{c(\overline{N}_i^Q(0))}\right\} + \sum_{\mu=1}^{2} \frac{\tilde{\psi}_\mu^K}{c_\mu^K(\tau)} \kappa_\mu^K Y^K(\tau)\right\},$$

$$E^{WWT}(\tau) = \tilde{\varepsilon}^{WWT}\left\{\ln\frac{c_Q^*(0)}{c_Q(0)} + \ln\frac{c_{\hat{j}}^*(0)}{c_{\hat{j}}(0)}\right\}\left[1 + r_{an}^{WWT}(0)\, e^{\varphi_{an}^{WWT} Y^Q(\tau)}\right] *$$

$$\ldots * \left[1 - r_{WW}^{WWC}(0)\, e^{\varphi_r^{WWC} Y^Q(\tau)}\right] *$$

$$\ldots * \left\{ Y^Q(\tau)\left\{\rho^{WW}(\tau)\psi_Q^{RP} + \sum_{i=1}^{2} \frac{\tilde{\psi}_i^Q \kappa_i^Q e^{\varphi_i^Q \kappa_i^Q \sum_{t=1}^{\tau-1} Y^Q(t)}}{c(\overline{N}_i^Q(0))}\right\} + \sum_{\mu=1}^{2} \frac{\tilde{\psi}_\mu^K}{c_\mu^K(\tau)} \kappa_\mu^K Y^K(\tau)\right\}.$$

9.5 Constraints for the Human Labour Input Amounts

The constraint (BM11) for the labour input amounts of the water use model

$$H_L(\tau) = L^P - \left\{\begin{matrix} L^Q(\tau) + L^K(\tau) + L_{i=1+2}^Q(\tau) + L_{\mu=1+2}^K(\tau) \\ + L^W(\tau) + L^{WWT}(\tau) + L^E(\tau) \end{matrix}\right\} \geq 0$$

Must be extended to

$$H_L(\tau) = L^P - \left\{\begin{matrix} L^Q(\tau) + L^K(\tau) + L_{i=1+2}^Q(\tau) + L_{\mu=1+2}^K(\tau) \\ + L^{CW}(\tau) + L^{ME}(\tau) + \sum_{\nu=1}^{2} L_\nu^{CW}(\tau) + \sum_{o=1}^{2} L_o^{ME}(\tau) \\ + L^{WT}(\tau) + L^{WD}(\tau) + L^{WWC}(\tau) + L^{WWT}(\tau) + L^E(\tau) \end{matrix}\right\} \geq 0 \qquad \text{(EM15)}$$

With the following detailed expressions

$$L^Q(\tau) = \tilde{l}^Q Y^Q(\tau) \overline{K}(0)\, e^{-\lambda^Q \sum_{t=1}^{\tau-1} Y_k^Q(t)} \quad \wedge \quad L^K(\tau) = l^K Y^K(\tau),$$

$$L^Q_{i=1+2}(\tau) = Y^Q(\tau) \sum_{i=1}^{2} \frac{l^Q_i \, \kappa^Q_i \, e^{+\varphi^Q_i \kappa^Q_i \sum_{t=1}^{\tau-1} Y^Q(t)}}{c(\overline{N}^Q_i(0))} \wedge$$
$$L^K_{\mu=1+2}(\tau) = Y^K(\tau) \sum_{\mu=1}^{2} \frac{l^K_\mu \, \kappa^K_\mu}{c(\overline{N}^K_\mu(0))} e^{+\varphi^K_\mu \kappa^K_\mu \sum_{t=1}^{\tau-1} Y^K(t)},$$

$$L^{CW}(\tau) = l^{CW} Y^{CW}(\tau) \wedge L^{ME}(\tau) = l^{ME} Y^{ME}(\tau),$$

$$L^{CW}_{\nu=1+2}(\tau) = Y^{CW}(\tau) \sum_{\nu=1}^{2} \frac{l_\nu \, \kappa^{CW}_\nu}{c(\overline{N}^{CW}_\nu(0))} \wedge L^{ME}_{o=1+2}(\tau) = Y^{ME}(\tau) \sum_{o=1}^{2} \frac{l_o \, \kappa^{ME}_o}{c(\overline{N}^{ME}_o(0))},$$

$$L^{WT} = l^{WT} \left[1 + r^{WD}_W(0)\, e^{\varphi^{WD}_r Y^Q(\tau)}\right] \left\{ \begin{array}{l} Y^Q(\tau) \left\{ \psi^{RP}_Q + \sum_{i=1}^{2} \frac{\tilde{\psi}^Q_i \kappa^Q_i e^{\varphi^Q_i \kappa^Q_i \sum_{t=1}^{\tau-1} Y^Q(t)}}{c(\overline{N}^Q_i(0))} \right\} \\ + Y^K(\tau) \sum_{\mu=1}^{2} \frac{\tilde{\psi}^K_\mu \kappa^K_\mu e^{\varphi^K_\mu \kappa^K_\mu \sum_{t=1}^{\tau-1} Y^K(t)}}{c(\overline{N}^K_\mu(0))} \end{array} \right\},$$

$$L^{WD} = l^{WD} \left[1 + r^{WD}_W(0)\, e^{\varphi^{WD}_r Y^Q(\tau)}\right] \left\{ \begin{array}{l} Y^Q(\tau) \left\{ \begin{array}{l} \psi^{RP}_Q \\ + \sum_{i=1}^{2} \frac{\tilde{\psi}^Q_i \kappa^Q_i e^{\varphi^Q_i \kappa^Q_i \sum_{t=1}^{\tau-1} Y^Q(t)}}{c(\overline{N}^Q_i(0))} \end{array} \right\} \\ + Y^K(\tau) \sum_{\mu=1}^{2} \frac{\tilde{\psi}^K_\mu \kappa^K_\mu e^{\varphi^K_\mu \kappa^K_\mu \sum_{t=1}^{\tau-1} Y^K(t)}}{c(\overline{N}^K_\mu(0))} \end{array} \right\},$$

$$L^{WWC}(\tau) = l^{WWC} \left\{ \begin{array}{l} Y^Q(\tau) \left\{ \begin{array}{l} \rho^{WW}(\tau) \psi^{RP}_Q \\ + \sum_{i=1}^{2} \frac{\tilde{\psi}^Q_i \kappa^Q_i e^{\varphi^Q_i \kappa^Q_i \sum_{t=1}^{\tau-1} Y^Q(t)}}{c(\overline{N}^Q_i(0))} \end{array} \right\} \\ + Y^K(\tau) \sum_{\mu=1}^{2} \frac{\tilde{\psi}^K_\mu \kappa^K_\mu e^{\varphi^K_\mu \kappa^K_\mu \sum_{t=1}^{\tau-1} Y^K(t)}}{c(\overline{N}^K_\mu(0))} \end{array} \right\},$$

$$L^{WWT}(\tau) = l^{WWT}\left[1 - r_{WW}^{WWC}(0)\, e^{\varphi_r^{WWC} Y^Q(\tau)}\right]\left\{\begin{array}{l} Y^Q(\tau)\left\{\begin{array}{l}\rho^{WW}(\tau)\psi_Q^{RP} \\ + \sum\limits_{i=1}^{2} \dfrac{\tilde{\psi}_i^Q \kappa_i^Q e^{\varphi_i^Q \kappa_i^Q \sum_{t=1}^{\tau-1} Y^Q(t)}}{c(\overline{N}_i^Q(0))}\end{array}\right\} \\ + Y^K(\tau) \sum\limits_{\mu=1}^{2} \dfrac{\tilde{\psi}_\mu^K \kappa_\mu^K e^{\varphi_\mu^K \kappa_\mu^K \sum_{t=1}^{\tau-1} Y^K(t)}}{c(\overline{N}_\mu^K(0))}\end{array}\right\},$$

$$L^E(\tau) = \frac{l^E}{c^E(\tau)\, \sigma^E \eta^E}\left\{\begin{array}{l} E^Q(\tau) + E^K(\tau) + \sum\limits_{i=1}^{2} E_i^Q(\tau) + \sum\limits_{\mu=1}^{2} E_\mu^K(\tau) \\ + E^{CW}(\tau) + E^{ME}(\tau) + \sum\limits_{\upsilon=1}^{2} E_\upsilon^{CW}(\tau) + \sum\limits_{o=1}^{2} E_o^{ME}(\tau) \\ + E^{WT}(\tau) + E^{WD}(\tau) + E^{WWC}(\tau) + E^{WWT}(\tau)\end{array}\right\} \geq 0.$$

with the details summarized in the context of the energy constraint (EM14)

The internal human labour input for producing free energy will also be ignored. Otherwise, there would be a large number of additional input terms, which would complicate the analysis. It would only be of special interest if the energy sector had infrastructure and characteristics similar to those introduced for the water sectors. Since we will focus on the water sector, this simplification is acceptable.

9.6 Constraints for the Capital Stocks

The capital good stock constraint for producing the consumption good is equal to

$$H_{\overline{K}^Q}(\tau) := K^Q(0) + \sum_{t=1}^{\tau-1} Y^K(t) - dk^Q \sum_{t=1}^{\tau-1} Y^Q(t) - k^Q Y^Q(\tau) \geq 0. \quad \text{(EM16)}$$

Since the water and wastewater systems require the civil work and mechanical & electrical capital goods, the following constraints must also be fulfilled

$$\begin{aligned} H_{C\bar{W}}(\tau) := &\sum_{g'''} K_{CW}^{g'''}(0) + \sum_{t=1}^{\tau-1} Y^{CW}(t) \\ &- d^{CW} \sum_{t=1}^{\tau-1} \sum_{g'''} k_{CW}^{g'''} W^{g'''}(t) - \sum_{g'''} k_{CW}^{g'''} W^{g'''}(\tau) \geq 0, \end{aligned} \quad \text{(EM17)}$$

$$\begin{aligned} H_{M\bar{E}}(\tau) := &\sum_{g'''} K_{ME}^{g'''}(0) + \sum_{t=1}^{\tau-1} Y^{ME}(t) \\ &- d^{ME} \sum_{t=1}^{\tau-1} \sum_{g'''} k_{ME}^{g'''} W^{g'''}(t) - \sum_{g'''} k_{ME}^{g'''} W^{g'''}(\tau) \geq 0. \end{aligned} \quad \text{(EM18)}$$

The indices $g''' \in \{^{WT}, ^{WD}, ^{WWC}, ^{WWT}\}$ symbolize the water treatment, water distribution, wastewater collection, and wastewater treatment processes. The civil work and mechanical & electrical equipment capital goods inputs amounts are equal to the sums

$$Y^{CW}(t) := \sum_{g'''} Y_{CW}^{g'''}(t) \wedge Y^{ME}(t) := \sum_{g'''} Y_{ME}^{g'''}(t)$$

And lead to the sub-constraints

$$H_{C\bar{W}}^{g'''}(\tau) := K_{CW}^{g'''}(0) + \sum_{t=1}^{\tau-1} Y_{CW}^{g'''}(t) - d^{CW} \sum_{t=1}^{\tau-1} k_{CW}^{g'''} W^{g'''}(t) - k_{CW}^{g'''} W^{g'''}(\tau) \geq 0, \tag{EM17g\#}$$

$$H_{M\bar{E}}^{g'''}(\tau) := K_{ME}^{g'''}(0) + \sum_{t=1}^{\tau-1} Y_{ME}^{g'''}(t) - d^{ME} \sum_{t=1}^{\tau-1} k_{ME}^{g'''} W^{g'''}(t) - k_{ME}^{g'''} W^{g'''}(\tau) \geq 0. \tag{EM18g\#}$$

The accumulated stocks for the civil work and mechanical & electrical equipment capital goods for water treatment and distribution, as well as for wastewater collection and treatment are equal to sums of the initial capital stocks plus the produced capital goods in the past time intervals. The constraints mean that these accumulated stocks must be larger than or equal to the depreciation of the capital stock applied in the past plus the newly acquired investments into the corresponding capital stocks in time interval τ.

9.7 Constraints for Reduced Variables

From the transformed system of constraints, only those for the human labour and energy inputs remain relevant for deriving the shadow prices for producing one unit of consumption good. The other constraints can be embedded into the following energy and human labour constraints.

9.7.1 Human Labour Constraint for Reduced Variables

After substituting the activity level of the capital good process by the activity level of the consumption good process, we obtain the human labour constraints

$$H_L(\tau) = L^P - \left\{ \begin{array}{l} L^Q(\tau) + L^K(\tau) + L^Q_{i=1+2}(\tau) + L^K_{\mu=1+2}(\tau) \\ +L^{CW}(\tau) + L^{ME}(\tau) + \sum_{\nu=1}^{2} L^{CW}_{\nu}(\tau) + \sum_{o=1}^{2} L^{ME}_{o}(\tau) \\ +L^{WT}(\tau) + L^{WD}(\tau) + L^{WWC}(\tau) + L^{WWT}(\tau) + L^E(\tau) \end{array} \right\} \geq 0 \tag{EM15*}$$

With the following detailed human labour inputs of the different processes that are needed to produce one unit of consumption good

$$L^Q(\tau) = \tilde{l}^Q Y^Q(\tau)\overline{K}(0)\, e^{-\lambda^Q k^Q \sum_{t=1}^{\tau-1} Y^Q(t)\left\{d^Q + \gamma_t^Q\right\}},$$

$$L^K(\tau) = l^K k^Q Y^Q(\tau)\left\{d^Q + \gamma_\tau^Q\right\},$$

$$L^Q_{i=1+2}(\tau) = Y^Q(\tau)\sum_{i=1}^{2} \frac{l_i^Q \kappa_i^Q e^{+\varphi_i^Q \kappa_i^Q \sum_{t=1}^{\tau-1} Y^Q(t)}}{c(\overline{N}_i^Q(0))},$$

$$L^K_{\mu=1+2}(\tau) = k^Q Y^Q(\tau)\left\{d^Q + \gamma_\tau^Q\right\}\sum_{\mu=1}^{2} \frac{l_\mu^K \kappa_\mu^K}{c(\overline{N}_\mu^K(0))}\, e^{+\varphi_\mu^K \kappa_\mu^K k^Q \left\{\sum_{t=1}^{\tau-1} Y^Q(t)\left\{d^Q + \gamma_t^Q\right\}\right\}},$$

$$L^{CW}(\tau) = \sum_{g'''} L^{g'''}_{CW}(\tau) = \sum_{g'''} l^{CW} k^{g'''}_{CW} W^{g'''}(\tau)\left\{d^{CW} + \gamma_\tau^{g'''}\right\}$$
$$\text{with } g''' \in \{WT, WD, WWC, WWT\},$$

$$L^{ME}(\tau) = \sum_{g'''} L^{g'''}_{ME}(\tau) = \sum_{g'''} l^{ME} k^{g'''}_{ME} W^{g'''}(\tau)\left\{d^{ME} + \gamma_\tau^{g'''}\right\},$$

$$L^{CW}_{\nu=1+2}(\tau) = \sum_{g'''} L^{g'''}_{CW_{\nu=1+2}}(\tau) = \sum_{g'''}\left\{\sum_{\nu=1}^{2} \frac{l_\nu \kappa_\nu^{CW}}{c(\overline{N}_\nu^{CW}(0))}\right\} k^{g'''}_{CW} W^{g'''}(\tau)\left\{d^{CW} + \gamma_\tau^{g'''}\right\}$$

$$L^{ME}_{o=1+2}(\tau) = \sum_{g'''} L^{g'''}_{ME_{o=1+2}}(\tau) = \sum_{g'''}\left\{\sum_{o=1}^{2} \frac{l_o \kappa_o^{ME}}{c(\overline{N}_o^{ME}(0))}\right\} k^{g'''}_{ME} W^{g'''}(\tau)\left\{d^{ME} + \gamma_\tau^{g'''}\right\},$$

$$L^{WT} = l^{WT} W^{WT}(\tau);\ L^{WD} = l^{WD} W^{WD}(\tau);$$
$$L^{WWC} = l^{WWC} W^{WWC}(\tau);\ L^{WWT} = l^{WWT} W^{WWT}(\tau);\ L^E = 0.$$

9.7.2 Energy Constraint for Reduced Variables

After substituting the activity level of the capital good process by the activity level of the consumption good process, we obtain the energy input constraint

$$H_{N^E}(\tau) = N^E(\tau) - \frac{1}{c^E(0)\,\eta^E \sigma^E}\left\{\begin{array}{l} E^Q(\tau) + E^K(\tau) + \sum_{i=1}^{2} E_i^Q(\tau) + \sum_{\mu=1}^{2} E_\mu^K(\tau) \\ +E^{CW}(\tau) + E^{ME}(\tau) + \sum_{\nu=1}^{2} E_\nu^{CW}(\tau) + \sum_{o=1}^{2} E_o^{ME}(\tau) \\ +E^{WT}(\tau) + E^{WD}(\tau) + E^{WWC}(\tau) + E^{WWT}(\tau) \end{array}\right\} \geq 0 \tag{EM14*}$$

with the following detailed energy input expressions for the different processes involved in producing one unit of consumption good

$$E^Q(\tau) = \frac{ex^Q}{\eta^Q} Y^Q(\tau) \wedge E^K(\tau) = \frac{ex^K}{\eta^K} k^Q Y^Q(\tau)\left\{d^Q + \gamma_\tau^Q\right\},$$

$$E^Q_{i=1+2}(\tau) = Y^Q(\tau) \sum_{i=1}^{2} \frac{\kappa_i^Q}{\eta_i^Q} \left\{\begin{array}{l} \dfrac{ex_i^{app}}{c(\overline{N}_i^Q(0))} e^{\varphi_i^Q \kappa_i^Q \sum_{t=1}^{\tau-1} Y^Q(t)} \\ + \varepsilon \left(- \ln c(\overline{N}_i^Q(0)) + \varphi_i^Q \kappa_i^Q \sum_{t=1}^{\tau-1} Y^Q(t)\right) \end{array}\right\},$$

$$E^K_{\mu=1+2}(\tau) = k^Q Y^Q(\tau)\left\{d^Q + \gamma_\tau^Q\right\} \sum_{\mu=1}^{2} \frac{\kappa_\mu^K}{\eta_\mu^K} \left\{\begin{array}{l} \dfrac{ex_\mu^{app}}{c(\overline{N}_\mu^K(0))} e^{\varphi_\mu^K \kappa_\mu^K k^Q \left\{\sum_{t=1}^{\tau-1} Y^Q(t)\left\{d^Q + \gamma_t^Q\right\}\right\}} \\ +\varepsilon \left(\begin{array}{l} - \ln c(\overline{N}_\mu^K(0)) \\ + \varphi_\mu^K \kappa_\mu^K k^Q \left\{\sum_{t=1}^{\tau-1} Y^Q(t)\left\{d^Q + \gamma_t^Q\right\}\right\} \end{array}\right) \end{array}\right\},$$

$$E^{CW}(\tau) = \sum_{g'''} E_{CW}^{g'''}(\tau) = \sum_{g'''} \frac{ex^{CW}}{\eta^{CW}} k_{CW}^{g'''} W^{g'''}(\tau)\left\{d^{CW} + \gamma_\tau^{g'''}\right\} \text{ with}$$

$g''' \in \{WT, WD, WWC, WWT\}$,

$$E^{ME}(\tau) = \sum_{g'''} E_{ME}^{g'''}(\tau) = \sum_{g'''} \frac{ex^{ME}}{\eta^{ME}} k_{ME}^{g'''} W^{g'''}(\tau)\left\{d^{ME} + \gamma_\tau^{g'''}\right\},$$

$$E^{CW}_{\nu=1+2}(\tau) = \sum_{g'''} E^{g'''}_{\substack{CW\\ \nu=1+2}}(\tau) = \sum_{g'''}\left\{\sum_{\nu=1}^{2} \frac{e_\nu \kappa_\nu^{CW}}{c(\overline{N}_\nu^{CW}(0))}\right\} k_{CW}^{g'''} W^{g'''}(\tau)\left\{d^{CW} + \gamma_\tau^{g'''}\right\},$$

$$E^{ME}_{o=1+2}(\tau) = \sum_{g'''} E^{g'''}_{\substack{ME\\ o=1+2}}(\tau) = \sum_{g'''}\left\{\sum_{o=1}^{2} \frac{e_o \kappa_o^{ME}}{c(\overline{N}_o^{ME}(0))}\right\} k_{ME}^{g'''} W^{g'''}(\tau)\left\{d^{ME} + \gamma_\tau^{g'''}\right\},$$

$$E^{WT}(\tau) = \tilde{\varepsilon}^{WT} \sum_{\omega} \ln \frac{c^*_{\omega}(0)}{c_{\omega}(0)} \left[1 + r^{WT}_{an}(0)\, e^{\varphi^{WT}_{an}\, Y^Q(\tau)}\right] W^{WT}(\tau),$$

$$E^{WD}(\tau) = ex^{WD}(0) \left\{1 + r^{WD}_{an}(0)\, e^{\varphi^{WD}_{an}\, Y^Q(\tau)}\right\} W^{WD}(\tau),$$

$$E^{WWC}(\tau) = ex^{WWC}(0) \left[1 + r^{WWC}_{an}(0)\, e^{\varphi^{WWC}_{an}\, Y^Q(\tau)})\right] W^{WWC}(\tau),$$

$$E^{WWT}(\tau) = \tilde{\varepsilon}^{WWT} \left\{\ln \frac{c^*_Q(0)}{c_Q(0)} + \ln \frac{c^*_j(0)}{c_j(0)}\right\} \left[1 + r^{WWT}_{an}(0)\, e^{\varphi^{WWT}_{an}\, Y^Q(\tau)}\right] W^{WWT}(\tau).$$

9.7.3 Water and Wastewater Constraints for Reduced Variables

The water and wastewater constraints include expressions for the water input amounts for water treatment, water distribution, wastewater collection, and wastewater treatment. The different factors are already in the form that allows the Leibniz rule for differentiation to be applied.

$$W^{WT}(\tau) = Y^Q(\tau) \left[1 + r^{WD}_W(0) e^{\varphi^{WD}_r Y^Q(\tau)}\right]$$

$$\ldots * \left\{ \begin{array}{l} \left\{ \psi^{RP}_Q + \sum\limits_{i=1}^{2} \dfrac{\tilde{\psi}^Q_i \kappa^Q_i e^{\varphi^Q_i \kappa^Q_i \sum_{t=1}^{\tau-1} Y^Q(t)}}{c(\overline{N}^Q_i(0))} \right\} \\ + \left\{ k^Q \left\{ d^Q + \gamma^Q_\tau \right\} \right\} \sum\limits_{\mu=1}^{2} \dfrac{\tilde{\psi}^K_\mu \kappa^K_\mu e^{\varphi^K_\mu \kappa^K_\mu k^Q \sum_{t=1}^{\tau-1} Y^Q(t) \left\{ d^Q + \gamma^Q_t \right\}}}{c(\overline{N}^K_\mu(0))} \end{array} \right\},$$

$$W^{WD}(\tau) = Y^Q(\tau) \left[1 + r^{WD}_W(0) e^{\varphi^{WD}_r Y^Q(\tau)}\right] *$$

$$\ldots * \left\{ \begin{array}{l} \left\{ \psi^{RP}_Q + \sum\limits_{i=1}^{2} \dfrac{\tilde{\psi}^Q_i \kappa^Q_i e^{\varphi^Q_i \kappa^Q_i \sum_{t=1}^{\tau-1} Y^Q(t)}}{c(\overline{N}^Q_i(0))} \right\} \\ + \left\{ k^Q \left\{ d^Q + \gamma^Q_\tau \right\} \right\} \sum\limits_{\mu=1}^{2} \dfrac{\tilde{\psi}^K_\mu \kappa^K_\mu e^{\varphi^K_\mu \kappa^K_\mu k^Q \left\{ \sum_{t=1}^{\tau-1} Y^Q(t) \left\{ d^Q + \gamma^Q_t \right\} \right\}}}{c(\overline{N}^K_\mu(0))} \end{array} \right\},$$

$$W^{WWC}(\tau) = Y^{Q}(\tau) *$$
$$\ldots * \left\{ \begin{array}{l} \left\{ \rho^{WW}(\tau)\psi_{Q}^{RP} + \sum\limits_{i=1}^{2} \dfrac{\tilde{\psi}_{i}^{Q}\kappa_{i}^{Q} e^{\varphi_{i}^{Q}\kappa_{i}^{Q}\sum\limits_{t=1}^{\tau-1} Y^{Q}(t)}}{c(\overline{N}_{i}^{Q}(0))} \right\} \\ + \left\{ k^{Q}\left\{ d^{Q} + \gamma_{\tau}^{Q} \right\} \right\} \sum\limits_{\mu=1}^{2} \dfrac{\tilde{\psi}_{\mu}^{K}\kappa_{\mu}^{K} e^{\varphi_{\mu}^{K}\kappa_{\mu}^{K}k^{Q}\left\{ \sum\limits_{t=1}^{\tau-1} Y^{Q}(t)\left\{ d^{Q}+\gamma_{t}^{Q} \right\} \right\}}}{c(\overline{N}_{\mu}^{K}(0))} \end{array} \right\},$$

$$W^{WWT}(\tau) = Y^{Q}(\tau)\left[1 - r_{WW}^{WWC}(0)\, e^{\varphi_{r}^{WWC} Y^{Q}(\tau)}\right]$$
$$\ldots * \left\{ \begin{array}{l} \left\{ \rho^{WW}(\tau)\psi_{Q}^{RP} + \sum\limits_{i=1}^{2} \dfrac{\tilde{\psi}_{i}^{Q}\kappa_{i}^{Q} e^{\varphi_{i}^{Q}\kappa_{i}^{Q}\sum\limits_{t=1}^{\tau-1} Y^{Q}(t)}}{c(\overline{N}_{i}^{Q}(0))} \right\} \\ + \left\{ k^{Q}\left\{ d^{Q} + \gamma_{\tau}^{Q} \right\} \right\} \sum\limits_{\mu=1}^{2} \dfrac{\tilde{\psi}_{\mu}^{K}\kappa_{\mu}^{K} e^{\varphi_{\mu}^{K}\kappa_{\mu}^{K}k^{Q}\left\{ \sum\limits_{t=1}^{\tau-1} Y^{Q}(t)\left\{ d^{Q}+\gamma_{t}^{Q} \right\} \right\}}}{c(\overline{N}_{\mu}^{K}(0))} \end{array} \right\}.$$

9.8 Aggregation of Process Inputs to Sector Inputs

It is more efficient to aggregate the different processes to sectors before the optimality conditions are derived.

9.8.1 Aggregation of Processes to the Production Sector

The set of processes $g \in \left\{ Q, K, \underset{i=1,2}{Q}, \underset{\nu=1,2}{K} \right\}$ represents the consumption and capital good processes and the corresponding extraction processes of the raw materials from the production sector. Since the production sector of the water use and water infrastructure models is the same as in the basic model, there is no need to derive the marginal human labour and energy costs for the production sector again. These marginal costs for the production sector are already available in Chap. 6. They will be combined with the marginal costs for the water and wastewater sectors, which now have a more complex structure.

As a matter of completeness, the total human labour and energy amounts of the production are equal to the sums

9.8.1.1 Aggregation of the Human Labour Input Amounts for the Production Sector

$$L^P_{\text{sector}} = L^Q(\tau) + L^K(\tau) + L^Q_{i=1+2}(\tau) + L^K_{\mu=1+2}(\tau) =$$
$$= Y^Q(\tau) \left\{ \begin{array}{l} \tilde{l}^Q\,\overline{K}(0)\,e^{-\lambda^Q k^Q \left\{ \sum\limits_{t=1}^{\tau-1} Y^Q(t)\left\{ d^Q + \gamma_t^Q \right\} \right\}} \\ + l^K k^Q \left\{ d^Q + \gamma_\tau^Q \right\} \\ + \sum\limits_{i=1}^{2} \dfrac{l_i^Q\,\kappa_i^Q\,e^{+\varphi_i^Q \kappa_i^Q \sum\limits_{t=1}^{\tau-1} Y^Q(t)}}{c(\overline{N}_i^Q(0))} \\ + k^Q \left\{ d^Q + \gamma_\tau^Q \right\} \sum\limits_{\mu=1}^{2} \dfrac{l_\mu^K\,\kappa_\mu^K}{c(\overline{N}_\mu^K(0))}\, e^{+\varphi_\mu^K \kappa_\mu^K k^Q \left\{ \sum\limits_{t=1}^{\tau-1} Y^Q(t)\left\{ d^Q + \gamma_t^Q \right\} \right\}} \end{array} \right\}. \tag{9.1}$$

9.8.1.2 Aggregation of the Energy Input Amounts for the Production Sector

$$E^P_{\text{sector}} = E^Q(\tau) + E^K(\tau) + E^Q_{i=1+2}(\tau) + E^K_{\mu=1+2}(\tau) =$$
$$Y^Q(\tau) \left\{ \begin{array}{l} \dfrac{ex^Q}{\eta^Q} \\ + \dfrac{ex^K}{\eta^K} k^Q \left\{ d^Q + \gamma_\tau^Q \right\} \\ + \sum\limits_{i=1}^{2} \dfrac{\kappa_i^Q}{\eta_i^Q} \left\{ \begin{array}{l} \dfrac{ex_i^{app}}{c(\overline{N}_i^Q(0))} e^{\varphi_i^Q \kappa_i^Q \sum\limits_{t=1}^{\tau-1} Y^Q(t)} + \\ \varepsilon \left(-\ln c(\overline{N}_i^Q(0)) + \varphi_i^Q \kappa_i^Q \sum\limits_{t=1}^{\tau-1} Y^Q(t) \right) \end{array} \right\} \\ + k^Q \left\{ d^Q + \gamma_\tau^Q \right\} \sum\limits_{\mu=1}^{2} \dfrac{\kappa_\mu^K}{\eta_\mu^K} \left\{ \begin{array}{l} \dfrac{ex_\mu^{app}}{c(\overline{N}_\mu^K(0))} e^{\varphi_\mu^K \kappa_\mu^K k^Q \left\{ \sum\limits_{t=1}^{\tau-1} Y^Q(t)\left\{ d^Q + \gamma_t^Q \right\} \right\}} + \\ \varepsilon \left(\begin{array}{l} -\ln c(\overline{N}_\mu^K(0)) \\ + \varphi_\mu^K \kappa_\mu^K k^Q \left\{ \sum\limits_{t=1}^{\tau-1} Y^Q(t) \left\{ d^Q + \gamma_t^Q \right\} \right\} \end{array} \right) \end{array} \right\} \end{array} \right\}. \tag{9.2}$$

9.8.2 Aggregation of Processes to the Water Sectors

The water amounts could be inserted separately into each of the energy and human labour expressions for the energy and human labour constraints for deriving the optimal conditions. However, we prefer to aggregate the processes that belong to the water and wastewater sectors. Later, we will differentiate the resulting expressions by the corresponding water amounts, which are determined by the development path of the consumption good amounts. Before we determine the non-profit conditions in the next step, we introduce the following sets of abbreviations for the essential components, which must be considered and clearly distinguished from each other.

The first set of abbreviations

$$U = \left\{ \begin{array}{l} U^{WT}_{Y^Q(\tau)}, U^{WD}_{Y^Q(\tau)}, U^{WWC}_{Y^Q(\tau)}, U^{WWT}_{Y^Q(\tau)} \\ with\ U^{WD}_{Y^Q(\tau)} = U^{WT}_{Y^Q(\tau)} \wedge U^{WWT}_{Y^Q(\tau)} = U^{WWC}_{Y^Q(\tau)} \end{array} \right\}$$

Summarizes the structural changes of the water and wastewater input coefficients for the water and wastewater infrastructure in the relation to the activity levels $Y^Q(\tau)$ and the water saving innovations of the water supply and wastewater system. The other sets of abbreviations

$$\begin{aligned} V^L = & \left[V^L_{\substack{WT \\ CW+ME_{dep}}}, V^L_{\substack{WD \\ CW+ME_{dep}}}, V^L_{\substack{WWC \\ CW+ME_{dep}}}, V^L_{\substack{WWT \\ CW+ME_{dep}}} \right] \\ & \cup \left[V^L_{\substack{WT \\ CW+ME_{dev}}}, V^L_{\substack{WD \\ CW+ME_{dev}}}, V^L_{\substack{WWC \\ CW+ME_{dev}}}, V^L_{\substack{WWT \\ CW+ME_{dev}}} \right], \\ V^E = & \left[V^E_{\substack{WT \\ CW+ME_{dep}}}, V^E_{\substack{WD \\ CW+ME_{dep}}}, V^E_{\substack{WWC \\ CW+ME_{dep}}}, V^E_{\substack{WWT \\ CW+ME_{dep}}} \right] \\ & \cup \left[V^E_{\substack{WT \\ CW+ME_{dev}}}, V^E_{\substack{WD \\ CW+ME_{dev}}}, V^E_{\substack{WWC \\ CW+ME_{dev}}}, V^E_{\substack{WWT \\ CW+ME_{dev}}} \right] \\ & \cup \left[V^E_{WT_{inn}}, V^E_{WD_{inn}}, V^E_{WWC_{inn}}, V^E_{WWT_{inn}} \right], \end{aligned}$$

Summarize the structural changes of the human labour, energy, and capital good input coefficients, where the effects of sustaining and developing the capital stocks and energy saving innovations of the water and wastewater system are taken into account. The detailed expressions for the sets of abbreviations $V^L \wedge V^E$ are given as part of the following aggregated human labour and energy input amounts for the water and wastewater sectors. The water saving innovations of the water and wastewater system are represented by the factors

$$\left[1 + r^{WD}_{W}(0) e^{\varphi^{WD}_r Y^Q(\tau)}\right] \wedge \left[1 - r^{WWC}_{WW}(0)\, e^{\varphi^{WWC}_r Y^Q(\tau)}\right]$$

And the energy saving innovations are represented by the factors

$$\left[1 + r^{\hat{g}}_{an}(0)\, e^{\varphi^{\hat{g}}_{an} Y^Q(\tau)}\right], \hat{g} \in \{WT, WD, WWC, WWT\}.$$

9.8.2.1 Aggregation of the Human Labour and Energy Input Amounts for the Water Treatment Sector

Human Labour Input Amounts for Water Treatment

$$L_{\text{sector}}^{WT}(\tau) = \left[1 + r_W^{WD}(0)e^{\varphi_r^{WD}Y^Q(\tau)}\right] Y^Q(\tau) * \\ \ldots * U^{WT}_{\substack{Q \\ N^Q_{i=1+2}+N^K_{\mu=1,2}}}(\tau)\left[l^{WT} + V^L_{\substack{WT \\ CW+ME_{dep}}} + V^L_{\substack{WT \\ CW+ME_{dev}}}(\tau)\right] \tag{9.3}$$

with

$$U^{WT}_{\substack{Q \\ N^Q_{i=1+2}+N^K_{\mu=1,2}}}(\tau) = \left\{ \begin{array}{l} \left\{ \psi_Q^{RP} + \sum\limits_{i=1}^{2} \dfrac{\tilde{\psi}_i^Q \kappa_i^Q e^{\varphi_i^Q \kappa_i^Q \sum_{t=1}^{\tau-1} Y^Q(t)}}{c(\overline{N}_i^Q(0))} \right\} + \\ \left\{ k^Q \left\{ d^Q + \gamma_\tau^Q \right\} \right\} \sum\limits_{\mu=1}^{2} \dfrac{\tilde{\psi}_\mu^K \kappa_\mu^K e^{\varphi_\mu^K \kappa_\mu^K k^Q \sum_{t=1}^{\tau-1} Y^Q(t)\left\{ d^Q + \gamma_t^Q \right\}}}{c(\overline{N}_\mu^K(0))} \end{array} \right\},$$

$$V^L_{\substack{WT \\ CW+MEdep}}(\tau) \\ = \left\{ \sum_{\nu=1}^{2} \frac{l_\nu \kappa_\nu^{CW}}{c(\overline{N}_\nu^{CW}(0))} + l^{CW} \right\} k_{CW}^{WT} d^{CW} + \left\{ \sum_{o=1}^{2} \frac{l_o \kappa_o^{ME}}{c(\overline{N}_o^{ME}(0))} + l^{ME} \right\} k_{ME}^{WT} d^{ME},$$

$$V^L_{\substack{WT \\ CW+ME_{dev}}}(\tau) \\ = \left\{ \left\{ \sum_{\nu=1}^{2} \frac{l_\nu \kappa_\nu^{CW}}{c(\overline{N}_\nu^{CW}(0))} + l^{CW} \right\} k_{CW}^{WT} + \left\{ \sum_{o=1}^{2} \frac{l_o \kappa_o^{ME}}{c(\overline{N}_o^{ME}(0))} + l^{ME} \right\} k_{ME}^{WT} \right\} \gamma_\tau^{WT}.$$

Energy Input Amounts for Water Treatment

$$E_{\text{sector}}^{WT}(\tau) = \left[1 + r_W^{WD}(0)e^{\varphi_r^{WD}Y^Q(\tau)}\right] Y^Q(\tau)* \\ \ldots * U^{WT}_{\substack{Q \\ N^Q_{i=1+2}+N^K_{\mu=1,2}}}(\tau)\left[V^E_{\substack{WT \\ CW+ME_{dep}}} + V^E_{\substack{WT \\ CW+ME_{dev}}}(\tau) + V^E_{WT_{innv}}(\tau)\right] \tag{9.4}$$

with

$$U^{WT}_{Q_{N^Q_{i=1+2}+N^K_{\mu=1,2}}}(\tau) := \left\{ \begin{array}{l} \left\{ \psi^{RP}_Q + \sum_{i=1}^{2} \frac{\tilde{\psi}^Q_i \kappa^Q_i e^{\varphi^Q_i \kappa^Q_i \sum_{t=1}^{\tau-1} Y^Q(t)}}{c(\overline{N}^Q_i(0))} \right\} + \\ \left\{ k^Q Y^Q(\tau) \left\{ d^Q + \gamma^Q_\tau \right\} \right\} \sum_{\mu=1}^{2} \frac{\tilde{\psi}^K_\mu \kappa^K_\mu e^{\varphi^K_\mu \kappa^K_\mu k^Q \sum_{t=1}^{\tau-1} Y^Q(t) \left\{ d^Q + \gamma^Q_t \right\}}}{c(\overline{N}^K_\mu(0))} \end{array} \right\},$$

$$V^E_{WT_{CW+ME_{dep}}} = \left\{ \sum_{\nu=1}^{2} \frac{e_\nu \kappa^{CW}_\nu}{c(\overline{N}^{CW}_\nu(0))} + \frac{ex^{CW}}{\eta^{CW}} \right\} k^{WT}_{CW} d^{CW} + \left\{ \sum_{o=1}^{2} \frac{e_o \kappa^{ME}_o}{c(\overline{N}^{ME}_o(0))} + \frac{ex^{ME}}{\eta^{ME}} \right\} k^{WT}_{ME} d^{ME},$$

$$V^E_{WT_{CW+ME_{dev}}}(\tau) = \left\{ \left\{ \sum_{\nu=1}^{2} \frac{e_\nu \kappa^{CW}_\nu}{c(\overline{N}^{CW}_\nu(0))} + \frac{ex^{CW}}{\eta^{CW}} \right\} k^{WT}_{CW} + \left\{ \sum_{o=1}^{2} \frac{e_o \kappa^{ME}_o}{c(\overline{N}^{ME}_o(0))} + \frac{ex^{ME}}{\eta^{ME}} \right\} k^{WT}_{ME} \right\} \gamma^{WT}_\tau,$$

$$V^E_{WT_{innv}}(\tau) = \tilde{\varepsilon}^{WT} \sum_{\omega} \ln \frac{c^*_\omega(0)}{c_\omega(0)} \left[1 + r^{WT}_{an}(0)\, e^{\varphi^{WT}_{an} Y^Q(\tau)} \right].$$

9.8.2.2 Aggregation of the Human Labour and Energy Input Amounts for the Water Distribution Sector

The Human Labour Input Amounts for Water Distribution

$$L^{WD}_{\text{sector}}(\tau) = \left[1 + r^{WD}_W(0) e^{\varphi^{WD}_r Y^Q(\tau)} \right] Y^Q(\tau)\, U^{WD}_{Q_{N^Q_{i=1+2}+N^K_{\mu=1,2}}}(\tau) \left[l^{WD} + V^L_{WD_{CW+ME_{dep}}} + V^L_{WD_{CW+ME_{dev}}}(\tau) \right] \tag{9.5}$$

with

$$U^{WD}_{Q_{N^Q_{i=1+2}+N^K_{\mu=1,2}}}(\tau) = \left\{ \begin{array}{l} \left\{ \psi^{RP}_Q + \sum_{i=1}^{2} \frac{\tilde{\psi}^Q_i \kappa^Q_i e^{\varphi^Q_i \kappa^Q_i \sum_{t=1}^{\tau-1} Y^Q(t)}}{c(\overline{N}^Q_i(0))} \right\} \\ + \left\{ k^Q \left\{ d^Q + \gamma^Q_\tau \right\} \right\} \sum_{\mu=1}^{2} \frac{\tilde{\psi}^K_\mu \kappa^K_\mu e^{\varphi^K_\mu \kappa^K_\mu k^Q \left\{ \sum_{t=1}^{\tau-1} Y^Q(t) \left\{ d^Q + \gamma^Q_t \right\} \right\}}}{c(\overline{N}^K_\mu(0))} \end{array} \right\},$$

$$V^{L}_{\substack{WD\\CW+ME_{dep}}} = \left\{ \sum_{\nu=1}^{2} \frac{l_\nu\, \kappa_\nu^{CW}}{c(\overline{N}_\nu^{CW}(0))} + l^{CW} \right\} k_{CW}^{WD} d^{CW} + \left\{ \sum_{o=1}^{2} \frac{l_o\, \kappa_o^{ME}}{c(\overline{N}_o^{ME}(0))} + l^{ME} \right\} k_{ME}^{WD} d^{ME},$$

$$V^{L}_{\substack{WD\\CW+ME_{dev}}}(\tau) = \left\{ \left\{ \sum_{\nu=1}^{2} \frac{l_\nu\, \kappa_\nu^{CW}}{c(\overline{N}_\nu^{CW}(0))} + l^{CW} \right\} k_{CW}^{WD} + \left\{ \sum_{o=1}^{2} \frac{l_o\, \kappa_o^{ME}}{c(\overline{N}_o^{ME}(0))} + l^{ME} \right\} k_{ME}^{WD} \right\} \gamma_\tau^{WD}.$$

The Energy Input Amounts for Water Distribution

$$\begin{aligned} E^{WD}_{\text{sector}}(\tau) &= \left[1 + r_W^{WD}(0) e^{\varphi_r^{WD} Y^Q(\tau)} \right] Y^Q(\tau) * \\ &\ldots * U^{WD}_{\substack{Q\\N^Q_{i=1+2}+N^K_{\mu=1,2}}}(\tau) \left[V^{E}_{\substack{WD\\CW+ME_{dep}}} + V^{E}_{\substack{WD\\CW+ME_{dev}}}(\tau) + V^{E}_{WD_{inn}}(\tau) \right] \end{aligned} \tag{9.6}$$

with

$$U^{WD}_{\substack{Q\\N^Q_{i=1+2}+N^K_{\mu=1,2}}}(\tau) = \left\{ \begin{aligned} &\left\{ \psi_Q^{RP} + \sum_{i=1}^{2} \frac{\tilde{\psi}_i^Q \kappa_i^Q e^{\varphi_i^Q \kappa_i^Q \sum_{t=1}^{\tau-1} Y^Q(t)}}{c(\overline{N}_i^Q(0))} \right\} \\ &+ \left\{ k^Q \left\{ d^Q + \gamma_\tau^Q \right\} \right\} \sum_{\mu=1}^{2} \frac{\tilde{\psi}_\mu^K \kappa_\mu^K e^{\varphi_\mu^K \kappa_\mu^K k^Q \left\{ \sum_{t=1}^{\tau-1} Y^Q(t) \left\{ d^Q + \gamma_t^Q \right\} \right\}}}{c(\overline{N}_\mu^K(0))} \end{aligned} \right\},$$

$$\begin{aligned} V^{E}_{\substack{WD\\CW+ME_{dep}}} &= \left\{ \sum_{\nu=1}^{2} \frac{e_\nu\, \kappa_\nu^{CW}}{c(\overline{N}_\nu^{CW}(0))} + \frac{ex^{CW}}{\eta^{CW}} \right\} k_{CW}^{WD} d^{CW} \\ &+ \left\{ \sum_{o=1}^{2} \frac{e_o\, \kappa_o^{ME}}{c(\overline{N}_o^{ME}(0))} + \frac{ex^{ME}}{\eta^{ME}} k_{ME}^{WD} \right\} k_{ME}^{WD} d^{ME}, \end{aligned}$$

$$V^{E}_{\substack{WD\\CW+ME_{dev}}}(\tau) = \left\{ \begin{aligned} &\left\{ \sum_{\nu=1}^{2} \frac{e_\nu\, \kappa_\nu^{CW}}{c(\overline{N}_\nu^{CW}(0))} + \frac{ex^{CW}}{\eta^{CW}} \right\} k_{CW}^{WD} \\ &+ \left\{ \sum_{o=1}^{2} \frac{e_o\, \kappa_o^{ME}}{c(\overline{N}_o^{ME}(0))} + \frac{ex^{ME}}{\eta^{ME}} k_{ME}^{WD} \right\} k_{ME}^{WD} \end{aligned} \right\} \gamma_\tau^{WD},$$

$$V^{E}_{WD_{inn}}(\tau) := ex^{WD}(0) \left\{ 1 + r_{an}^{WD}(0)\, e^{\varphi_{an}^{WD} Y^Q(\tau)} \right\}.$$

9.8.2.3 Aggregation of the Human Labour and Energy Input Amounts for the Wastewater Collection Sector

The Human Labour Input Amounts for Wastewater Collection

$$L_{\text{sector}}^{WWC}(\tau) = Y^{Q}(\tau)\, U_{\substack{Q \\ N_{i=1+2}^{Q}+N_{\mu=1,2}^{K}}}^{WWC}(\tau)\left[l^{WWC} + V_{\substack{WWC \\ CW+ME_{dep}}}^{L} + V_{\substack{WWC \\ CW+ME_{dev}}}^{L}(\tau)\right] \quad (9.7)$$

with

$$U_{\substack{Q \\ N_{i=1+2}^{Q}+N_{\mu=1,2}^{K}}}^{WWC}(\tau) = \left\{ \begin{array}{l} \left\{ \rho^{WW}(\tau)\psi_{Q}^{RP} + \sum_{i=1}^{2} \dfrac{\tilde{\psi}_{i}^{Q}\kappa_{i}^{Q} e^{\varphi_{i}^{Q}\kappa_{i}^{Q}\sum_{t=1}^{\tau-1} Y^{Q}(t)}}{c(\overline{N}_{i}^{Q}(0))} \right\} \\ + \left\{ k^{Q}\left\{ d^{Q} + \gamma_{\tau}^{Q} \right\}\right\} \sum_{\mu=1}^{2} \dfrac{\tilde{\psi}_{\mu}^{K}\kappa_{\mu}^{K} e^{\varphi_{\mu}^{K}\kappa_{\mu}^{K}k^{Q}\left\{\sum_{t=1}^{\tau-1} Y^{Q}(t)\left\{ d^{Q}+\gamma_{t}^{Q}\right\}\right\}}}{c(\overline{N}_{\mu}^{K}(0))} \end{array} \right\},$$

$$V_{\substack{WWC \\ CW+ME_{dep}}}^{L} = \left\{ \sum_{\nu=1}^{2} \frac{l_{\nu}\kappa_{\nu}^{CW}}{c(\overline{N}_{\nu}^{CW}(0))} + l^{CW} \right\} k_{CW}^{WWC} d^{CW} + \left\{ \sum_{o=1}^{2} \frac{l_{o}\kappa_{o}^{ME}}{c(\overline{N}_{o}^{ME}(0))} + l^{ME} \right\} k_{ME}^{WWC} d^{ME},$$

$$V_{\substack{WWC \\ CW+ME_{dev}}}^{L}(\tau) = \left\{ \left\{ \sum_{\nu=1}^{2} \frac{l_{\nu}\kappa_{\nu}^{CW}}{c(\overline{N}_{\nu}^{CW}(0))} + l^{CW} \right\} k_{CW}^{WWC} + \left\{ \sum_{o=1}^{2} \frac{l_{o}\kappa_{o}^{ME}}{c(\overline{N}_{o}^{ME}(0))} + l^{ME} \right\} k_{ME}^{WWC} \right\} \gamma_{\tau}^{WWC}.$$

The Energy Input Amount for Wastewater Collection

$$E_{\text{sector}}^{WWC}(\tau) = Y^{Q}(\tau)\, U_{\substack{Q \\ N_{i=1+2}^{Q}+N_{\mu=1,2}^{K}}}^{WWC}(\tau)\left[V_{\substack{WWC \\ CW+ME_{dep}}}^{E} + V_{\substack{WWC \\ CW+ME_{dev}}}^{E}(\tau) + V_{WWC_{inn}}^{E}\right] \quad (9.8)$$

with

$$U_{Q\atop N_{i=1+2}^{Q}+N_{\mu=1,2}^{K}}^{WWC}(\tau) = \left\{ \begin{matrix} \left\{ \rho^{WW}(\tau)\psi_Q^{RP} + \sum_{i=1}^{2} \frac{\tilde{\psi}_i^Q \kappa_i^Q e^{\varphi_i^Q \kappa_i^Q \sum_{t=1}^{\tau-1} Y^Q(t)}}{c(\overline{N}_i^Q(0))} \right\} \\ + \left\{ k^Q \left\{ d^Q + \gamma_\tau^Q \right\} \right\} \sum_{\mu=1}^{2} \frac{\tilde{\psi}_\mu^K \kappa_\mu^K e^{\varphi_\mu^K \kappa_\mu^K k^Q \left\{ \sum_{t=1}^{\tau-1} Y^Q(t) \left\{ d^Q + \gamma_t^Q \right\} \right\}}}{c(\overline{N}_\mu^K(0))} \end{matrix} \right\},$$

$$V_{WWC\atop CW+ME_{dep}}^{E} = \left\{ \sum_{\nu=1}^{2} \frac{e_\nu \kappa_\nu^{CW}}{c(\overline{N}_\nu^{CW}(0))} + \frac{ex^{CW}}{\eta^{CW}} \right\} k_{CW}^{WWC} d^{CW} + \left\{ \sum_{o=1}^{2} \frac{e_o \kappa_o^{ME}}{c(\overline{N}_o^{ME}(0))} + \frac{ex^{ME}}{\eta^{ME}} \right\} k_{ME}^{WWC} d^{ME},$$

$$V_{WWC\atop CW+ME_{dev}}^{E}(\tau) = \left\{ \left\{ \sum_{\nu=1}^{2} \frac{e_\nu \kappa_\nu^{CW}}{c(\overline{N}_\nu^{CW}(0))} + \frac{ex^{CW}}{\eta^{CW}} \right\} k_{CW}^{WWC} + \left\{ \sum_{o=1}^{2} \frac{e_o \kappa_o^{ME}}{c(\overline{N}_o^{ME}(0))} + \frac{ex^{ME}}{\eta^{ME}} \right\}_{ME}^{WWC} \right\} \gamma_\tau^{WWC},$$

$$V_{WWC_{inn}}^{E}(\tau) := ex^{WWC}(0) \left[1 + r_{an}^{WWC}(0)\, e^{\varphi_{an}^{WWC} Y^Q(\tau)} \right].$$

9.8.2.4 Aggregation of the Human Labour and Energy Input Amounts for the Wastewater Treatment Sector

The Human Labour Input Amounts for Wastewater Treatment

$$\begin{aligned} L_{\text{sector}}^{WWT}(\tau) &= \left[1 - r_{WW}^{WWC}(0)\, e^{\varphi_r^{WWC} Y^Q(\tau)} \right] Y^Q(\tau)^* \\ &\dots * U_{Q\atop N_{i=1+2}^{Q}+N_{\mu=1,2}^{K}}^{WWT}(\tau) \left[l^{WWT} + V_{WWT\atop CW+ME_{dep}}^{L} + V_{WWT\atop CW+ME_{dev}}^{L}(\tau) \right] \end{aligned} \tag{9.9}$$

with

$$U_{Q\atop N_{i=1+2}^{Q}+N_{\mu=1,2}^{K}}^{WWT}(\tau) = \left\{ \begin{matrix} \left\{ \rho^{WW}(\tau)\psi_Q^{RP} + \sum_{i=1}^{2} \frac{\tilde{\psi}_i^Q \kappa_i^Q e^{\varphi_i^Q \kappa_i^Q \sum_{t=1}^{\tau-1} Y^Q(t)}}{c(\overline{N}_i^Q(0))} \right\} \\ + \left\{ k^Q \left\{ d^Q + \gamma_\tau^Q \right\} \right\} \sum_{\mu=1}^{2} \frac{\tilde{\psi}_\mu^K \kappa_\mu^K e^{\varphi_\mu^K \kappa_\mu^K k^Q \left\{ \sum_{t=1}^{\tau-1} Y^Q(t) \left\{ d^Q + \gamma_t^Q \right\} \right\}}}{c(\overline{N}_\mu^K(0))} \end{matrix} \right\},$$

$$V^{L}_{\substack{WWT\\ CW+ME_{dep}}} = \left\{\sum_{\nu=1}^{2} \frac{l_\nu\, \kappa_\nu^{CW}}{c(\overline{N}_\nu^{CW}(0))} + l^{CW}\right\} k_{CW}^{WWT} d^{CW} + \left\{\sum_{o=1}^{2} \frac{l_o\, \kappa_o^{ME}}{c(\overline{N}_o^{ME}(0))} + l^{ME}\right\} k_{ME}^{WWT} d^{ME},$$

$$V^{L}_{\substack{WWT\\ CW+ME_{dev}}}(\tau) = \left\{\left\{\sum_{\nu=1}^{2} \frac{l_\nu\, \kappa_\nu^{CW}}{c(\overline{N}_\nu^{CW}(0))} + l^{CW}\right\} k_{CW}^{WWT} + \left\{\sum_{o=1}^{2} \frac{l_o\, \kappa_o^{ME}}{c(\overline{N}_o^{ME}(0))} + l^{ME}\right\} k_{ME}^{WWT}\right\} \gamma_\tau^{WWT}.$$

The Energy Input Amounts for Wastewater Treatment

$$\begin{aligned} E^{WWT}_{\text{sector}}(\tau) &= \left[1 - r_{WW}^{WWC}(0)\, e^{\varphi_r^{WWC} Y^Q(\tau)}\right] Y^Q(\tau)^* \\ &\dots * U^{WWT}_{\substack{Q\\ N^Q_{i=1+2}+N^K_{\mu=1,2}}}(\tau) \left[V^{E}_{\substack{WWT\\ CW+ME_{dep}}} + V^{E}_{\substack{WWT\\ CW+ME_{dev}}}(\tau) + V^{E}_{WWT_{inn}}(\tau)\right] \end{aligned} \tag{9.10}$$

with

$$U^{WWT}_{\substack{Q\\ N^Q_{i=1+2}+N^K_{\mu=1,2}}}(\tau) = \left\{ \begin{array}{l} \left\{ \rho^{WW}(\tau)\psi_Q^{RP} + \sum_{i=1}^{2} \dfrac{\tilde{\psi}_i^Q \kappa_i^Q e^{\varphi_i^Q \kappa_i^Q \sum_{t=1}^{\tau-1} Y^Q(t)}}{c(\overline{N}_i^Q(0))} \right\} \\ + \left\{k^Q \left\{d^Q + \gamma_\tau^Q\right\}\right\} \sum_{\mu=1}^{2} \dfrac{\tilde{\psi}_\mu^K \kappa_\mu^K e^{\varphi_\mu^K \kappa_\mu^K k^Q \left\{\sum_{t=1}^{\tau-1} Y^Q(t) \left\{d^Q + \gamma_t^Q\right\}\right\}}}{c(\overline{N}_\mu^K(0))} \end{array} \right\},$$

$$V^{E}_{\substack{WWT\\ CW+MEdep}}(\tau) = \left\{\sum_{\nu=1}^{2} \frac{e_\nu\, \kappa_\nu^{CW}}{c(\overline{N}_\nu^{CW}(0))} + \frac{ex^{CW}}{\eta^{CW}}\right\} k_{CW}^{WWT} d^{CW} + \left\{\sum_{o=1}^{2} \frac{e_o\, \kappa_o^{ME}}{c(\overline{N}_o^{ME}(0))} + \frac{ex^{ME}}{\eta^{ME}}\right\} k_{ME}^{WWT} d^{ME},$$

$$V^{E}_{\substack{WWT\\ CW+ME_{dev}}}(\tau) = \left\{\left\{\sum_{\nu=1}^{2} \frac{e_\nu \kappa_\nu^{CW}}{c(\overline{N}_\nu^{CW}(0))} + \frac{ex^{CW}}{\eta^{CW}}\right\} k_{CW}^{WWT} + \left\{\sum_{o=1}^{2} \frac{e_o \kappa_o^{ME}}{c(\overline{N}_o^{ME}(0))} + \frac{ex^{ME}}{\eta^{ME}}\right\} k_{ME}^{WWC}\right\} \gamma_\tau^{WWT},$$

$$V^{E}_{WWT_{inn}}(\tau) := \tilde{\varepsilon}^{WWT} \left\{\ln \frac{c_Q^*(0)}{c_Q(0)} + \ln \frac{c_{\tilde{j}}^*(0)}{c_{\tilde{j}}(0)}\right\} \left[1 + r_{an}^{WWT}(0)\, e^{\varphi_{an}^{WWT} Y^Q(\tau)}\right].$$

Chapter 10
Optimality Conditions of the Water Infrastructure Model

Abstract The optimality conditions are formulated as non-profit conditions. They are expressed as actual and inter-temporal marginal costs for human labour and energy inputs for the production sector and for the water and wastewater sectors. These marginal costs depend on the development path of the consumption goods, which are accompanied by structural changes that are caused by innovation effects within the different sectors. The reduction of water losses from the water distribution system, for example, leads to a decrease in the required water quantities. The reduction of wastewater leakages, on the other hand, results in an increase in the quantities of wastewater to be treated. These changes in the actual and inter-temporal marginal costs for wastewater treatment can be interpreted as beneficial increases of reduced wastewater leakages, which generate social costs in the form of health risks and damages to the water infrastructure. As a result, the actual and inter-temporal marginal costs for using non-renewable resources for production are combined with those costs for water and its infrastructure.

Within the framework of the model constraints, we will now derive the optimality conditions for the water infrastructure model. The objective is once again to maximize the quasi-concave, inter-temporal welfare function, where the amounts of the consumption good vector $\vec{Q}(1)$ of the first activity interval determine the initial conditions of the model

$$W\left(\vec{Q}(1),...,\vec{Q}(T)\right) = \sum\nolimits_{\tau=1}^{T} (1+\delta)^{1-\tau} W_\tau \left(\vec{Q}(\tau)\right).$$

Because the consumption good vector has only one component, the following welfare function must be maximized

$$W\left(\bar{Q}(1), Q(2),...,Q(T)\right) = W_1(\bar{Q}(1) + \sum_{\tau=2}^{T} (1+\delta)^{1-\tau} W_\tau(Q(\tau)). \tag{10.1}$$

A feasible production program is called optimal when it maximizes this welfare function. According to the procedure of Kuhn-Tucker (for more details, see Hadley

H. Niemes, M. Schirmer, *Entropy, Water and Resources*,
DOI 10.1007/978-3-7908-2416-2_10, © Springer-Verlag Berlin Heidelberg 2010

1964:185–212), this means forming the LAGRANGE function (Eq. 10.2) from the welfare function and the system of constraints of the extended model.

$$\begin{aligned} V &= W_1(\bar{Q}(1) + \sum_{\tau=2}^{T} (1+\delta)^{1-\tau} W_\tau(Q(\tau)) + \sum_{\tau=1}^{T} \left\{ \sum_{g} p_g(\tau) H_g(\tau) \right\} \Leftrightarrow \\ \tilde{V} &:= V - W_1(\bar{Q}(1) = \sum_{\tau=2}^{T} (1+\delta)^{1-\tau} W_\tau(Q(\tau)) + \sum_{\tau=1}^{T} \left\{ \sum_{g} p_g(\tau) H_g(\tau) \right\}. \end{aligned} \tag{10.2}$$

With the indices

$$g \in \left\{ \begin{array}{l} Q, N^{Q}_{i=1,2}, \bar{N}^{Q}_{i=1,2}, N^{Q}_{ii}, N^{K}_{\mu=1,2}, \bar{N}^{K}_{\mu=1,2}, N^{Q}_{\mu\mu}, N^{CW}_{\nu=1,2}, N^{CW}_{\nu\nu}, N^{ME}_{o=1,2}, N^{ME}_{oo}, \\ \genfrac{}{}{0pt}{}{WT}{C\bar{W}}, \genfrac{}{}{0pt}{}{WD}{C\bar{W}}, \genfrac{}{}{0pt}{}{WWC}{C\bar{W}}, \genfrac{}{}{0pt}{}{WWT}{C\bar{W}}, \genfrac{}{}{0pt}{}{WT}{M\bar{E}}, \genfrac{}{}{0pt}{}{WD}{M\bar{E}}, \genfrac{}{}{0pt}{}{WWC}{M\bar{E}}, \genfrac{}{}{0pt}{}{WWT}{M\bar{E}}, WT, WD, WWC, WWT, N^{E}, L, \bar{K}^{Q} \end{array} \right\}.$$

When using $Y_g(\bar{\tau})$ for the variables of the water infrastructure model, an optimal feasible production program must fulfil the general conditions for each of the selected activity intervals $\bar{\tau} \in \{1, ..., T\}$, where the LAGRANGE multipliers $p_g(\bar{\tau})$ are interpreted as shadow prices of the variables $Y_g(\bar{\tau})$

$$\begin{aligned} &\left\{ \frac{\vartheta \tilde{V}}{\vartheta Y_g(\bar{\tau})} = 0 \; for \; Y_g(\bar{\tau}) > 0 \right\} \vee \left\{ \frac{\vartheta \tilde{V}}{\vartheta Y_g(\bar{\tau})} \leq 0 \; for \; Y_g(\bar{\tau}) = 0 \right\} \\ &with \\ &\left\{ H_g(\bar{\tau}) = 0 \; for \; p_g(\bar{\tau}) \geq 0 \right\} \vee \left\{ H_g(\bar{\tau}) > 0 \; for \; p_g(\bar{\tau}) = 0 \right\} \wedge p_g(\bar{\tau}+1) \geq 0. \end{aligned} \tag{10.3}$$

When using the no-storage assumption of the different process input and output amounts, the substitution of the different variables $Y_g(\bar{\tau})$ allows the activity level of the production process for the consumption good amount $Y^Q(\bar{\tau})$ to remain a variable of the production side. The consumption good amounts $Q(\bar{\tau})$ represent the variable of the demand side. These simplifications mean that all production processes are always in operation. There is, therefore, no need to complete case distinctions for interpreting the conditions of optimal feasible production programs. The following parts of the LAGRANGE function will be the subject of our analysis

$$\begin{aligned} &\left\{ \frac{\vartheta \tilde{V}}{\vartheta Y_g(\bar{\tau})} = 0 \text{ for } Y_g(\bar{\tau}) > 0 \right\} \\ &\text{with } \left\{ H_g(\bar{\tau}) = 0 \text{ for } p_g(\bar{\tau}) \geq 0 \right\} \wedge p_g(\bar{\tau}) > 0 \end{aligned} \tag{10.4}$$

10.1 Optimality Conditions for the Demand Side

The consumption good constraint describes how the demand side is connected with the production side. In our case with no durable consumption goods, we use the equivalent constraints to describe how both sides are connected

$$H_Q(\tau) = Y^Q(\tau) - Q(\tau+1) \geq 0 \Leftrightarrow H_{YQ}(\tau) = Y^Q(\tau) - Q(\tau+1) \geq 0$$

If we had durable consumption goods that were not consumed within one activity interval, another inter-temporal connection would have to be used for the demand side.

We obtain the following identity relation between the shadow prices for the consumption good and for producing the consumption good:

$$p_Q(\bar{\tau}+1) = (1+\delta)^{-\bar{\tau}} \frac{\vartheta W_{\bar{\tau}+1}(Q(\bar{\tau}+1)}{\vartheta Q(\bar{\tau}+1)} = p_{YQ}(\bar{\tau}) \tag{10.5}$$

10.2 Optimality Conditions for the Production Side

If the optimality conditions are fulfilled, we also achieve the so-called non-profit condition. The non-profit condition states that the shadow price of the consumption good, $p_{YQ}(\bar{\tau})$, must be equal to the sum of the actual and inter-temporal marginal costs of the different cost components for producing one unit of consumption good. The other constraints can be considered as embedded into the human labour and energy constraints. The shadow price of the consumption good is determined by the sum of the marginal costs of the direct and indirect human labour and energy inputs of the different processes that are required to produce one unit of the consumption good.

When differentiating the LAGRANGE function, the product, or Leibniz, rule for a collection of functions $f_1, \ldots, f_n$, will be applied

$$\frac{d}{dx}\prod_{i=1}^{k} f_i(x) = \left(\sum_{i=1}^{k} \frac{\frac{d}{dx} f_i(x)}{f_i(x)}\right)\prod_{i=1}^{k} f_i(x). \tag{10.6}$$

The water sectors of the extended model have a more complex structure. Differentiations of functions with three or four products must be developed because two additional capital goods are required. Furthermore, water and energy innovation aspects will also be evaluated.

The detailed expressions for the marginal costs of the human labour and energy inputs of the processes involved in producing one unit of consumption good will first be derived for the production sector and secondly for the water supply and wastewater sectors. These derivations will be based on the two essential constraints in Eqs. (EM14) and (EM15).

10.2.1 Non-profit Conditions for the Production Sector

The total marginal costs of the energy input of the production sector are equal to the sum

$$MCE^{P}_{sector}(\bar{\tau} \mapsto T) = MCE^{Q}_{Q}(\bar{\tau} \mapsto T) + MCE^{Q}_{K}(\bar{\tau} \mapsto T) + \\ \ldots + MCE^{Q}_{N_{i=1+2}}(\bar{\tau} \mapsto T) + MCE^{Q}_{K_{N_{\mu=1+2}}}(\bar{\tau} \mapsto T).$$

The actual and inter-temporal marginal human labour and energy costs for the production sector processes are already known from the basic model. They have been summarized in the form of the matrix derived and interpreted in Chap. 6. We therefore continue to derive the marginal costs for the water and wastewater sector.

process $\hat{g}$	actual m. lab. costs	intertemp. lab costs $(\bar{\tau} \to T)$	actual m. energy costs	intertemporal m. energy costs $(\bar{\tau} \to T)$
$\overset{Q}{Q}$	$MCL^{Q}_{Q}(\bar{\tau})$	$ITCL^{Q}_{K}$	$MCE^{Q}_{Q}(\bar{\tau})$	
$\overset{Q}{N_{i=1+2}}$	$MCL^{Q}_{N_{i=1+2}}(\bar{\tau})$	$ITCL^{Q}_{N_{i=1+2}}$	$MCE^{Q}_{N_{i=1+2}}(\bar{\tau})$	$ITCE^{Q}_{N_{i=1+2}}$
$\overset{Q}{K}$	$MCL^{Q}_{K}(\bar{\tau})$		$MCE^{Q}_{K}(\bar{\tau})$	
$\overset{Q}{K_{dep}}$ $N_{\mu=1+2}$	$MCL^{Q}_{K_{dep}\,N_{\mu=1+2}}(\bar{\tau})$	$ITCL^{Q}_{K_{dep}\,N_{\mu=1+2}}$	$MCE^{Q}_{K_{dep}\,N_{\mu=1+2}}(\bar{\tau})$	$ITCE^{Q}_{K_{dep}\,N_{\mu=1+2}}$
$\overset{Q}{K_{dev}}$ $N_{\mu=1+2}$	$MCL^{Q}_{K_{dev}\,N_{\mu=1+2}}(\bar{\tau})$	$ITCL^{Q}_{K_{dev}\,N_{\mu=1+2}}$	$MCE^{Q}_{K_{dev}\,N_{\mu=1+2}}(\bar{\tau})$	$ITCE^{Q}_{K_{dev}\,N_{\mu=1+2}}$
TMC^{p}_{sec}	$\sum_{\hat{g}} MCL^{p}_{\hat{g}}(\bar{\tau})$	$\sum_{\hat{g}} ITCL^{p}_{\hat{g}}$	$\sum_{\hat{g}} MCE^{p}_{\hat{g}}(\bar{\tau})$	$\sum_{\hat{g}} ITCE^{p}_{\hat{g}}$

10.2.2 Non-profit Conditions for the Water Sectors

The water sector of the extended model has a more complex structure than the water use model because the additional capital stocks of civil work and mechanical equipment capital goods carry water and energy saving innovations. The abbreviations introduced in the context of the aggregation of human labour and energy input amounts allow the optimality conditions (or marginal costs) for the water treatment, water distribution, and for the wastewater collection and treatment to be derived more easily. However, this is because only the so-called non-profit conditions are considered.

The derivation of the marginal costs for the water and wastewater sectors requires a large number of mathematical steps. The derivation and interpretation of the marginal costs are presented in detail as a mathematical appendix to this section. This allows us to continue with a general interpretation of the actual and inter-temporal human labour and energy costs of the water and wastewater sectors, which are summarized in the following matrix.

For each group in this summary matrix on page 244, we distinguish row by row between the different effects on the actual and inter-temporal marginal human labour and energy costs of the water and wastewater amounts that are required to produce one unit of consumption good. In the first row of the water treatment sector, for example, the reduction of the water losses leads to lower marginal costs for water treatment. The second row represents the effect of changing the activity levels or consumption good amounts. Both effects influence the actual marginal human labour and energy costs.

The change of activity level is combined with structural changes to the required raw materials and the corresponding water amounts that transport the undesired by-products to the wastewater treatment sector by the wastewater collection system. The raw materials that are required for producing the desired consumption and capital goods are extracted from the natural environment. The concentrations of these raw materials are decreasing functions when there is no ability or artificial recycling processes. These structural changes are represented by the next two rows and they lead to inter-temporal marginal human labour and energy costs.

The development path of the consumption good amounts is also combined with structural changes for sustaining and developing the capital stocks of the production and water treatment sector. The fifth row for each of the water and wastewater sectors shows the growth rate of the development path, which also influences the development of the capital stocks of the water and wastewater sectors. These additional actual and inter-temporal marginal cost effects are also accompanied by energy saving innovations.

Reducing the water losses leads to a reduction in the amounts of water that need to be treated. Reducing the wastewater leakage increases the marginal costs for wastewater treatment. This effect is interpreted as additional social benefits, or social costs, for preventing health risks and environmental damages caused by otherwise untreated wastewater.

10.3 Conclusions and Perspectives

The natural sciences consistent models guided by water uses and water infrastructure confirm that, in the case where producing desired consumption goods mainly relies on extracting raw materials from non-renewable natural resources, the actual and inter-temporal marginal costs occur for the production sector. Additional water amounts for the production and reproduction activities are required because of the overall decrease in concentrations of raw materials extracted from the natural environment. As a result, actual and inter-temporal marginal costs for the water and wastewater sectors are also generated. The past and current economic activities generate inter-temporal marginal costs even in the case where the water resources are renewable.

Sector effects $\hat{g}$	actual m. labour costs	intertemp. m. labour costs $\bar{\tau} \rightarrow T$	actual m. energy costs	intertemp m. energy costs $\bar{\tau} \rightarrow T$
$Q_{\substack{WT\\W_{inn}}}$	$MCL^{Q}_{\substack{WT\\W_{inn}}}(\bar{\tau})$		$MCE^{Q}_{\substack{WT\\W_{inn}}}(\bar{\tau})$	
$Q_{\substack{WT\\Y^{Q}}}$	$MCL^{Q}_{\substack{WT\\Y^{Q}}}(\bar{\tau})$		$MCE^{Q}_{\substack{WT\\Y^{Q}}}(\bar{\tau})$	
$Q_{\substack{WT\\N^{Q}_{i=1+2}}}$		$ITMLC^{Q}_{\substack{WT\\N^{Q}_{i=1+2}}}$		$ITMEC^{Q}_{\substack{WT\\N^{Q}_{i=1+2}}}$
$Q_{\substack{WT\\N^{K}_{\mu=1+2}}}$		$ITMLC^{Q}_{\substack{WT\\N^{K}_{\mu=1+2}}}$		$ITMEC^{Q}_{\substack{WT\\N^{K}_{\mu=1+2}}}$
$Q_{\substack{WT\\\gamma_{\tau}+E_{inn}}}$		$ITMLC^{Q}_{\gamma_{\gamma}^{WT}}$	$MEC^{Q}_{\substack{WT\\E_{inn}}}(\bar{\tau})$	$ITMEC^{Q}_{\substack{WT\\\gamma_{\tau}+E_{inn}}}$
$Q_{\substack{WD\\W_{inn}}}$	$MCL^{Q}_{\substack{WD\\W_{inn}}}(\bar{\tau})$		$MCE^{Q}_{\substack{WD\\W_{inn}}}(\bar{\tau})$	
$Q_{\substack{WD\\Y^{Q}}}$	$MCL^{Q}_{\substack{WD\\Y^{Q}}}(\bar{\tau})$		$MCE^{Q}_{\substack{WD\\Y^{Q}}}(\bar{\tau})$	
$Q_{\substack{WD\\N^{Q}_{i=1+2}}}$		$ITMLC^{Q}_{\substack{WD\\N^{Q}_{i=1+2}}}$		$ITMEC^{Q}_{\substack{WD\\N^{Q}_{i=1+2}}}$
$Q_{\substack{WD\\N^{K}_{\mu=1+2}}}$		$ITMLC^{Q}_{\substack{WD\\N^{K}_{\mu=1+2}}}$		$ITMEC^{Q}_{\substack{WD\\N^{K}_{\mu=1+2}}}$
$Q_{\substack{WD\\\gamma_{\tau}+E_{inn}}}$		$ITMLC^{Q}_{\gamma_{\gamma}^{WD}}$	$MEC^{Q}_{\substack{WD\\E_{inn}}}(\bar{\tau})$	$ITMEC^{Q}_{\substack{WD\\\gamma_{\tau}+E_{inn}}}$
$Q_{\substack{WWC\\W_{inn}}}$	$MCL^{Q}_{\substack{WWC\\W_{inn}}}(\bar{\tau})$		$MCE^{Q}_{\substack{WWC\\W_{inn}}}(\bar{\tau})$	
$Q_{\substack{WWC\\Y^{Q}}}$	$MCL^{Q}_{\substack{WWC\\Y^{Q}}}(\bar{\tau})$		$MCE^{Q}_{\substack{WWC\\Y^{Q}}}(\bar{\tau})$	
$Q_{\substack{WWC\\N^{Q}_{i=1+2}}}$		$ITMLC^{Q}_{\substack{WWC\\N^{Q}_{i=1+2}}}$		$ITMEC^{Q}_{\substack{WWC\\N^{Q}_{i=1+2}}}$
$Q_{\substack{WWC\\N^{K}_{\mu=1+2}}}$		$ITMLC^{Q}_{\substack{WWC\\N^{K}_{\mu=1+2}}}$		$ITMEC^{Q}_{\substack{WWC\\N^{K}_{\mu=1+2}}}$
$Q_{\substack{WT\\\gamma_{\tau}+E_{inn}}}$		$ITMLC^{Q}_{\gamma_{\gamma}^{WWC}}$	$MEC^{Q}_{\substack{WWC\\E_{inn}}}(\bar{\tau})$	$ITMEC^{Q}_{\substack{WWC\\\gamma_{\tau}+E_{inn}}}$
$Q_{\substack{WWT\\W_{inn}}}$	$MCL^{Q}_{\substack{WWT\\W_{inn}}}(\bar{\tau})$		$MCE^{Q}_{\substack{WWT\\W_{inn}}}(\bar{\tau})$	
$Q_{\substack{WWT\\Y^{Q}}}$	$MCL^{Q}_{\substack{WWT\\Y^{Q}}}(\bar{\tau})$		$MCE^{Q}_{\substack{WWT\\Y^{Q}}}(\bar{\tau})$	
$Q_{\substack{WWT\\N^{Q}_{i=1+2}}}$		$ITMLC^{Q}_{\substack{WWT\\N^{Q}_{i=1+2}}}$		$ITMEC^{Q}_{\substack{WWT\\N^{Q}_{i=1+2}}}$
$Q_{\substack{WWT\\N^{K}_{\mu=1+2}}}$		$ITMLC^{Q}_{\substack{WWT\\N^{K}_{\mu=1+2}}}$		$ITMEC^{Q}_{\substack{WWT\\N^{K}_{\mu=1+2}}}$
$Q_{\substack{WWT\\\gamma_{\tau}+E_{inn}}}$		$ITMLC^{Q}_{\gamma_{\gamma}^{WWT}}$	$MEC^{Q}_{\substack{WWT\\E_{inn}}}(\bar{\tau})$	$ITMEC^{Q}_{\substack{WWT\\\gamma_{\tau}+E_{inn}}}$
TMC^{W+WW}_{sector}	$\sum_{\hat{g}} MCL^{Q}_{\hat{g}}(\bar{\tau})$	$\sum_{\hat{g}} ITMCL^{Q}_{\hat{g}}$	$\sum_{\hat{g}} MCE^{Q}_{\hat{g}}(\bar{\tau})$	$\sum_{\hat{g}} ITMCE^{Q}_{\hat{g}}$

The extended dynamic model proved that the water and energy saving innovations can partly compensate for the negative trend of increasing actual and inter-temporal marginal costs for water treatment, water distribution, wastewater collection, and wastewater treatment. Nevertheless, we come to the conclusion that a society is confronted with two crises when the production and reproduction activities mainly rely on non-renewable resources. The first is the inter-temporal marginal costs for production and reproduction activities and the second is the inter-temporal marginal costs of the water supply and wastewater sectors.

An essential surplus of information about the structure of the marginal costs for producing one unit of the consumption good can be derived because the water uses and water and wastewater infrastructure are integrated into the models. We are convinced that our approach closes the gap between economic theory and natural science and engineering concepts.

Apart from the human labour innovations of the production sector, we have introduced additional innovation effects, especially for the water supply and wastewater sectors. Our models have the flexibility to be extended in different directions. Some refinements of the model, for example, could be to replace some of the assumed time-invariant coefficients for the different sectors by time-variant coefficients or to introduce recycling processes into the models.

Another interesting way to refine our model would be to replace the exogenously given water quality standards by an interrelation between the residuals after wastewater treatment (or emissions) and environmental goods. The amounts or qualities of the environmental goods would be additional components of the inter-temporal welfare function. These kinds of interrelations between the emissions and the welfare function, however, require the introduction of so-called diffusion, self-purification, and damage functions in our model. For river basin water quality management strategies, for example, the extension of the model in this direction would be of special interest. In this context, the model could also be used to compare the efficiency of alternative environmental instruments.

Apart from such refinements, our perspectives on developing and applying our extended model will be to combine climate change problems with water and energy scarcity problems. First, our model has the flexibility to take into account either additionally, or separately, the production of free energy and to introduce the use of additional environmental goods such as air and soil. We only have to release the assumption that the natural resources for producing free energy are not a limiting factor. Second, an infrastructure similar to the one introduced for the water and wastewater sectors could be established for the energy production or air production sectors. The model can therefore be adapted to address the three problems of water, energy, and global warming, which will confront mankind in the near future.

Aside from analysing these environmental and resource problems, we plan to revise the welfare function so that durable consumption goods are also taken into consideration. This means that the reproduction activities will also be based on a capital stock, which is a better representation of the learning and innovation effects for the production sector. The number of consumption goods will be increased and therefore the production side must accordingly be generalized. The production of the capital and durable good amounts required for sustaining and developing the capital stocks for the production and reproduction activities might be revised to strengthen the information aspects of production and reproduction activities.

These generalizations might imply that it is necessary to modify the concept of the capital theory or to replace it by other concepts. The loss of biodiversity by the degradation of natural resources, for example, needs natural science consistent economic models where the information components of the natural and economical activities are taken into consideration.

Our proposal to analyse the dynamic interrelations of the energy, water, and global warming crises that will face mankind in the future are in line with an impressive number of contributions to this subject. One example is the recently published United Nations World Development Report 3: "Water in a changing world" (2009). In this context, we will try to apply our theoretical work to analyse select geographical region, such as the Alps of Switzerland. This region has significant fresh water storage capacity, which is influenced by global warming, as analysed by Jasper et al. (2004, 2006).

Mathematical Appendix and Detailed Explanations to Sect. 10.2.2

A Marginal Costs for Water Treatment

A.1 Marginal Human Labour Costs for Water Treatment

$$
\begin{aligned}
MCL^{Q}_{WT}(\bar{\tau} \mapsto T) &= \frac{\vartheta}{\vartheta Y^{Q}(\bar{\tau})} \sum_{\tau=1}^{T} p_L(\tau) \left[1 + r_W^{WD}(0) e^{\varphi_r^{WD} Y^{Q}(\tau)}\right] \\
&\quad Y^{Q}(\tau)\, U^{WT}_{\substack{Q \\ N^{Q}_{i=1+2}+N^{K}_{\mu=1,2}+^{W}_{RP}}}(\tau) \left[1^{WT} + V^{L}_{\substack{WT \\ CW+ME_{dep}}} + V^{L}_{\substack{WT \\ CW+ME_{dev}}}(\tau)\right] \\
&= \sum_{\tau=1}^{T} p_L(\tau) \left\{ \frac{\vartheta \left[1 + r_W^{WD}(0) e^{\varphi_r^{WD} Y^{Q}(\tau)}\right]}{\vartheta Y^{Q}(\bar{\tau})} \right\} \\
&\quad Y^{Q}(\tau)\, U^{WT}_{\substack{Q \\ N^{Q}_{i=1+2}+N^{K}_{\mu=1,2}+^{W}_{RP}}}(\tau) \left[1^{WT} + V^{L}_{\substack{WT \\ CW+ME_{dep}}} + V^{L}_{\substack{WT \\ CW+ME_{dev}}}(\tau)\right] \\
&\quad + \sum_{\tau=1}^{T} p_L(\tau) \left[1 + r_W^{WD}(0) e^{\varphi_r^{WD} Y^{Q}(\tau)}\right] \left\{ \frac{\vartheta Y^{Q}(\tau)}{\vartheta Y^{Q}(\bar{\tau})} \right\} \\
&\quad U^{WT}_{\substack{Q \\ N^{Q}_{i=1+2}+N^{K}_{\mu=1,2}+^{W}_{RP}}}(\tau) \left[1^{WT} + V^{L}_{\substack{WT \\ CW+ME_{dep}}} + V^{L}_{\substack{WT \\ CW+ME_{dev}}}(\tau)\right]
\end{aligned}
$$

$$+\sum_{\tau=1}^{T} p_L(\tau)\left[1+\ r_W^{WD}(0)e^{\varphi_r^{WD}Y^Q(\tau)}\right] Y^Q(\tau)\left\{\frac{\vartheta\, U^{WT}_{Q_{N^Q_{i=1+2}+N^K_{\mu=1,2}+{}^W_{RP}}}(\tau)}{\vartheta Y^Q(\bar{\tau})}\right\}$$

$$\left[l^{WT}+\ V^L_{WT_{CW+ME_{dep}}}+V^L_{WT_{CW+ME_{dev}}}(\tau)\right]$$

$$+\sum_{\tau=1}^{T} p_L(\tau)\left[1+\ r_W^{WD}(0)e^{\varphi_r^{WD}Y^Q(\tau)}\right]\ Y^Q(\tau)\ U^{WT}_{Q_{N^Q_{i=1+2}+N^K_{\mu=1,2}+{}^W_{RP}}}(\tau)$$

$$\left\{\frac{\vartheta}{\vartheta Y^Q(\bar{\tau})}\left[l^{WT}+\ V^L_{WT_{CW+ME_{dep}}}+V^L_{WT_{CW+ME_{dev}}}(\tau)\right]\right\}$$

with

$$U^{WT}_{Q_{N^Q_{i=1+2}+N^K_{\mu=1,2}+{}^W_{RP}}}(\tau):=\left\{\begin{array}{l}\left\{\psi_Q^{RP}+\sum_{i=1}^{2}\frac{\tilde{\psi}_i^Q\kappa_i^Q e^{\varphi_i^Q\kappa_i^Q\sum_{t=1}^{\tau-1}Y^Q(t)}}{c(\bar{N}_i^Q(0))}\right\}+\\ \left\{k^Q\left\{d^Q+\ \gamma_\tau^Q\right\}\right\}\sum_{\mu=1}^{2}\frac{\tilde{\psi}_\mu^K\kappa_\mu^K e^{\varphi_\mu^K\kappa_\mu^K k^Q\sum_{t=1}^{\tau-1}Y^Q(t)\left\{d^Q+\gamma_t^Q\right\}}}{c(\bar{N}_\mu^K(0))}\end{array}\right\}$$

$$V^L_{WT_{CW+ME_{dep}}}\ :=\ \left\{\sum_{\nu=1}^{2}\frac{l_\nu\,\kappa_\nu^{CW}}{c(\bar{N}_\nu^{CW}(0))}+l^{CW}\right\}k^{WT}_{CW}d^{CW}+\left\{\sum_{o=1}^{2}\frac{l_o\,\kappa_o^{ME}}{c(\bar{N}_o^{ME}(0))}+l^{ME}\right\}k^{WT}_{ME}d^{ME}$$

$$V^L_{WT_{CW+ME_{dev}}}(\tau):=\left\{\left\{\sum_{\nu=1}^{2}\frac{l_\nu\,\kappa_\nu^{CW}}{c(\bar{N}_\nu^{CW}(0))}+l^{CW}\right\}k^{WT}_{CW}+\left\{\sum_{o=1}^{2}\frac{l_o\,\kappa_o^{ME}}{c(\bar{N}_o^{ME}(0))}+l^{ME}\right\}k^{WT}_{ME}\right\}\gamma_\tau^{WT}.$$

When applying the Leibniz rule for differentiation of a function of several product factors, we obtain the following marginal human labour and energy cost components for water treatment.

A.1.1 Actual Marginal Human Labour Costs for Water Treatment Caused by Water Saving Innovation Effects

The water amounts for water treatment and distribution are identical – because water losses for water treatment are ignored. Therefore, the water saving innovation effects on the water supply sector are expressed by the partial differentiation factors

$$\left\{\frac{\vartheta\left[1+\ r_W^{WD}(0)e^{\varphi_r^{WD}Y^Q(\tau)}\right]}{\vartheta Y^Q(\bar{\tau})}\right\}=\left\{\begin{array}{l}\varphi_r^{WD}r_W^{WD}(0)e^{\varphi_r^{WD}Y^Q(\bar{\tau})}\ =\ \varphi_r^{WD}r_W^{WD}(\bar{\tau})\ if\ \tau\ =\ \bar{\tau}\\ 0\ if\ \tau\ \neq\ \bar{\tau}\end{array}\right\}.$$

When considering only the first row of the total marginal human labour for water treatment, we obtain the following expression for the actual marginal costs (or benefits) component generated by the water saving innovation effect on the water supply sector

$$MCL^{Q}_{\substack{WT\\W_{inn}}}(\bar{\tau}) = p_L(\bar{\tau})\varphi_r^{WD} r_W^{WD}(\bar{\tau}) Y^Q(\bar{\tau})\, U^{WT}_{\substack{Q\\N^Q_{i=1+2}+N^K_{\mu=1,2}+^W_{RP}}}(\bar{\tau}) \begin{bmatrix} l^{WT} \\ + V^L_{\substack{WT\\CW+ME_{dep}}} \\ + V^L_{\substack{WT\\CW+ME_{dev}}}(\bar{\tau}) \end{bmatrix} \tag{10.7}$$

Since the structural change of the specific water input amounts $U^{WT}_{\substack{Q\\N^Q_{i=1+2}+N^K_{\mu=1,2}+^W_{RP}}}(\bar{\tau})$ and the specific labour inputs coefficient for water treatment $\left[1^{WT} + V^L_{\substack{WT\\CW+ME_{dep}}} + V^L_{\substack{WT\\CW+ME_{dev}}}(\bar{\tau})\right]$

Are aggregations of different sub-components, the actual marginal human labour costs (or benefits) could accordingly be disaggregated.

A.1.2 The Dependence of the Actual Marginal Human Labour Costs for Water Treatment on the Activity Level or Consumption Good Amounts

Since the additional information received by such disaggregations is less relevant, we prefer to continue with the second component of the marginal human labour costs for water treatment. The change of the total activity levels of our system of constraints are expressed by the partial differentiation factors $\left\{\frac{\vartheta\, Y(\tau)}{\vartheta Y^Q(\bar{\tau})}\right\} = \left\{\begin{matrix} 1 \text{ if } \tau = \bar{\tau} \\ 0 \text{ if } \tau \neq \bar{\tau} \end{matrix}\right\}$. Therefore, the actual marginal human labour costs for water treatment in relation to the change of the consumption good amounts of a selected period are equal to

$$\begin{aligned} &MCL^{Q}_{\substack{WT\\Y^Q}}(\bar{\tau}) \\ &= p_L(\bar{\tau}) \left[1 + r_W^{WD}(\bar{\tau})\right] U^{WT}_{\substack{Q\\N^Q_{i=1+2}+N^K_{\mu=1,2}+^W_{RP}}}(\bar{\tau}) \left[l^{WT} + V^L_{\substack{WT\\CW+ME_{dep}}} + V^L_{\substack{WT\\CW+ME_{dev}}}(\bar{\tau})\right]. \end{aligned} \tag{10.8}$$

While the actual marginal costs caused by the water-saving innovation effect depend on the water saving coefficient of the water distribution system φ_r^{WD}, another kind of actual marginal human labour costs for water treatment occurs, which depend on the change of the activity level.

In the context of the water saving innovation effect, we might have expected to also get its inter-temporal effects directly. Because of the simplified function $1 + r_W^{WD}(0)e^{\varphi_r^{WD} Y^Q(\tau)}$, the inter-temporal water saving effects are a function of the development path of the required water amounts for producing one unit of the consumption good amounts.

A.1.3 Inter-Temporal Marginal Human Labour Costs for Water Treatment Caused by the Structural Change of the Water Amounts

The expression for the inter-temporal marginal human labour input costs caused by the structural change of the water input amounts for the reproduction and production activities is partly known from the water use model. The net water amounts in the extended model are also determined by the water input amounts. The structural changes of the water input amounts are expressed by the partial differentiation factors

$$\frac{\vartheta}{\vartheta Y^Q(\bar{\tau})} U^{WT}_{Q \; N^Q_{i=1+2}+N^K_{\mu=1,2}+W_{RP}} (\tau)$$

$$= \frac{\vartheta}{\vartheta Y^Q(\bar{\tau})} \left\{ \begin{array}{l} \left\{ \psi^{RP}_Q + \sum_{i=1}^{2} \frac{\tilde{\psi}^Q_i \kappa^Q_i e^{\varphi^Q_i \kappa^Q_i \sum_{t=1}^{\tau-1} Y^Q(t)}}{c(\bar{N}^Q_i(0))} \right\} + \\ \left\{ k^Q \left\{ d^Q + \gamma^Q_\tau \right\} \right\} \sum_{\mu=1}^{2} \frac{\tilde{\psi}^K_\mu \kappa^K_\mu e^{\varphi^K_\mu \kappa^K_\mu k^Q \sum_{t=1}^{\tau-1} Y^Q(t) \left\{ d^Q + \gamma^Q_t \right\}}}{c(\bar{N}^K_\mu(0))} \end{array} \right\}$$

$$= \left\{ \begin{array}{l} \sum_{i=1}^{2} \frac{\tilde{\psi}^Q_i \kappa^Q_i \varphi^Q_i \kappa^Q_i}{c(\bar{N}^Q_i(\tau^f))} \\ + k^Q \left\{ d^Q + \gamma^Q_{\tau^f} \right\} k^Q \left\{ d^Q + \gamma^Q_{\tau^f} \right\} \sum_{\mu=1}^{2} \frac{\tilde{\psi}^K_\mu \kappa^K_\mu \varphi^K_\mu \kappa^K_\mu}{c(\bar{N}^K_\mu(\tau^f))} \; if \; \tau^f \geq \bar{\tau} + 1 \\ 0 \; if \; \tau^f < \bar{\tau} + 1 \end{array} \right\}.$$

There are two terms. The first term represents the inter-temporal structural change caused by the raw materials needed for the consumption good process. The second term represents the structural change caused by the raw materials required for sustaining and developing the capital stock for producing one unit of the consumption good. The water input coefficient for the reproduction activities is time invariant so there is no influence on the inter-temporal structural change of the water input coefficient occurs.

The total inter-temporal marginal human labour costs caused by the structural change of the water input coefficients for the production sector and the reproduction activities are equal to

$$ITMLC^{Q}_{\substack{WT\\N^{Q}_{i=1+2}}}\ {}^{+WT}_{N^{K}_{\mu=1+2}}(\bar{\tau} \rightarrow T) =$$

$$= \sum_{\tau^f=\bar{\tau}+1}^{T} \left\{ \begin{array}{l} p_L(\tau^f)\left[1 + r_W^{WD}(\tau^f\right]\ Y^Q(\tau^f)^* \\ \ldots^* \left\{ \begin{array}{l} \sum\limits_{i=1}^{2} \dfrac{\tilde{\psi}_i^Q \kappa_i^Q \varphi_i^Q \kappa_i^Q}{c(\bar{N}_i^Q(\tau^f)} \\ +k^Q\left\{d^Q + \gamma^Q_{\tau^f}\right\}\ k^Q\left\{d^Q + \gamma^Q_{\tau^f}\right\} \sum\limits_{\mu=1}^{2} \dfrac{\tilde{\psi}_\mu^K \kappa_\mu^K \varphi_\mu^K \kappa_\mu^K}{c(\bar{N}_\mu^K(\tau^f)} \end{array} \right\}^* \\ \ldots^* \left[1^{WT} + V^L_{\substack{WT\\CW+ME_{dep}}} + V^L_{\substack{WT\\CW+ME_{dev}}}(\tau^f)\right] \end{array} \right\}$$

This sum can be divided into two terms: the inter-temporal marginal human labour costs caused by the structural change of the water amounts for extracting the raw materials for the consumption good; and the inter-temporal marginal human labour costs caused by the water amounts for extracting the raw materials to sustain and to develop the capital stock.

$$ITMLC^{Q}_{\substack{WT\\N^{Q}_{i=1+2}}}(\bar{\tau} \rightarrow T) =$$

$$\sum_{\tau^f=\bar{\tau}+1}^{T} \left\{ \begin{array}{l} p_L(\tau^f)\left[1 + r_W^{WD}(\tau^f\right]\ Y^Q(\tau^f)^* \\ \ldots^* \left\{ \sum\limits_{i=1}^{2} \dfrac{\tilde{\psi}_i^Q \kappa_i^Q \varphi_i^Q \kappa_i^Q}{c(\bar{N}_i^Q(\tau^f)} \right\} \left[1^{WT} + V^L_{\substack{WT\\CW+ME_{dep}}} + V^L_{\substack{WT\\CW+ME_{dev}}}(\tau^f)\right] \end{array} \right\} \tag{10.9}$$

$$ITMLC^{Q}_{\substack{WT\\N^{K}_{\mu=1+2}}}(\bar{\tau} \rightarrow T) =$$

$$\sum_{\tau^f=\bar{\tau}+1}^{T} \left\{ \begin{array}{l} p_L(\tau^f)\left[1 + r_W^{WD}(\tau^f\right]\ Y^Q(\tau^f)^* \\ \ldots^* \left\{ k^Q\left\{d^Q + \gamma^Q_{\tau^f}\right\}\ k^Q\left\{d^Q + \gamma^Q_{\tau^f}\right\} \sum\limits_{\mu=1}^{2} \dfrac{\tilde{\psi}_\mu^K \kappa_\mu^K \varphi_\mu^K \kappa_\mu^K}{c(\bar{N}_\mu^K(\tau^f)} \right\}^* \\ \ldots^* \left[1^{WT} + V^L_{\substack{WT\\CW+ME_{dep}}} + V^L_{\substack{WT\\CW+ME_{dev}}}(\tau^f)\right] \end{array} \right\} \tag{10.10}$$

A.1.4 The Dependence of the Inter-Temporal Marginal Human Labour Costs for Water Treatment on the Development Path for Consumption Good Amounts

From the fourth partial differentiation factor

$$\left\{\frac{\vartheta}{\vartheta Y^{Q}(\bar{\tau})}\left[l^{WT} + V^{L}_{\substack{WT\\CW+ME_{\text{dep}}}} + V^{L}_{\substack{WT\\CW+ME_{dev}}}(\tau)\right]\right\} = \frac{\vartheta}{\vartheta Y^{Q}(\bar{\tau})}\left[V^{L}_{\substack{WT\\CW+ME_{dev}}}(\tau)\right],$$

we obtain the information that dynamic changes of the specific human labour input amounts are only determined by the dynamics of the development path of the water demand for production

$$\frac{\vartheta}{\vartheta Y^{Q}(\bar{\tau})}\left[V^{L}_{\substack{WT\\CW+ME_{dev}}}(\tau)\right] =$$
$$\frac{\vartheta}{\vartheta Y^{Q}(\bar{\tau})}\left\{\left\{\sum_{\nu=1}^{2}\frac{l_{\nu}\,\kappa_{\nu}^{CW}}{c(\bar{N}_{\nu}^{CW}(0))} + l^{CW}\right\} k^{WT}_{CW} + \left\{\sum_{o=1}^{2}\frac{l_{o}\,\kappa_{o}^{ME}}{c(\bar{N}_{o}^{ME}(0))} + l^{ME}\right\} k^{WT}_{ME}\right\}\gamma_{\tau}^{WT}$$
$$= l_{k^{WT}_{CW+ME}}\frac{\vartheta\,\gamma_{\tau}^{WT}}{\vartheta Y^{Q}(\bar{\tau})}$$

with

$$l_{k^{WT}_{CW+ME}} := \left\{\left\{\sum_{\nu=1}^{2}\frac{l_{\nu}\,\kappa_{\nu}^{CW}}{c(\bar{N}_{\nu}^{CW}(0))} + l^{CW}\right\} k^{WT}_{CW} + \left\{\sum_{o=1}^{2}\frac{l_{o}\,\kappa_{o}^{ME}}{c(\bar{N}_{o}^{ME}(0))} + l^{ME}\right\} k^{WT}_{ME}\right\}.$$

Let us take into consideration how the water treatment amount and its growth rates are determined and defined by

$$\gamma_{\tau}^{WT} := \frac{W^{WT}(\tau+1) - W^{WT}(\tau)}{W^{WT}(\tau)}.$$

We learn that development path of the consumption good amounts determine the growth rate and the development path of the water amounts for water treatment. Even for the special case of a time-invariant consumption good level, for example, the water amounts would not be constant. Since the concentrations of the raw materials for production are a declining function, the water amounts required to transport the un-avoidable by-products to the wastewater treatment sector will increase as

$$
\begin{aligned}
&W^{WT}(\tau) \\
&= Y^Q(\tau) \left\{ \begin{aligned} &\left\{ \psi_Q^{RP} + \sum_{i=1}^{2} \frac{\tilde{\psi}_i^Q \kappa_i^Q e^{\varphi_i^Q \kappa_i^Q \sum_{t=1}^{\tau-1} Y^Q(t)}}{c(\bar{N}_i^Q(0))} \right\} \\ &+ \left\{ k^Q \left\{ d^Q + \gamma_\tau^Q \right\} \right\} \sum_{\mu=1}^{2} \frac{\tilde{\psi}_\mu^K \kappa_\mu^K e^{\varphi_\mu^K \kappa_\mu^K k^Q \sum_{t=1}^{\tau-1} Y^Q(t) \left\{ d^Q + \gamma_t^Q \right\}}}{c(\bar{N}_\mu^K(0))} \end{aligned} \right\}^* \\
&\ldots^* \left[1 + r_W^{WD}(0) e^{\varphi_r^{WD} Y^Q(\tau)} \right].
\end{aligned}
$$

The inter-temporal differential change of the marginal human labour costs depends on the differential change of the development path for the water treatment and the consumption good amounts are equal to

$$
\begin{aligned}
&ITMLC^Q_{\gamma_Y^{WT}}(\bar{\tau} \rightarrow T) \\
&= \sum_{\tau^f=\bar{\tau}+1}^{T} \left\{ p_L(\tau^f) \left[1 + r_W^{WD}(\tau^f \right] Y^Q(\tau^f)\, U^{WT}_{Q\, N^Q_{i=1+2} + N^K_{\mu=1,2}}(\tau^f)\, l_{k^{WT}_{CW+ME}} \frac{\vartheta \gamma^{WT}_{\tau^f}}{\vartheta Y^Q(\bar{\tau})} \right\}
\end{aligned} \tag{10.11}
$$

A.2 Marginal Energy Costs for Water Treatment

$$
\begin{aligned}
MCE^Q_{WT}(\bar{\tau} \mapsto T) &= \frac{\vartheta}{\vartheta Y^Q(\bar{\tau})} \sum_{\tau=1}^{T} p_E(\tau) \left[1 + r_W^{WD}(0) e^{\varphi_r^{WD} Y^Q(\tau)} \right] \\
&\quad Y^Q(\tau)\, U^{WT}_{Q\, N^Q_{i=1+2} + N^K_{\mu=1,2} + {}^W_{RP}}(\tau) \left[V^E_{\substack{WT \\ CW+ME_{dep}}} + V^E_{\substack{WT \\ CW+ME_{dev}}}(\tau) + V^E_{WT_{innv}}(\tau) \right] \\
&= \sum_{\tau=1}^{T} p_E(\tau) \left\{ \frac{\vartheta \left[1 + r_W^{WD}(0) e^{\varphi_r^{WD} Y^Q(\tau)} \right]}{\vartheta Y^Q(\bar{\tau})} \right\} \\
&\quad Y^Q(\tau)\, U^{WT}_{Q\, N^Q_{i=1+2} + N^K_{\mu=1,2} + {}^W_{RP}}(\tau) \left[V^E_{\substack{WT \\ CW+ME_{dep}}} + V^E_{\substack{WT \\ CW+ME_{dev}}}(\tau) + V^E_{WT_{innv}}(\tau) \right] \\
&\quad + \sum_{\tau=1}^{T} p_E(\tau) \left[1 + r_W^{WD}(0) e^{\varphi_r^{WD} Y^Q(\tau)} \right] \left\{ \frac{\vartheta Y^Q(\tau)}{\vartheta Y^Q(\bar{\tau})} \right\} \\
&\quad U^{WT}_{Q\, N^Q_{i=1+2} + N^K_{\mu=1,2} + {}^W_{RP}}(\tau) \left[V^E_{\substack{WT \\ CW+ME_{dep}}} + V^E_{\substack{WT \\ CW+ME_{dev}}}(\tau) + V^E_{WT_{innv}}(\tau) \right]
\end{aligned}
$$

$$+\sum_{\tau=1}^{T} p_E(\tau)\left[1+ r_W^{WD}(0)e^{\varphi_r^{WD}Y^Q(\tau)}\right] Y^Q(\tau)\left\{\frac{\vartheta U_{Q\; N^Q_{i=1+2}+N^K_{\mu=1,2}+^W_{RP}}^{WT}(\tau)}{\vartheta Y^Q(\bar{\tau})}\right\}$$

$$\left[V^E_{\substack{WT\\CW+ME_{dep}}} + V^E_{\substack{WT\\CW+ME_{dev}}}(\tau) + V^E_{WT_{innv}}(\tau)\right]$$

$$+\sum_{\tau=1}^{T} p_E(\tau)\left[1+ r_W^{WD}(0)e^{\varphi_r^{WD}Y^Q(\tau)}\right] Y^Q(\tau)\; U_{Q\; N^Q_{i=1+2}+N^K_{\mu=1,2}+^W_{RP}}^{WT}(\tau)$$

$$\left\{\frac{\vartheta\left[V^E_{\substack{WT\\CW+ME_{dep}}} + V^E_{\substack{WT\\CW+ME_{dev}}}(\tau) + V^E_{WT_{innv}}(\tau)\right]}{\vartheta Y^Q(\bar{\tau})}\right\}$$

with

$$U_{Q\; N^Q_{i=1+2}+N^K_{\mu=1,2}+^W_{RP}}^{WT}(\tau) = \left\{\begin{array}{l}\left\{\psi_Q^{RP} + \sum_{i=1}^{2}\frac{\tilde{\psi}_i^Q\kappa_i^Q e^{\varphi_i^Q\kappa_i^Q\sum_{t=1}^{\tau-1}Y^Q(t)}}{c(\bar{N}_i^Q(0))}\right\}+\\ \left\{k^Q Y^Q(\tau)\left\{d^Q+\gamma_\tau^Q\right\}\right\}\sum_{\mu=1}^{2}\frac{\tilde{\psi}_\mu^K\kappa_\mu^K e^{\varphi_\mu^K\kappa_\mu^K k^Q\sum_{t=1}^{\tau-1}Y^Q(t)\left\{d^Q+\gamma_t^Q\right\}}}{c(\bar{N}_\mu^K(0))}\end{array}\right\},$$

$$V^E_{\substack{WT\\CW+ME_{dep}}} = \left\{\sum_{\nu=1}^{2}\frac{e_\nu\kappa_\nu^{CW}}{c(\bar{N}_\nu^{CW}(0))}+\frac{ex^{CW}}{\eta^{CW}}\right\} k_{CW}^{WT}\, d^{CW} + \left\{\sum_{o=1}^{2}\frac{e_o\kappa_o^{ME}}{c(\bar{N}_o^{ME}(0))}+\frac{ex^{ME}}{\eta^{ME}}\right\} k_{ME}^{WT} d^{ME},$$

$$V^E_{\substack{WT\\CW+ME_{dev}}}(\tau) = \left\{\left\{\sum_{\nu=1}^{2}\frac{e_\nu\kappa_\nu^{CW}}{c(\bar{N}_\nu^{CW}(0))}+\frac{ex^{CW}}{\eta^{CW}}\right\} k_{CW}^{WT} + \left\{\sum_{o=1}^{2}\frac{e_o\kappa_o^{ME}}{c(\bar{N}_o^{ME}(0))}+\frac{ex^{ME}}{\eta^{ME}}\right\} k_{ME}^{WT}\right\}\gamma_\tau^{WT}$$

$$V^E_{WT_{innv}}(\tau) = \tilde{\varepsilon}^{WT}\sum_{\omega}\ln\frac{c_\omega^*(0)}{c_\omega(0)}\left[1+ r_{an}^{WT}(0)\, e^{\varphi_{an}^{WT}Y^Q(\tau)}\right].$$

A.2.1 Marginal Energy Costs for Water Treatment Caused by the Water Saving Innovation Effect

Let us consider the first row of the total marginal energy costs for water treatment. We obtain the following expression for the actual marginal energy costs (or benefits) component generated by the water saving innovation effect on the water supply sector

$$
\begin{aligned}
& MCE^{Q}_{\substack{WT \\ W_{inn}}} (\bar{\tau}) \\
& = p_E(\bar{\tau}) \varphi_r^{WD} r_W^{WD}(\bar{\tau}) Y^Q(\bar{\tau})\, U^{WT}_{\substack{Q \\ N^Q_{i=1+2}+N^K_{\mu=1,2}+^W_{RP}}} (\bar{\tau})^* \\
& \ldots * \left[V^E_{\substack{WT \\ CW+ME_{dep}}} + V^E_{\substack{WT \\ CW+ME_{dev}}} (\bar{\tau}) + V^E_{WT_{innv}} (\bar{\tau}) \right]
\end{aligned}
\tag{10.12}
$$

The structural changes of the specific water input amounts represented by the factor $U^{WT}_{\substack{Q \\ N^Q_{i=1+2}+N^K_{\mu=1,2}+^W_{RP}}} (\bar{\tau})$ and the structural change of the specific energy inputs for water treatment represented by the factor $\left[V^E_{\substack{WT \\ CW+ME_{dep}}} + V^E_{\substack{WT \\ CW+ME_{dev}}} (\bar{\tau}) + V^E_{WT_{innv}} (\bar{\tau}) \right]$ influence the actual marginal energy costs for wastewater treatment because of the reduction of water losses in the water distribution system.

The first two terms for the energy input coefficients for sustaining and developing of the civil work and electrical equipment capital stocks are

$$
V^E_{\substack{WT \\ CW+ME_{dep}}} = \left\{ \sum_{\nu=1}^{2} \frac{e_\nu \kappa_\nu^{CW}}{c(N_\nu^{CW}(0))} + \frac{ex^{CW}}{\eta^{CW}} \right\} k^{WT}_{CW} d^{CW} + \left\{ \sum_{o=1}^{2} \frac{e_o \kappa_o^{ME}}{c(N_o^{ME}(0))} + \frac{ex^{ME}}{\eta^{ME}} \right\} k^{WT}_{ME} d^{ME}
$$

and

$$
V^E_{\substack{WT \\ CW+ME_{dev}}} (\bar{\tau}) = \left\{ \left\{ \sum_{\nu=1}^{2} \frac{e_\nu \kappa_\nu^{CW}}{c(N_\nu^{CW}(0))} + \frac{ex^{CW}}{\eta^{CW}} \right\} k^{WT}_{CW} + \left\{ \sum_{o=1}^{2} \frac{e_o \kappa_o^{ME}}{c(N_o^{ME}(0))} + \frac{ex^{ME}}{\eta^{ME}} \right\} k^{WT}_{ME} \right\} \gamma^{WT}_{\bar{\tau}}.
$$

The third term for the energy saving innovation effect of the selected time interval

$$
V^E_{\substack{WT \\ E_{inn}}} (\bar{\tau}) = \tilde{\varepsilon}^{WT} \sum_{\omega} \ln \frac{c^*_\omega(0)}{c_\omega(0)} \left[1 + r^{WT}_{an}(0)\, e^{\varphi^{WT}_{an} Y^Q(\bar{\tau})} \right]
$$

also shows that the actual marginal energy costs (or benefits) caused by the water saving innovations are closely connected with the energy saving innovation effect. Therefore, the two innovation effects are connected with each other and saving water strengthens the energy saving innovation.

The actual marginal energy costs could once again be disaggregated into a large number of sub-components. The additional information received by the disaggregation is less useful and therefore, we will not do any additional disaggregation.

A.2.2 Actual Marginal Energy Costs for Water Treatment Determined by the Activity Level

We now continue with the second component of the total marginal energy costs for water treatment. With the expressions for the change of the total activity levels of our system of constraints

$$\left\{\frac{\vartheta\, Y(\tau)}{\vartheta Y^Q(\bar\tau)}\right\} = \left\{\begin{matrix} 1 \; if\, \tau = \bar\tau \\ 0 \; if\, \tau \neq \bar\tau \end{matrix}\right\},$$

we obtain that the actual marginal energy labour costs for water treatment in relation to the change of the activity level (or consumption good amount) of a selected period are equal to

$$\begin{aligned} MCE^{Q}_{\substack{WT\\ Y^Q}}(\bar\tau) &= p_E(\bar\tau)\left[1+\ r^{WD}_W(\bar tau)\right]\ U^{WT}_{\substack{Q\\ N^Q_{i=1+2}+N^K_{\mu=1,2}}}\ {}_{+^W_{RP}}(\bar\tau)^* \\ \ldots^* &\left[V^E_{\substack{WT\\ CW+ME_{dep}}} + V^E_{\substack{WT\\ CW+ME_{dev}}}(\bar\tau) + V^E_{WT_{innv}}(\bar\tau)\right] \end{aligned} \tag{10.13}$$

A.2.3 Inter-Temporal Marginal Energy Costs for Water Treatment Caused by the Structural Change of the Water Amounts

The expression of the inter-temporal marginal energy costs caused by the structural change of the water input amounts for the reproduction and production activities is taken partly from the basic model. The net water amounts in the extended model are also determined by the water input amounts required to produce the raw materials for the consumption good and capital good amounts plus the water input amounts for the reproduction activities.

The total inter-temporal marginal energy costs caused by the structural change of the water input coefficient for the production sector and the reproduction activities are equal to

$$ITMEC^{Q}_{\substack{WT\\ N^Q_{i=1+2}}}{}_{+^{WT}_{N^K_{\mu=1+2}}}(\bar\tau \to T):$$

$$= \sum_{\tau^f=\bar\tau+1}^{T} \left\{\begin{matrix} p_E(\tau^f)\left[1+\ r^{WD}_W(\tau^f)\right]\ Y^Q(\tau^f)^* \\ \ldots^* \left\{\begin{matrix} \sum\limits_{i=1}^{2} \frac{\tilde\psi^Q_i \kappa^Q_i \varphi^Q_i \kappa^Q_i}{c(\bar N^Q_i(\tau^f))} \\ +k^Q\left\{d^Q+\gamma^Q_{\tau^f}\right\}\ k^Q\left\{d^Q+\gamma^Q_{\tau^f}\right\} \sum\limits_{\mu=1}^{2} \frac{\tilde\psi^K_\mu \kappa^K_\mu \varphi^K_\mu \kappa^K_\mu}{c(\bar N^K_\mu(\tau^f))} \end{matrix}\right\}^* \\ \ldots^* \left[V^E_{\substack{WT\\ CW+ME_{dep}}} + V^E_{\substack{WT\\ CW+ME_{dev}}}(\tau) + V^E_{WT_{innv}}(\tau)\right] \end{matrix}\right\}.$$

This expression can also be divided into two terms. The first term is the inter-temporal marginal energy costs caused by the structural change of the water amounts for extracting the raw materials of the consumption good. The second term is the water amounts for extracting the raw materials to sustain and to develop the capital stock. No energy inputs are taken into consideration for the reproduction activities. Therefore, reproduction has no influence on the inter-temporal marginal energy costs caused by the dynamic structural change of the water amounts. The inter-temporal marginal energy costs for water treatment caused by the structural change of the water input amounts are, however, accompanied by inter-temporal energy saving innovation effects. This means that the inter-temporal effects of structural changes to the water input amounts are closely connected with the inter-temporal energy innovations for water treatment as follows.

$$ITMEC^{Q}_{WT} \underset{i=1+2}{N^{Q}} (\bar{\tau} \to T) = \sum_{\tau^f=\bar{\tau}+1}^{T} \left\{ \begin{array}{l} p_E(\tau^f)\left[1+\ r^{WD}_{W}(\tau^f)\right]\ Y^Q(\tau^f)\left\{\sum_{i=1}^{2}\frac{\tilde{\psi}^Q_i\kappa^Q_i\varphi^Q_i\kappa^Q_i}{c(\bar{N}^Q_i(\tau^f))}\right\}^* \\ \ldots^*\left[V^E_{\underset{CW+ME_{dep}}{WT}} + V^E_{\underset{CW+ME_{dev}}{WT}}(\tau^f) + V^E_{WT_{innv}}(\tau^f)\right] \end{array} \right\} \tag{10.14}$$

$$ITMEC^{Q}_{WT} \underset{\mu=1+2}{N^{K}} (\bar{\tau} \to T) = \sum_{\tau^f=\bar{\tau}+1}^{T} \left\{ \begin{array}{l} p_E(\tau^f)\left[1+\ r^{WD}_{W}(\tau^f\right]\ Y^Q(\tau^f)^* \\ \ldots^*\left\{k^Q\left\{d^Q+\gamma^Q_{\tau^f}\right\}\ k^Q\left\{d^Q+\gamma^Q_{\tau^f}\right\}\ \sum_{\mu=1}^{2}\frac{\tilde{\psi}^K_\mu\kappa^K_\mu\varphi^K_\mu\kappa^K_\mu}{c(\bar{N}^K_\mu(\tau^f))}\right\}^* \\ \ldots^*\left[V^E_{\underset{CW+ME_{dep}}{WT}} + V^E_{\underset{CW+ME_{dev}}{WT}}(\tau^f) + V^E_{WT_{innv}}(\tau^f)\right] \end{array} \right\} \tag{10.15}$$

A.2.4 Actual and Inter-Temporal Marginal Energy Costs for Water Treatment Determined by the Development Path of the Consumption Good Amounts

From the fourth partial differentiation factor

$$\left\{\frac{\vartheta}{\vartheta Y^Q(\bar{\tau})}\left[V^E_{\underset{CW+ME_{dep}}{WT}} + V^E_{\underset{CW+ME_{dev}}{WT}}(\tau) + V^E_{WT_{innv}}(\tau)\right]\right\} = \frac{\vartheta\, V^E_{\underset{CW+ME_{dev}}{WT}}(\tau)}{\vartheta Y^Q(\bar{\tau})} + \frac{\vartheta\, V^E_{WT_{innv}}(\tau)}{\vartheta Y^Q(\bar{\tau})},$$

we observe that dynamic changes in the specific energy input amounts are determined by two effects. First, they are determined by the dynamics of the development path of the water demand for production

$$\frac{\vartheta}{\vartheta Y^{Q}(\bar{\tau})}\left[V^{E}_{\substack{WT\\CW+ME_{dev}}}(\tau)\right] = \frac{\vartheta}{\vartheta Y^{Q}(\bar{\tau})}\left\{\begin{array}{l}\left\{\sum_{\nu=1}^{2}\frac{e_{\nu}\,\kappa_{\nu}^{CW}}{c(\bar{N}_{\nu}^{CW}(0))}+\frac{ex^{CW}}{\eta^{CW}}\right\}k_{CW}^{WT}+\\ +\left\{\sum_{o=1}^{2}\frac{e_{o}\,\kappa_{o}^{ME}}{c(\bar{N}_{o}^{ME}(0))}+\frac{ex^{ME}}{\eta^{ME}}\right\}k_{ME}^{WT}\end{array}\right\}\gamma_{\bar{\tau}}^{WT}$$

$$= e_{k_{CW+ME}^{WT}}\frac{\vartheta\,\gamma_{\tau}^{WT}}{\vartheta Y^{Q}(\bar{\tau})}$$

with

$$e_{k_{CW+ME}^{WT}} := \left\{\left\{\sum_{\nu=1}^{2}\frac{l_{\nu}\,\kappa_{\nu}^{CW}}{c(\bar{N}_{\nu}^{CW}(0))}+l^{CW}\right\}k_{CW}^{WT}+\left\{\sum_{o=1}^{2}\frac{l_{o}\,\kappa_{o}^{ME}}{c(\bar{N}_{o}^{ME}(0))}+l^{ME}\right\}k_{ME}^{WT}\right\}.$$

The second effect is the energy saving innovation that is generated within the selected time interval, which can be called the actual marginal energy innovation effect for water treatment

$$\begin{aligned}\frac{\vartheta\,V^{E}_{\substack{WT\\E_{inn}}}(\tau)}{\vartheta Y^{Q}(\bar{\tau})} &= \frac{\vartheta}{\vartheta Y^{Q}(\bar{\tau})}\tilde{\varepsilon}^{WT}\sum_{\omega}\ln\frac{c_{\omega}^{*}(0)}{c_{\omega}(0)}\left[1+r_{an}^{WT}(0)\,e^{\varphi_{an}^{WT}\,Y^{Q}(\tau)}\right]\\ &= MEC^{Q}_{\substack{WT\\E_{inn}}}(\bar{\tau}) = \varphi_{an}^{WT}\tilde{\varepsilon}^{WT}\,r_{an}^{WT}(0)\,e^{\varphi_{an}^{WT}\,Y^{Q}(\bar{\tau})}\sum_{\omega}\ln\frac{c_{\omega}^{*}(0)}{c_{\omega}(0)}\\ &= \varphi_{an}^{WT}\tilde{\varepsilon}^{WT}\,r_{an}^{WT}(\bar{\tau})\sum_{\omega}\ln\frac{c_{\omega}^{*}(0)}{c_{\omega}(0)}.\end{aligned}\tag{10.16}$$

The development path of the inter-temporal marginal energy costs determined by the development path of the consumption good amounts are shifted slightly to a lower level by the energy innovation effect generated by the consumption good amount of the selected time interval. In the case of an increasing economy, this innovation effect would be accelerated because of the increasing consumption good amounts.

The total inter-temporal marginal energy costs as a function of the development path for the consumption good amounts, including the energy saving innovation effect for water treatment, are equal to

$$\begin{aligned}&ITMEC^{Q}_{\substack{WT\\\gamma_{\tau}+E_{inn}}}(\bar{\tau}\rightarrow T) =\\ &\sum_{\tau^{f}=\bar{\tau}+1}^{T}\left\{\begin{array}{l}p_{E}(\tau^{f})\left[1+r_{W}^{WD}(\tau^{f})\right]\,Y^{Q}(\tau^{f})U^{WT}_{\substack{Q\\N_{i=1+2}^{Q}+N_{\mu=1,2}^{K}+_{RP}^{W}}}(\tau^{f})^{*}\\ \ldots^{*}\left[e_{k_{CW+ME}^{WT}}\frac{\vartheta\,\gamma_{\tau^{f}}^{WT}}{\vartheta Y^{Q}(\bar{\tau})}+\varphi_{an}^{WT}\tilde{\varepsilon}^{WT}r_{an}^{WT}(\tau^{f})\sum_{\omega}\ln\frac{c_{\omega}^{*}(0)}{c_{\omega}(0)}\right]\end{array}\right\}.\end{aligned}\tag{10.17}$$

Direct innovation effects for water treatment and the development path of the consumption good amounts determine the total energy saving innovation effect. The development path of the consumption good amounts is also influenced by the structural change of the production sector and by the water saving innovations.

Wasting water means wasting energy and vice versa, especially in the case where the economic activities are based on non-renewable resources.

If the concentrations of the residuals $c_{\omega}(0)$, which are to be reduced to exogenously given water quality standards, are not time invariant then the energy input for water treatment would change accordingly. The wastewater discharged into the natural environment, for example, might be the reason for polluting the water resources. This leads to an increase of energy costs for water treatment or to the decision, as seen in the Adana case study, to switch from a groundwater to a surface water supply system. An interrelation between the water concentrations and the concentrations of the wastewater pollution stock could be introduced in our model. If this were done, the optimal conditions would be achieved by balancing the actual and intertemporal marginal costs of water treatment with those for wastewater treatment. Our model can easily be altered in this way and this would be of special interest when comparing alternative environmental instruments.

B Marginal Costs for Water Distribution

B.1 Marginal Human Labour Costs for Water Distribution

$$
\begin{aligned}
MCL^{Q}_{WD}(\bar{\tau} \mapsto T) &= \\
&\frac{\vartheta}{\vartheta Y^{Q}(\bar{\tau})} \sum_{\tau=1}^{T} p_L(\tau) \left[1 + r^{WD}_{W}(0) e^{\varphi^{WD}_{r} Y^{Q}(\tau)} \right] Y^{Q}(\tau)\; U^{WD}_{\substack{Q \\ N^{Q}_{i=1+2} + N^{K}_{\mu=1,2} + {}^{W}_{RP}}}(\tau) \\
&\left[1^{WD} + V^{L}_{\substack{WD \\ CW+ME_{dep}}} + V^{L}_{\substack{WD \\ CW+ME_{dev}}}(\tau) \right] \\
&= \sum_{\tau=1}^{T} p_L(\tau) \left\{ \frac{\vartheta \left[1 + r^{WD}_{W}(0) e^{\varphi^{WD}_{r} Y^{Q}(\tau)} \right]}{\vartheta Y^{Q}(\bar{\tau})} \right\} Y^{Q}(\tau)\; U^{WD}_{\substack{Q \\ N^{Q}_{i=1+2} + N^{K}_{\mu=1,2} + {}^{W}_{RP}}}(\tau) \\
&\left[1^{WD} + V^{L}_{\substack{WD \\ CW+ME_{dep}}} + V^{L}_{\substack{WD \\ CW+ME_{dev}}}(\tau) \right] \\
&+ \sum_{\tau=1}^{T} p_L(\tau) \left[1 + r^{WD}_{W}(0) e^{\varphi^{WD}_{r} Y^{Q}(\tau)} \right] \left\{ \frac{\vartheta Y^{Q}(\tau)}{\vartheta Y^{Q}(\bar{\tau})} \right\} U^{WD}_{\substack{Q \\ N^{Q}_{i=1+2} + N^{K}_{\mu=1,2} + {}^{W}_{RP}}}(\tau) \\
&\left[1^{WD} + V^{L}_{\substack{WD \\ CW+ME_{dep}}} + V^{L}_{\substack{WD \\ CW+ME_{dev}}}(\tau) \right] \\
&+ \sum_{\tau=1}^{T} p_L(\tau) \left[1 + r^{WD}_{W}(0) e^{\varphi^{WD}_{r} Y^{Q}(\tau)} \right] Y^{Q}(\tau) \left\{ \frac{\vartheta U^{WD}_{\substack{Q \\ N^{Q}_{i=1+2} + N^{K}_{\mu=1,2} + {}^{W}_{RP}}}(\tau)}{\vartheta Y^{Q}(\bar{\tau})} \right\} \\
&\left[1^{WD} + V^{L}_{\substack{WD \\ CW+ME_{dep}}} + V^{L}_{\substack{WD \\ CW+ME_{dev}}}(\tau) \right]
\end{aligned}
$$

$$+\sum_{\tau=1}^{T} p_L(\tau)\left[1+ r_W^{WD}(0)e^{\varphi_r^{WD} Y^Q(\tau)}\right] Y^Q(\tau)$$

$$U^{WD}_{Q\; N^Q_{i=1+2}+N^K_{\mu=1,2}+{}^W_{RP}} \left\{ \frac{\vartheta\left[l^{WD} + V^L_{\substack{WT\\ CW+ME_{dep}}} + V^L_{\substack{WD\\ CW+ME_{dev}}}(\tau)\right]}{\vartheta Y^Q(\bar{\tau})} \right\}$$

with

$$U^{WD}_{Q\; N^Q_{i=1+2}+N^K_{\mu=1,2}+{}^W_{RP}}(\tau) = \left\{ \begin{array}{l} \left\{ \psi_Q^{RP} + \sum_{i=1}^{2} \frac{\tilde{\psi}_i^Q \kappa_i^Q e^{\varphi_i^Q \kappa_i^Q \sum_{t=1}^{\tau-1} Y^Q(t)}}{c(\bar{N}_i^Q(0))} \right\} \\ + \left\{ k^Q\left\{ d^Q + \gamma_\tau^Q \right\}\right\} \sum_{\mu=1}^{2} \frac{\tilde{\psi}_\mu^K \kappa_\mu^K e^{\varphi_\mu^K \kappa_\mu^K k^Q \left\{ \sum_{t=1}^{\tau-1} Y^Q(t)\left\{ d^Q + \gamma_t^Q \right\}\right\}}}{c(\bar{N}_\mu^K(0))} \end{array} \right\}$$

$$V^L_{\substack{WD\\ CW+ME_{dep}}} = \left\{ \sum_{\nu=1}^{2} \frac{l_\nu \kappa_\nu^{CW}}{c(\bar{N}_\nu^{CW}(0))} + l^{CW} \right\} k^{WD}_{CW} d^{CW} + \left\{ \sum_{o=1}^{2} \frac{l_o \kappa_o^{ME}}{c(\bar{N}_o^{ME}(0))} + l^{ME} \right\} k^{WD}_{ME} d^{ME}$$

$$V^L_{\substack{WD\\ CW+ME_{dev}}}(\tau) := \left\{ \left\{ \sum_{\nu=1}^{2} \frac{l_\nu \kappa_\nu^{CW}}{c(\bar{N}_\nu^{CW}(0))} + l^{CW} \right\} k^{WD}_{CW} + \left\{ \sum_{o=1}^{2} \frac{l_o \kappa_o^{ME}}{c(\bar{N}_o^{ME}(0))} + l^{ME} \right\} k^{WD}_{ME} \right\} \gamma_\tau^{WD}.$$

The first three partial differentiation factors represent the water saving innovation effect, the influence of the activity level, and the structural change of the water input amounts for the production and reproduction sector. These factors do not differ for the water treatment and distribution sectors. The difference of marginal costs between the water treatment and water distribution sector is only determined by the differences in the fourth partial differentiation factor. Because the water amounts of the water treatment and distribution sector have the same value, the growth rates of the two sectors are also identical $\gamma_\tau^{WT} = \gamma_\tau^{WD}$. As a result, the two sectors are mainly differentiated by their human labour input coefficients and their civil work and mechanical & electrical equipment capital good input coefficients for the processes involved in the water treatment and distribution sectors. The depreciation rates of these two capital goods are the same for the water treatment and distribution sector.

For a less complex water distribution system, the civil work input coefficient is larger than the one for the water treatment sector, where the mechanical & electrical equipment input coefficient is larger than the one for the water distribution system. Furthermore, it can be expected that the human labour input for maintaining the water distribution system is larger than the one for maintaining the water treatment system. Therefore, a much less complex water distribution system with the following relations can be expected

$$k^{WD}_{CW} > k^{WT}_{CW} \text{ and } k^{WD}_{ME} < k^{WT}_{CW} \text{ and } l^{WD} > l^{WT}.$$

This leads to a different, but closely connected, development path for the two sectors. In the case where topographical conditions require a more complex water distribution system with a large number of pumping stations or water storage tanks, these relations could be the opposite.

As a matter of completeness, the expressions for the different marginal human labour costs of the water distribution follow.

B.1.1 Actual Marginal Human Labour Costs for Water Distribution Generated by Water Saving Innovation Effects

$$\begin{aligned} &MCL^{Q}_{\substack{WD\\W_{inn}}}(\bar{\tau}) \\ &= p_L(\bar{\tau})\varphi_r^{WD} r_W^{WD}(\bar{\tau}) Y^Q(\bar{\tau})^* \\ &\ldots^* U^{WD}_{\substack{Q\\ N^Q_{i=1+2}+N^K_{\mu=1,2}+{}^W_{RP}}}(\bar{\tau}) \left[l^{WD} + V^L_{\substack{WD\\CW+ME_{dep}}} + V^L_{\substack{WD\\CW+ME_{dev}}}(\bar{\tau}) \right]. \end{aligned} \tag{10.18}$$

B.1.2 Actual Marginal Human Labour Costs for Water Distribution as a Function of the Consumption Good Amounts

$$\begin{aligned} &MCL^{Q}_{\substack{WD\\Y^Q}}(\bar{\tau}) \\ &= p_L(\bar{\tau}) \left[1 + r_W^{WD}(\bar{\tau})\right] U^{WD}_{\substack{Q\\ N^Q_{i=1+2}+N^K_{\mu=1,2}+{}^W_{RP}}}(\bar{\tau}) \left[l^{WD} + V^L_{\substack{WD\\CW+ME_{dep}}} + V^L_{\substack{WD\\CW+ME_{dev}}}(\bar{\tau}) \right]. \end{aligned} \tag{10.19}$$

B.1.3 Inter-Temporal Marginal Human Labour Costs for Water Distribution Caused by the Structural Change of the Water Input Amounts

$$\begin{aligned} &ITMLC^{Q}_{\substack{WD\\ N^Q_{i=1+2}+{}^{WD}_{N^K_{\mu=1+2}}}}(\bar{\tau} \to T) = \\ &\sum_{\tau^f=\bar{\tau}+1}^{T} \left\{ \begin{aligned} &p_L(\tau^f)\left[1 + r_W^{WD}(\tau^f)\right] Y^Q(\tau^f)^* \\ &\ldots^* \left\{ \begin{aligned} &\sum_{i=1}^{2} \frac{\tilde{\psi}_i^Q \kappa_i^Q \varphi_i^Q \kappa_i^Q}{c(\bar{N}_i^Q(\tau^f))} \\ &+ k^Q \left\{ d^Q + \gamma^Q_{\tau^f} \right\} k^Q \left\{ d^Q + \gamma^Q_{\tau^f} \right\} \sum_{\mu=1}^{2} \frac{\tilde{\psi}_\mu^K \kappa_\mu^K \varphi_\mu^K \kappa_\mu^K}{c(\bar{N}_\mu^K(\tau^f))} \end{aligned} \right\}^* \\ &\ldots^* \left[l^{WD} + V^L_{\substack{WD\\CW+ME_{dep}}} + V^L_{\substack{WD\\CW+ME_{dev}}}(\tau^f) \right] \end{aligned} \right\}. \end{aligned}$$

$$ITMLC^{Q}_{WD_{N^Q_{i=1+2}}}(\bar{\tau} \to T) = \sum_{\tau^f=\bar{\tau}+1}^{T} \left\{ \begin{array}{l} p_L(\tau^f)\left[1 + r^{WD}_{W}(\tau^f)\right] Y^Q(\tau^f)^* \\ \ldots^* \left\{ \sum_{i=1}^{2} \frac{\tilde{\psi}^Q_i \kappa^Q_i \varphi^Q_i \kappa^Q_i}{c(\bar{N}^Q_i(\tau^f))} \right\} \left[l^{WD} + V^L_{WD_{CW+ME_{dep}}} + V^L_{WD_{CW+ME_{dev}}}(\tau^f) \right] \end{array} \right\}. \tag{10.20}$$

$$ITMLC^{Q}_{WT_{N^K_{\mu=1+2}}}(\bar{\tau} \to T) = \sum_{\tau^f=\bar{\tau}+1}^{T} \left\{ \begin{array}{l} p_L(\tau^f)\left[1 + r^{WD}_{W}(\tau^f)\right] Y^Q(\tau^f)^* \\ \ldots^* \left\{ k^Q\left\{d^Q + \gamma^Q_{\tau^f}\right\} k^Q\left\{d^Q + \gamma^Q_{\tau^f}\right\} \sum_{\mu=1}^{2} \frac{\tilde{\psi}^K_\mu \kappa^K_\mu \varphi^K_\mu \kappa^K_\mu}{c(\bar{N}^K_\mu(\tau^f))} \right\} \\ \ldots^* \left[l^{WD} + V^L_{WD_{CW+ME_{dep}}} + V^L_{WD_{CW+ME_{dev}}}(\tau^f) \right] \end{array} \right\}. \tag{10.21}$$

B.1.4 Inter-Temporal Marginal Human Labour Costs for Water Distribution as a Function of the Development Path of the Water Amounts

$$ITMLC^{Q}_{\gamma^{WD}_{\bar{\tau}}}(\bar{\tau} \to T) = \sum_{\tau^f=\bar{\tau}+1}^{T} \left\{ p_L(\tau^f)\left[1 + r^{WD}_{W}(\tau^f)\right] Y^Q(\tau^f) U^{WD}_{Q_{N^Q_{i=1+2}+N^K_{\mu=1,2}+{}^{W}_{RP}}}(\tau^f)\, l_{k^{WD}_{CW+ME}} \frac{\vartheta\, \gamma^{WD}_{\tau^f}}{\vartheta\, Y^Q(\bar{\tau})} \right\} \tag{10.22}$$

B.2 Marginal Human Energy Costs for Water Distribution

$$MCE^{Q}_{WT}(\bar{\tau} \mapsto T) =$$

$$\frac{\vartheta}{\vartheta Y^Q(\bar{\tau})} \sum_{\tau=1}^{T} p_E(\tau)\left[1 + r^{WD}_{W}(0) e^{\varphi^{WD}_r Y^Q(\tau)}\right] Y^Q(\tau)\, U^{WD}_{Q_{N^Q_{i=1+2}+N^K_{\mu=1,2}}}(\tau) \left[\begin{array}{l} V^E_{WD_{CW+ME_{dep}}} \\ + V^E_{WD_{CW+ME_{dev}}}(\tau) \\ + V^E_{WD_{inn}}(\tau) \end{array} \right]$$

$$= \sum_{\tau=1}^{T} p_E(\tau) \left\{ \frac{\vartheta\left[1 + r^{WD}_{W}(0) e^{\varphi^{WD}_r Y^Q(\tau)}\right]}{\vartheta Y^Q(\bar{\tau})} \right\} Y^Q(\tau)\, U^{WD}_{Q_{N^Q_{i=1+2}+N^K_{\mu=1,2}}}(\tau) \left[\begin{array}{l} V^E_{WD_{CW+ME_{dep}}} \\ + V^E_{WD_{CW+ME_{dev}}}(\tau) \\ + V^E_{WD_{inn}}(\tau) \end{array} \right]$$

$$+\sum_{\tau=1}^{T} p_E(\tau)\left[1+\,r_W^{WD}(0)e^{\varphi_r^{WD}Y^Q(\tau)}\right]\left\{\frac{\vartheta Y^Q(\tau)}{\vartheta Y^Q(\bar{\tau})}\right\} U^{WD}_{\substack{Q\\ N^Q_{i=1+2}+N^K_{\mu=1,2}}}(\tau)\begin{bmatrix} V^E_{\substack{WD\\ CW+ME_{dep}}}\\ +\,V^E_{\substack{WD\\ CW+ME_{dev}}}(\tau)\\ +\,V^E_{WD_{inn}}(\tau)\end{bmatrix}$$

$$+\sum_{\tau=1}^{T} p_E(\tau)\left[1+\,r_W^{WD}(0)e^{\varphi_r^{WD}Y^Q(\tau)}\right] Y^Q(\tau)\left\{\frac{\vartheta}{\vartheta Y^Q(\bar{\tau})}\,U^{WD}_{\substack{Q\\ N^Q_{i=1+2}+N^K_{\mu=1,2}}}(\tau)\right\}\begin{bmatrix} V^E_{\substack{WD\\ CW+ME_{dep}}}\\ +\,V^E_{\substack{WD\\ CW+ME_{dev}}}(\tau)\\ +\,V^E_{WD_{inn}}(\tau)\end{bmatrix}$$

$$+\sum_{\tau=1}^{T} p_E(\tau)\left[1+\,r_W^{WD}(0)e^{\varphi_r^{WD}Y^Q(\tau)}\right] Y^Q(\tau)\,U^{WD}_{\substack{Q\\ N^Q_{i=1+2}+N^K_{\mu=1,2}}}(\tau)\left\{\frac{\vartheta}{\vartheta Y^Q(\bar{\tau})}\begin{bmatrix} V^E_{\substack{WD\\ CW+ME_{dep}}}\\ +\,V^E_{\substack{WD\\ CW+ME_{dev}}}(\tau)\\ +\,V^E_{WD_{inn}}(\tau)\end{bmatrix}\right\}$$

with

$$U^{WD}_{\substack{Q\\ N^Q_{i=1+2}+N^K_{\mu=1,2}}}(\tau):=\left\{\begin{array}{l}\left\{\psi_Q^{RP}+\sum\limits_{i=1}^{2}\dfrac{\tilde{\psi}_i^Q\kappa_i^Q e^{\varphi_i^Q\kappa_i^Q\sum_{t=1}^{\tau-1}Y^Q(t)}}{c(\bar{N}_i^Q(0))}\right\}\\ +\left\{k^Q\left\{d^Q+\gamma_\tau^Q\right\}\right\}\sum\limits_{\mu=1}^{2}\dfrac{\tilde{\psi}_\mu^K\kappa_\mu^K e^{\varphi_\mu^K\kappa_\mu^K k^Q\left\{\sum_{t=1}^{\tau-1}Y^Q(t)\left\{d^Q+\gamma_t^Q\right\}\right\}}}{c(\bar{N}_\mu^K(0))}\end{array}\right\}$$

$$V^E_{\substack{WD\\ CW+ME_{dep}}}=\left\{\sum_{\nu=1}^{2}\frac{e_\nu\,\kappa_\nu^{CW}}{c(\bar{N}_\nu^{CW}(0))}+\frac{ex^{CW}}{\eta^{CW}}\right\}k_{CW}^{WD}\,d^{CW}+\left\{\sum_{o=1}^{2}\frac{e_o\,\kappa_o^{ME}}{c(\bar{N}_o^{ME}(0))}+\frac{ex^{ME}}{\eta^{ME}}k_{ME}^{WD}\right\}k_{ME}^{WD}d^{ME},$$

$$V^E_{\substack{WD\\ CW+ME_{dev}}}(\tau)=\left\{\left\{\sum_{\nu=1}^{2}\frac{e_\nu\,\kappa_\nu^{CW}}{c(\bar{N}_\nu^{CW}(0))}+\frac{ex^{CW}}{\eta^{CW}}\right\}k_{CW}^{WD}+\left\{\sum_{o=1}^{2}\frac{e_o\,\kappa_o^{ME}}{c(\bar{N}_o^{ME}(0))}+\frac{ex^{ME}}{\eta^{ME}}k_{ME}^{WD}\right\}k_{ME}^{WD}\right\}\gamma_\tau^{WD},$$

$$V^E_{WD_{inn}}(\tau):=\;ex^{WD}(0)\left\{1\,+\,r_{an}^{WD}(0)\,e^{\varphi_{an}^{WD}\,Y^Q(\tau)}\right\}.$$

The first three partial differentiation factors represent the water saving innovation effect, the influence of the activity level, and the structural change of the water input amounts for the production and reproduction sector. These three factors do not differ between the water treatment and distribution sectors. The difference in marginal energy costs between the water treatment and water distribution sector is only determined by the differences of the fourth partial differentiation factor.

Because the water amounts of the water treatment and distribution sector have the same values, the growth rates of the two sectors are also identical $\gamma_\tau^{WT}=\gamma_\tau^{WD}$. As a result, the differences between the two sectors are mainly characterized by the energy input coefficients and civil work and mechanical & electrical equipment capital good input coefficients for the processes involved in the water treatment and distribution sectors. The depreciation rates of these two capital goods are the same for the water treatment and distribution sectors.

The expressions for the different marginal energy costs of the water distribution sector follow below.

B.2.1 Actual Marginal Energy Costs for Water Distribution Generated by Water Saving Innovation Effects

$$MCE^{Q}_{\substack{WD\\W_{inn}}}(\bar{\tau}) = p_E(\bar{\tau})\varphi_r^{WD} r_W^{WD}(\bar{\tau}) Y^Q(\bar{\tau})\, U^{WD}_{\substack{Q\\N^Q_{i=1+2}+N^K_{\mu=1,2}+^W_{RP}}}(\bar{\tau}) \begin{bmatrix} V^E_{\substack{WD\\CW+ME_{dep}}} \\ +V^E_{\substack{WD\\CW+ME_{dev}}}(\bar{\tau}) \\ +V^E_{WD_{innv}}(\bar{\tau}) \end{bmatrix}. \tag{10.23}$$

B.2.2 Actual Marginal Energy Costs for Water Distribution as a Function of the Activity Level

$$\begin{aligned} MCE^{Q}_{\substack{WD\\Y^Q}}(\bar{\tau}) &= p_E(\bar{\tau})\left[1 + r_W^{WD}(\bar{\tau})\right] * \\ &\dots * U^{WD}_{\substack{Q\\N^Q_{i=1+2}+N^K_{\mu=1,2}+^W_{RP}}}(\bar{\tau}) \left[V^E_{\substack{WD\\CW+ME_{dep}}} + V^E_{\substack{WD\\CW+ME_{dev}}}(\bar{\tau}) + V^E_{WD_{innv}}(\bar{\tau})\right]. \end{aligned} \tag{10.24}$$

B.2.3 Inter-Temporal Marginal Energy Costs for Water Distribution Caused by the Structural Change of the Water Input Amounts

$$ITMEC^{Q}_{\substack{WD\\N^Q_{i=1+2}}}{}_{+^{WD}_{N^K_{\mu=1+2}}}(\bar{\tau}\to T) = \sum_{\tau^f=\bar{\tau}+1}^{T} \left\{ \begin{array}{l} p_E(\tau^f)\left[1 + r_W^{WD}(\tau^f)\right] Y^Q(\tau^f) * \\ \dots * \left\{ \begin{array}{l} \sum\limits_{i=1}^{2} \frac{\tilde{\psi}_i^Q \kappa_i^Q \varphi_i^Q \kappa_i^Q}{c(\bar{N}_i^Q(\tau^f))} \\ + k^Q\left\{d^Q + \gamma^Q_{\tau^f}\right\} k^Q\left\{d^Q + \gamma^Q_{\tau^f}\right\} \sum\limits_{\mu=1}^{2} \frac{\tilde{\psi}_\mu^K \kappa_\mu^K \varphi_\mu^K \kappa_\mu^K}{c(\bar{N}_\mu^K(\tau^f))} \end{array} \right\} \begin{bmatrix} V^E_{\substack{WD\\CW+ME_{dep}}} \\ +V^E_{\substack{WD\\CW+ME_{dev}}}(\tau) \\ +V^E_{WD_{innv}}(\tau) \end{bmatrix} \end{array} \right\}$$

with

$$ITMEC^{Q}_{\substack{WD\\N^Q_{i=1+2}}}(\bar{\tau}\to T) = \sum_{\tau^f=\bar{\tau}+1}^{T} \left\{ \begin{array}{l} p_E(\tau^f)\left[1 + r_W^{WD}(\tau^f)\right] * \\ \dots * Y^Q(\tau^f) \left\{ \sum\limits_{i=1}^{2} \frac{\tilde{\psi}_i^Q \kappa_i^Q \varphi_i^Q \kappa_i^Q}{c(\bar{N}_i^Q(\tau^f))} \right\} \begin{bmatrix} V^E_{\substack{WD\\CW+ME_{dep}}} \\ +V^E_{\substack{WD\\CW+ME_{dev}}}(\tau^f) \\ +V^E_{WD_{innv}}(\tau^f) \end{bmatrix} \end{array} \right\}, \tag{10.25}$$

$$ITMEC^{Q}_{WD} \quad (\bar{\tau} \to T) = \sum_{\tau^f=\bar{\tau}+1}^{T} \left\{ \begin{array}{l} p_E(\tau^f)\left[1 + r_W^{WD}(\tau^f)\right] * \\ \ldots * Y^Q(\tau^f) \left\{ \begin{array}{l} k^Q \left\{ d^Q + \gamma^Q_{\tau^f} \right\} k^Q \left\{ d^Q + \gamma^Q_{\tau^f} \right\} * \\ \ldots * \sum_{\mu=1}^{2} \frac{\tilde{\psi}^K_\mu \kappa^K_\mu \varphi^K_\mu \kappa^K_\mu}{c(\bar{N}^K_\mu(\tau^f))} \end{array} \right\} \left[\begin{array}{l} V^E_{\substack{WD \\ CW+ME_{dep}}} \\ + V^E_{\substack{WD \\ CW+ME_{dev}}}(\tau^f) \\ + V^E_{WD_{innv}}(\tau^f) \end{array} \right] \end{array} \right\}_{N^K_{\mu=1+2}}. \tag{10.26}$$

B.2.4 The Dependence of the Actual and Inter-Temporal Marginal Energy Costs for Water Distribution on the Development Path of Consumption

From the fourth partial differentiation factor

$$\left\{ \frac{\vartheta}{\vartheta Y^Q(\bar{\tau})} \left[V^E_{\substack{WD \\ CW+ME_{dep}}} + V^E_{\substack{WD \\ CW+ME_{dev}}}(\tau) + V^E_{WD_{innv}}(\tau) \right] \right\} = \frac{\vartheta V^E_{\substack{WD \\ CW+ME_{dev}}}(\tau)}{\vartheta Y^Q(\bar{\tau})} + \frac{\vartheta V^E_{WD_{innv}}(\tau)}{\vartheta Y^Q(\bar{\tau})}$$

we see that the dynamic changes of the specific energy input amounts are determined by two effects. First, they are determined by the dynamics of the development path of the water demand for production

$$\frac{\vartheta}{\vartheta Y^Q(\bar{\tau})} \left[V^E_{\substack{WD \\ CW+ME_{dev}}}(\tau) \right] = \frac{\vartheta}{\vartheta Y^Q(\bar{\tau})} \left\{ \begin{array}{l} \left\{ \sum_{\nu=1}^{2} \frac{e_\nu \kappa^{CW}_\nu}{c(\bar{N}^{CW}_\nu(0))} + \frac{ex^{CW}}{\eta^{CW}} \right\} k^{WD}_{CW} \\ + \left\{ \sum_{o=1}^{2} \frac{e_o \kappa^{ME}_o}{c(\bar{N}^{ME}_o(0))} + \frac{ex^{ME}}{\eta^{ME}} \right\} k^{WD}_{ME} \end{array} \right\} \gamma^{WD}_{\bar{\tau}} = e_{k^{WD}_{CW+ME}} \frac{\vartheta \gamma^{WD}_{\tau}}{\vartheta Y^Q(\bar{\tau})}$$

with

$$e_{k^{WD}_{CW+ME}} = \left\{ \left\{ \sum_{\nu=1}^{2} \frac{l_\nu \kappa^{CW}_\nu}{c(\bar{N}^{CW}_\nu(0))} + l^{CW} \right\} k^{WD}_{CW} + \left\{ \sum_{o=1}^{2} \frac{l_o \kappa^{ME}_o}{c(\bar{N}^{ME}_o(0))} + l^{ME} \right\} k^{WD}_{ME} \right\}.$$

Secondly, they are determined by the energy saving innovation effect generated within the selected time interval, which can be called the actual marginal energy innovation effect for water treatment

$$\frac{\vartheta V^E_{\substack{WD \\ E_{inn}}}(\tau)}{\vartheta Y^Q(\bar{\tau})} = \frac{\vartheta}{\vartheta Y^Q(\bar{\tau})} ex^{WD}(0) \left\{ 1 + r^{WD}_{an}(0)\, e^{\varphi^{WD}_{an} Y^Q(\tau)} \right\} = MEC^{Q}_{\substack{WD \\ E_{inn}}}(\bar{\tau})$$
$$= \varphi^{WD}_{an} ex^{WD}(0)\, r^{WD}_{an}(0)\, e^{\varphi^{WT}_{an} Y^Q(\bar{\tau})} = \varphi^{WD}_{an} ex^{WD}(0)\, r^{WD}_{an}(\bar{\tau}). \tag{10.27}$$

The total inter-temporal marginal energy costs as a function of the development path for the consumption good amounts, including the energy saving innovation effect for water treatment, are equal to

$$ITMEC_{WD}^{Q}{}_{\gamma\tau+E_{inn}}(\bar{\tau}\rightarrow T) = \sum_{\tau^f=\bar{\tau}+1}^{T}\left\{\begin{array}{l} p_E(\tau^f)\left[1+\ r_W^{WD}(\tau^f)\right]\ Y^Q(\tau^f)U_Q^{WD}{}_{N_{i=1+2}^{Q}+N_{\mu=1,2}^{K}+{}_{RP}^{W}}(\tau^f)^{*} \\ \ldots^{*}\left[e_{k_{CW+ME}^{WD}}\dfrac{\vartheta\,\gamma_{\tau^f}^{WD}}{\vartheta Y^Q(\bar{\tau})}+\varphi_{an}^{WD}ex^{WD}(0)\,r_{an}^{WD}(\tau^f)\right]\end{array}\right\}. \quad (10.28)$$

This means that the total energy saving innovation effect is not only determined by the direct innovation effect for water distribution, but also by the development path of the consumption good amounts. The development path of the consumption good amounts is influenced by the structural change of the production sector and by the water saving innovations. Wasting water means wasting energy, especially if the economic activities are based on non-renewable resources.

C Marginal Costs for Wastewater Collection

Before we derive the marginal costs for the wastewater collection sector, it is useful to discuss the relation between the water and wastewater amounts to gain a better understanding of the non-profit conditions of the wastewater sectors.

Let us consider the case where the water supply and wastewater services for the production and reproduction activities are arranged by a unique public system. The amounts of wastewater to be collected and treated are lower than the water amounts because the amounts of water lost from the water distribution and the water amounts generated by the reproduction activities are excluded. If additional water amounts were produced by other individual water supply systems, such as private wells for production and reproduction activities, it would be necessary to include these wastewater amounts for dimensioning the wastewater collection and treatment sectors. This possibility is excluded from the water infrastructure model and the wastewater amounts are lower than the water amounts.

As discussed in the context of the Adana case study, the water losses destabilize the foundations of the other infrastructure, especially the wastewater collection systems in urban centres. This is because water losses can destroy the foundation of sewer lines and generate wastewater leakage. This leakage might pollute groundwater resources and contaminate the water by infiltrating wastewater into the water supply system. Therefore, reducing the water and wastewater losses improves both the health and environmental conditions. Reducing wastewater leakage leads to higher wastewater treatment amounts and these additional costs for wastewater treatment can be interpreted as social benefits for the improvement of the health and environmental conditions.

C.1 Marginal Human Labour Costs for Wastewater Collection

The wastewater collection system must be dimensioned to transport the total wastewater amounts generated by the production and reproduction activities. Therefore, the innovations for reducing wastewater leakage have no influence on the dimensions of the wastewater collection sector. As a result, the partial differentiation factor of the leakage innovation is only relevant for the wastewater treatment sector. The marginal human labour costs for the wastewater collection sector are equal to

$$
\begin{aligned}
& MCL^{Q}_{WWC}(\bar{\tau} \mapsto T) \\
&= \frac{\vartheta}{\vartheta Y^{Q}(\bar{\tau})} \sum_{\tau=1}^{T} p_L(\tau)\, Y^{Q}(\tau)\, U^{WWC}_{Q\; N^{Q}_{i=1+2}+N^{K}_{\mu=1,2}+{}^{WW}_{RP}}(\tau) \left[l^{WWC} + V^{L}_{\substack{WWC\\CW+ME_{dep}}} + V^{L}_{\substack{WWC\\CW+ME_{dev}}}(\tau) \right] \\
&+ \sum_{\tau=1}^{T} p_L(\tau) \left\{ \frac{\vartheta Y^{Q}(\tau)}{\vartheta Y^{Q}(\bar{\tau})} \right\} U^{WWC}_{Q\; N^{Q}_{i=1+2}+N^{K}_{\mu=1,2}+{}^{WW}_{RP}}(\tau) \left[l^{WWC} + V^{L}_{\substack{WWC\\CW+ME_{dep}}} + V^{L}_{\substack{WWC\\CW+ME_{dev}}}(\tau) \right] \\
&+ \sum_{\tau=1}^{T} p_L(\tau)\, Y^{Q}(\tau) \left\{ \frac{\vartheta U^{WWC}_{Q\; N^{Q}_{i=1+2}+N^{K}_{\mu=1,2}+{}^{WW}_{RP}}(\tau)}{\vartheta Y^{Q}(\bar{\tau})} \left[l^{WWC} + V^{L}_{\substack{WWC\\CW+ME_{dep}}} + V^{L}_{\substack{WWC\\CW+ME_{dev}}}(\tau) \right] \right\} \\
&+ \sum_{\tau=1}^{T} p_L(\tau)\, Y^{Q}(\tau)\, U^{WWC}_{Q\; N^{Q}_{i=1+2}+N^{K}_{\mu=1,2}+{}^{WW}_{RP}}(\tau) \left\{ \frac{\vartheta}{\vartheta Y^{Q}(\bar{\tau})} \left[l^{WWC} + V^{L}_{\substack{WWC\\CW+ME_{dep}}} + V^{L}_{\substack{WWC\\CW+ME_{dev}}}(\tau) \right] \right\}
\end{aligned}
$$

with

$$
U^{WWC}_{Q\; N^{Q}_{i=1+2}+N^{K}_{\mu=1,2}+{}^{WW}_{RP}}(\tau) = \left\{ \begin{array}{l} \left\{ \rho^{WW}(\tau)\psi^{RP}_{Q} + \sum_{i=1}^{2} \dfrac{\tilde{\psi}^{Q}_{i} \kappa^{Q}_{i} e^{\varphi^{Q}_{i} \kappa^{Q}_{i} \sum_{t=1}^{\tau-1} Y^{Q}(t)}}{c(\bar{N}^{Q}_{i}(0))} \right\} \\ + \left\{ k^{Q} \left\{ d^{Q} + \gamma^{Q}_{\tau} \right\} \right\} \sum_{\mu=1}^{2} \dfrac{\tilde{\psi}^{K}_{\mu} \kappa^{K}_{\mu} e^{\varphi^{K}_{\mu} \kappa^{K}_{\mu} k^{Q} \left\{ \sum_{t=1}^{\tau-1} Y^{Q}(t) \left\{ d^{Q} + \gamma^{Q}_{t} \right\} \right\}}}{c(\bar{N}^{K}_{\mu}(0))} \end{array} \right\},
$$

$$
\begin{aligned}
& V^{L}_{\substack{WWC\\CW+ME_{dep}}} \\
&= \left\{ \sum_{\upsilon=1}^{2} \frac{l_{\upsilon}\, \kappa^{CW}_{\upsilon}}{c(\bar{N}^{CW}_{\upsilon}(0))} + l^{CW} \right\} k^{WWC}_{CW} d^{CW} + \left\{ \sum_{o=1}^{2} \frac{l_{o}\, \kappa^{ME}_{o}}{c(\bar{N}^{ME}_{o}(0))} + l^{ME} \right\} k^{WWC}_{ME} d^{ME} \\
& V^{L}_{\substack{WWC\\CW+ME_{dev}}}(\tau) \\
&= \left\{ \left\{ \sum_{\upsilon=1}^{2} \frac{l_{\upsilon}\, \kappa^{CW}_{\upsilon}}{c(\bar{N}^{CW}_{\upsilon}(0))} + l^{CW} \right\} k^{WWC}_{CW} + \left\{ \sum_{o=1}^{2} \frac{l_{o}\, \kappa^{ME}_{o}}{c(\bar{N}^{ME}_{o}(0))} + l^{ME} \right\} k^{WWC}_{ME} \right\} \gamma^{WWC}_{\tau}.
\end{aligned}
$$

In detail we obtain the following marginal human labour costs for wastewater collection.

C.1.1 Actual Marginal Human Labour Costs for Wastewater Collection Caused by the Reduction of Wastewater Leakage

The dimensioning of the wastewater collection system includes the wastewater leakage and therefore reducing this leakage has no influence on the actual marginal human labour costs.

C.1.2 Actual Marginal Human Labour Costs for Wastewater Collection as a Function of the Consumption Good Amounts

$$\underset{Y^Q}{MCL^Q_{WWC}(\bar{\tau})} = p_L(\bar{\tau})\, \underset{N^Q_{i=1+2}+N^K_{\mu=1,2}+WW_{RP}}{U_Q^{WWC}(\bar{\tau})} \left[l^{WWC} + V^L_{\substack{WCD\\CW+ME_{dep}}} + V^L_{\substack{WWC\\CW+ME_{dev}}}(\bar{\tau}) \right]. \tag{10.29}$$

As with the water distribution system, the actual marginal human labour costs caused by the change in activity level for the wastewater collection sector also depends on the wastewater generation rate $0 < \rho^{WW} \leq 1$, which is estimated at 0.8 (or 80%). If this value is time invariant, the following inter-temporal marginal human labour costs are not influenced by the wastewater generation rate. Water saving measures for the reproduction activities could be integrated into our model using a generation rate and water input coefficient for reproduction activities. Both of these could be assumed to be time variant by introducing an innovation factor.

C.1.3 Inter-Temporal Marginal Human Labour Costs for Wastewater Collection Caused by the Structural Change of the Wastewater Amounts

$$\underset{N^Q_{i=1+2}+N^K_{\mu=1+2}+WW_{RP}}{ITMLC^Q_{WWC}{}^{+WWC}}(\bar{\tau} \to T) = \sum_{\tau^f=\bar{\tau}+1}^{T} \left\{ \begin{array}{l} p_L(\tau^f)\, Y^Q(\tau^f)* \\ \ldots * \left\{ \begin{array}{l} \sum_{i=1}^{2} \frac{\tilde{\psi}^Q_i \kappa^Q_i \varphi^Q_i \kappa^Q_i}{c(\bar{N}^Q_i(\tau^f)} \\ +k^Q \left\{ d^Q + \gamma^Q_{\tau^f} \right\} k^Q \left\{ d^Q + \gamma^Q_{\tau^f} \right\} \sum_{\mu=1}^{2} \frac{\tilde{\psi}^K_\mu \kappa^K_\mu \varphi^K_\mu \kappa^K_\mu}{c(\bar{N}^K_\mu(\tau^f)} \end{array} \right\} * \\ \ldots * \left[l^{WWC} + V^L_{\substack{WWC\\CW+ME_{dep}}} + V^L_{\substack{WWC\\CW+ME_{dev}}}(\tau^f) \right] \end{array} \right\}.$$

$$\underset{N^Q_{i=1+2}}{ITMLC^Q_{WWC}}(\bar{\tau} \to T) = \sum_{\tau^f=\bar{\tau}+1}^{T} \left\{ \begin{array}{l} p_L(\tau^f)\, Y^Q(\tau^f)* \\ \ldots * \left\{ \sum_{i=1}^{2} \frac{\tilde{\psi}^Q_i \kappa^Q_i \varphi^Q_i \kappa^Q_i}{c(\bar{N}^Q_i(\tau^f)} \right\} \left[l^{WWC} + V^L_{\substack{WCD\\CW+ME_{dep}}} + V^L_{\substack{WWC\\CW+ME_{dev}}}(\tau^f) \right] \end{array} \right\}. \tag{10.30}$$

$$ITMLC^{Q}_{\substack{WWC \\ N^K_{\mu=1+2}}} \quad (\bar{\tau} \rightarrow T)$$

$$= \sum_{\tau^f=\bar{\tau}+1}^{T} \left\{ \begin{array}{l} p_L(\tau^f)\; Y^Q(\tau^f)^* \\ \ldots^* \left\{ k^Q \left\{ d^Q + \gamma^Q_{\tau^f} \right\} \; k^Q \left\{ d^Q + \gamma^Q_{\tau^f} \right\} \sum_{\mu=1}^{2} \frac{\tilde{\psi}^K_{\mu} \kappa^K_{\mu} \varphi^K_{\mu} \kappa^K_{\mu}}{c(N^K_{\mu}(\tau^f)} \right\}^* \\ \ldots^* \left[l^{WWC} + V^L_{\substack{WWC \\ CW+ME_{dep}}} + V^L_{\substack{WWC \\ CW+ME_{dev}}} (\tau^f) \right] \end{array} \right\}. \tag{10.31}$$

C.1.4 The Dependence of Inter-Temporal Marginal Human Labour Costs for Wastewater on the Development Path of the Consumption Good Amounts

$$ITMLC^{Q}_{\substack{WWC \\ \gamma_{\bar{\tau}}}} (\bar{\tau} \rightarrow T)$$

$$= \sum_{\tau^f=\bar{\tau}+1}^{T} \left\{ p_L(\tau^f)\; Y^Q(\tau^f) U^{WWC}_{\substack{Q \\ N^Q_{i=1+2}+N^K_{\mu=1,2}}+WW_{RP}} (\tau^f)\, l_{k^{WWC}_{CW+ME}} \frac{\vartheta\, \gamma^{WWC}_{\tau^f}}{\vartheta Y^Q(\bar{\tau})} \right\}. \tag{10.32}$$

The inter-temporal change in the marginal human labour costs depends on the development path for wastewater collection amounts. The wastewater collection amounts are higher because of the wastewater generation rate of the reproduction activities.

C.2 Marginal Energy Costs for Wastewater Collection

$$MCE^{Q}_{WWC} (\bar{\tau} \mapsto T)$$

$$= \frac{\vartheta}{\vartheta Y^Q(\bar{\tau})} \sum_{\tau=1}^{T} p_E(\tau) Y^Q(\tau)\, U^{WWC}_{\substack{Q \\ N^Q_{i=1+2}+N^K_{\mu=1,2}+^{WW}_{RP}}} (\tau) \left[V^E_{\substack{WWC \\ CW+ME_{dep}}} + V^E_{\substack{WWC \\ CW+ME_{dev}}} (\tau) + V^E_{WWC_{inn}} \right]$$

$$= \sum_{\tau=1}^{T} p_E(\tau) \left\{ \frac{\vartheta Y^Q(\tau)}{\vartheta Y^Q(\bar{\tau})} \right\} U^{WWC}_{\substack{Q \\ N^Q_{i=1+2}+N^K_{\mu=1,2}+^{WW}_{RP}}} (\tau) \left[V^E_{\substack{WWC \\ CW+ME_{dep}}} + V^E_{\substack{WWC \\ CW+ME_{dev}}} (\tau) + V^E_{WWC_{inn}}(\tau) \right]$$

$$+ \sum_{\tau=1}^{T} p_E(\tau)\, Y^Q(\tau) \left\{ \frac{\vartheta U^{WWC}_{\substack{Q \\ N^Q_{i=1+2}+N^K_{\mu=1,2}+^{WW}_{RP}}} (\tau)}{\vartheta Y^Q(\bar{\tau})} \right\} \left[V^E_{\substack{WWC \\ CW+ME_{dep}}} + V^E_{\substack{WWC \\ CW+ME_{dev}}} (\tau) + V^E_{WWC_{inn}}(\tau) \right]$$

$$+ \sum_{\tau=1}^{T} p_E(\tau)\; Y^Q(\tau)\; U^{WWC}_{\substack{Q \\ N^Q_{i=1+2}+N^K_{\mu=1,2}+^{WW}_{RP}}} (\tau) \left\{ \frac{\vartheta}{\vartheta Y^Q(\bar{\tau})} \left[V^E_{\substack{WWC \\ CW+ME_{dep}}} + V^E_{\substack{WWC \\ CW+ME_{dev}}} (\tau) + V^E_{WWC_{inn}}(\tau) \right] \right\}$$

with

$$U^{WWC}_{\substack{Q\\ N^Q_{i=1+2}+N^K_{\mu=1,2}}}(\tau) = \left\{ \begin{array}{l} \left\{ \rho^{WW}(\tau)\psi^{RP}_Q + \sum_{i=1}^{2} \frac{\tilde{\psi}^Q_i \kappa^Q_i e^{\varphi^Q_i \kappa^Q_i \sum_{t=1}^{\tau-1} Y^Q(t)}}{c(\bar{N}^Q_i(0))} \right\} \\ + \left\{ k^Q \left\{ d^Q + \gamma^Q_\tau \right\} \right\} \sum_{\mu=1}^{2} \frac{\tilde{\psi}^K_\mu \kappa^K_\mu e^{\varphi^K_\mu \kappa^K_\mu k^Q \left\{ \sum_{t=1}^{\tau-1} Y^Q(t) \left\{ d^Q + \gamma^Q_t \right\} \right\}}}{c(\bar{N}^K_\mu(0))} \end{array} \right\},$$

$$V^E_{\substack{WWC\\ CW+ME_{dep}}} = \left\{ \sum_{\nu=1}^{2} \frac{e_\nu \kappa^{CW}_\nu}{c(\bar{N}^{CW}_\nu(0))} + \frac{ex^{CW}}{\eta^{CW}} \right\} k^{WWC}_{CW} d^{CW} + \left\{ \sum_{o=1}^{2} \frac{e_o \kappa^{ME}_o}{c(\bar{N}^{ME}_o(0))} + \frac{ex^{ME}}{\eta^{ME}} \right\} k^{WWC}_{ME} d^{ME},$$

$$V^E_{\substack{WWC\\ CW+ME_{dev}}}(\tau) = \left\{ \left\{ \sum_{\nu=1}^{2} \frac{e_\nu \kappa^{CW}_\nu}{c(\bar{N}^{CW}_\nu(0))} + \frac{ex^{CW}}{\eta^{CW}} \right\} k^{WWC}_{CW} + \left\{ \sum_{o=1}^{2} \frac{e_o \kappa^{ME}_o}{c(\bar{N}^{ME}_o(0))} + \frac{ex^{ME}}{\eta^{ME}} \right\} k^{WWC}_{ME} \right\} \gamma^{WWC}_\tau,$$

$$V^E_{WWC_{inn}}(\tau) = ex^{WWC}(0) \left[1 + r^{WWC}_{an}(0)\, e^{\varphi^{WWC}_{an} Y^Q(\tau)}) \right].$$

C.2.1 Actual Marginal Energy Costs for Wastewater Collection Caused by Reducing Wastewater Leakage

Because the dimensioning of the wastewater collection system includes wastewater leakage, reducing the leakage has no influence on this actual marginal energy costs.

C.2.2 Actual Marginal Energy Costs for Wastewater Collection as a Function of the Activity Level or Consumption Good Amounts

$$MCE^Q_{\substack{WWC\\ Y^Q}}(\bar{\tau}) = p_E(\bar{\tau})\, U^{WWC}_{\substack{Q\\ N^Q_{i=1+2}+N^K_{\mu=1,2}+^{WW}_{RP}}}(\bar{\tau}) \left[V^E_{\substack{WWC\\ CW+ME_{dep}}} + V^E_{\substack{WWC\\ CW+ME_{dev}}}(\tau) + V^E_{WWC_{inn}}(\tau) \right]. \tag{10.33}$$

C.2.3 Inter-Temporal Marginal Energy Costs for Wastewater Collection Caused by the Structural Change of the Wastewater Amounts

$$ITMEC^{Q}_{\substack{WWC \\ N^{Q}_{i=1+2}} \; \substack{+WWC \\ N^{K}_{\mu=1+2}}}(\bar{\tau} \to T) = \sum_{\tau^f=\bar{\tau}+1}^{T} \left\{ \begin{array}{l} p_E(\tau^f)\ Y^Q(\tau^f)^* \\ \ldots^* \left\{ \begin{array}{l} \sum_{i=1}^{2} \frac{\tilde{\psi}^Q_i \kappa^Q_i \varphi^Q_i \kappa^Q_i}{c(\bar{N}^Q_i(\tau^f))} \\ +k^Q \left\{ d^Q + \gamma^Q_{\tau^f} \right\} k^Q \left\{ d^Q + \gamma^Q_{\tau^f} \right\} \sum_{\mu=1}^{2} \frac{\tilde{\psi}^K_\mu \kappa^K_\mu \varphi^K_\mu \kappa^K_\mu}{c(N^K_\mu(\tau^f))} \end{array} \right\}^* \\ \ldots^* \left[V^E_{\substack{WWC \\ CW+ME_{dep}}} + V^E_{\substack{WWC \\ CW+ME_{dev}}}(\tau) + V^E_{WWC_{inn}}(\tau) \right] \end{array} \right\}.$$

with

$$ITMEC^{Q}_{\substack{WWC \\ N^{Q}_{i=1+2}}}(\bar{\tau} \to T) = \sum_{\tau^f=\bar{\tau}+1}^{T} \left\{ \begin{array}{l} p_E(\tau^f)\ Y^Q(\tau^f) \left\{ \sum_{i=1}^{2} \frac{\tilde{\psi}^Q_i \kappa^Q_i \varphi^Q_i \kappa^Q_i}{c(\bar{N}^Q_i(\tau^f))} \right\}^* \\ \ldots^* \left[V^E_{\substack{WWC \\ CW+ME_{dep}}} + V^E_{\substack{WWC \\ CW+ME_{dev}}}(\tau) + V^E_{WWC_{inn}}(\tau) \right] \end{array} \right\}, \tag{10.34}$$

$$ITMEC^{Q}_{\substack{WWC \\ N^{K}_{\mu=1+2}}}(\bar{\tau} \to T) = \sum_{\tau^f=\bar{\tau}+1}^{T} \left\{ \begin{array}{l} p_E(\tau^f)\ Y^Q(\tau^f)^* \\ \ldots^* \left\{ k^Q \left\{ d^Q + \gamma^Q_{\tau^f} \right\} k^Q \left\{ d^Q + \gamma^Q_{\tau^f} \right\} \sum_{\mu=1}^{2} \frac{\tilde{\psi}^K_\mu \kappa^K_\mu \varphi^K_\mu \kappa^K_\mu}{c(\bar{N}^K_\mu(\tau^f))} \right\} \\ \ldots^* \left[V^E_{\substack{WWC \\ CW+ME_{dep}}} + V^E_{\substack{WWC \\ CW+ME_{dev}}}(\tau) + V^E_{WWC_{inn}}(\tau) \right] \end{array} \right\}. \tag{10.35}$$

C.2.4 The Dependence of Actual and Inter-Temporal Marginal Energy Costs for Wastewater Collection on the Development Path of the Wastewater Amounts

From the partial differentiation factor

$$\left\{ \frac{\vartheta}{\vartheta Y^Q(\bar{\tau})} \left[V^E_{\substack{WWC \\ CW+ME_{dep}}} + V^E_{\substack{WWC \\ CW+ME_{dev}}}(\tau) + V^E_{WWC_{inn}}(\tau) \right] \right\} = \frac{\vartheta\, V^E_{\substack{WWC \\ CW+ME_{dev}}}(\tau)}{\vartheta Y^Q(\bar{\tau})} + \frac{\vartheta\, V^E_{WWC_{inn}}(\tau)}{\vartheta Y^Q(\bar{\tau})},$$

we observe that the dynamic changes of the specific energy input amounts are determined by two effects. The first effect is the dynamics of the development path of the water demand for production

$$\frac{\vartheta}{\vartheta Y^Q(\bar{\tau})}\left[V^E_{\substack{WWC\\CW+ME_{dev}}}(\tau)\right] =$$

$$\frac{\vartheta}{\vartheta Y^Q(\bar{\tau})}\left\{\begin{array}{l}\left\{\sum\limits_{\nu=1}^{2}\frac{e_\nu\,\kappa_\nu^{CW}}{c(N_\nu^{CW}(0))}+\frac{ex^{CW}}{\eta^{CW}}\right\}k_{CW}^{WWC}\\+\left\{\sum\limits_{o=1}^{2}\frac{e_o\,\kappa_o^{ME}}{c(N_o^{ME}(0))}+\frac{ex^{ME}}{\eta^{ME}}\right\}k_{ME}^{WWC}\end{array}\right\}\gamma_\tau^{WWC} = e_{k^{WWC}_{CW+ME}}\frac{\vartheta\,\gamma_\tau^{WWC}}{\vartheta Y^Q(\bar{\tau})}$$

with

$$e_{k^{WWC}_{CW+ME}} := \left\{\begin{array}{l}\left\{\sum\limits_{\nu=1}^{2}\frac{e_\nu\,\kappa_\nu^{CW}}{c(N_\nu^{CW}(0))}+\frac{ex^{CW}}{\eta^{CW}}\right\}k_{CW}^{WWC}\\+\left\{\sum\limits_{o=1}^{2}\frac{e_o\,\kappa_o^{ME}}{c(N_o^{ME}(0))}+\frac{ex^{ME}}{\eta^{ME}}\right\}k_{ME}^{WWC}\end{array}\right\}.$$

The second effect is the energy saving innovation effect generated within the selected time interval, which can be called the actual marginal energy innovation effect for wastewater collection.

$$\begin{aligned}\frac{\vartheta\,V^E_{\substack{WWC\\E_{inn}}}(\tau)}{\vartheta Y^Q(\bar{\tau})} &= \frac{\vartheta}{\vartheta Y^Q(\bar{\tau})}ex^{WWC}(0)\left\{1\,+\,r_{an}^{WWC}(0)\,e^{\varphi_{an}^{WWC}\,Y^Q(\tau)}\right\}\\ = MEC^Q_{\substack{WWC\\E_{inn}}}(\bar{\tau}) &= \varphi_{an}^{WWC}ex^{WWC}(0)\,r_{an}^{WWC}(0)\,e^{\varphi_{an}^{WT}\,Y^Q(\bar{\tau})}\\ &= \varphi_{an}^{WWC}ex^{WWC}(0)\,r_{an}^{WWC}(\bar{\tau}).\end{aligned} \tag{10.36}$$

The total inter-temporal marginal energy costs as a function of the development path for the consumption good amounts, including the energy saving innovation effect for water treatment, are equal to

$$ITMEC^Q_{\substack{WWC\\\gamma_\tau+E_{inn}}}(\bar{\tau}\to T) = \sum_{\tau^f=\bar{\tau}+1}^{T}\left\{\begin{array}{l}p_E(\tau^f)\,Y^Q(\tau^f)U^{WWC}_{\substack{Q\\N^Q_{i=1+2}+N^K_{\mu=1,2}+WW_{RP}}}(\tau^f)^*\\ \ldots^*\left[e_{k^{WWC}_{CW+ME}}\frac{\vartheta\,\gamma^{WWC}_{\tau^f}}{\vartheta Y^Q(\bar{\tau})}+\varphi_{an}^{WWC}ex^{WWC}(0)\,r_{an}^{WWC}(\tau^f)\right]\end{array}\right\}. \tag{10.37}$$

This means that the total energy saving innovation effect is determined by the direct innovation effect for the wastewater collection system and also by the development path of the consumption good amounts, which is influenced by the structural change of the production sector. Wastewater leakage means wasting energy for wastewater collection.

D Marginal Costs for Wastewater Treatment

The reduction of wastewater leakage leads to an increase in wastewater amounts to be treated. Therefore, the marginal costs for wastewater treatment include the benefits, or social costs, caused by the inefficiency of the wastewater collection system.

D.1 Marginal Human Labour Costs for Wastewater Treatment

The marginal human labour costs for wastewater treatment are equal to

$$MCL^{Q}_{WWT}(\bar{\tau} \mapsto T) =$$

$$\frac{\vartheta}{\vartheta Y^{Q}(\bar{\tau})} \sum_{\tau=1}^{T} \left\{ \begin{array}{l} p_L(\tau) \left[1 - r^{WWC}_{WW}(0)\, e^{\varphi^{WWC}_{r} Y^{Q}(\tau)} \right] Y^{Q}(\tau)\, U^{WWT}_{Q \atop N^{Q}_{i=1+2} + N^{K}_{\mu=1,2} + RP^{WW}}(\tau)^{*} \\ \ldots^{*} \left[l^{WWT} + V^{L}_{WWT \atop CW+ME_{dep}} + V^{L}_{WWT \atop CW+ME_{dev}}(\tau) \right] \end{array} \right\}.$$

$$MCL^{Q}_{WWT}(\bar{\tau} \mapsto T)$$

$$= \sum_{\tau=1}^{T} \left\{ \begin{array}{l} p_L(\tau) \left\{ \dfrac{\vartheta \left[1 - r^{WWC}_{WW}(0)\, e^{\varphi^{WWC}_{r} Y^{Q}(\tau)} \right]}{\vartheta Y^{Q}(\bar{\tau})} \right\}^{*} \\ \ldots^{*} Y^{Q}(\tau)\, U^{WWT}_{Q \atop N^{Q}_{i=1+2} + N^{K}_{\mu=1,2} + RP^{WW}}(\tau) \left[l^{WWT} + V^{L}_{WWT \atop CW+ME_{dep}} + V^{L}_{WWT \atop CW+ME_{dev}}(\tau) \right] \end{array} \right\}$$

$$+ \sum_{\tau=1}^{T} \left\{ \begin{array}{l} p_L(\tau) \left[1 - r^{WWC}_{WW}(0)\, e^{\varphi^{WWC}_{r} Y^{Q}(\tau)} \right] \left\{ \dfrac{\vartheta Y^{Q}(\tau)}{\vartheta Y^{Q}(\bar{\tau})} \right\}^{*} \\ \ldots^{*} U^{WWT}_{Q \atop N^{Q}_{i=1+2} + N^{K}_{\mu=1,2} + RP^{WW}}(\tau) \left[l^{WWT} + V^{L}_{WWT \atop CW+ME_{dep}} + V^{L}_{WWT \atop CW+ME_{dev}}(\tau) \right] \end{array} \right\}$$

$$+ \sum_{\tau=1}^{T} \left\{ \begin{array}{l} p_L(\tau) \left[1 - r^{WWC}_{WW}(0)\, e^{\varphi^{WWC}_{r} Y^{Q}(\tau)} \right] Y^{Q}(\tau) \left\{ \dfrac{\vartheta U^{WWT}_{Q \atop N^{Q}_{i=1+2} + N^{K}_{\mu=1,2} + RP^{WW}}(\tau)}{\vartheta Y^{Q}(\bar{\tau})} \right\}^{*} \\ \ldots^{*} \left[l^{WWT} + V^{L}_{WWT \atop CW+ME_{dep}} + V^{L}_{WWT \atop CW+ME_{dev}}(\tau) \right] \end{array} \right\}$$

$$+ \sum_{\tau=1}^{T} \left\{ \begin{array}{l} p_L(\tau) \left[1 - r^{WWC}_{WW}(0)\, e^{\varphi^{WWC}_{r} Y^{Q}(\tau)} \right] Y^{Q}(\tau)\, U^{WWT}_{Q \atop N^{Q}_{i=1+2} + N^{K}_{\mu=1,2} + RP^{WW}}(\tau) \\ \ldots^{*} \left\{ \dfrac{\vartheta \left[l^{WWT} + V^{L}_{WWT \atop CW+ME_{dep}} + V^{L}_{WWT \atop CW+ME_{dev}}(\tau) \right]}{\vartheta Y^{Q}(\bar{\tau})} \right\} \end{array} \right\}$$

with

$$U_{Q\;N^{Q}_{i=1+2}+N^{K}_{\mu=1,2}}^{WWT}(\tau) = \left\{ \begin{array}{l} \left\{ \rho^{WW}(\tau)\psi_{Q}^{RP} + \sum_{i=1}^{2} \frac{\tilde{\psi}_i^{Q}\kappa_i^{Q} e^{\varphi_i^{Q}\kappa_i^{Q}\sum_{t=1}^{\tau-1} Y^{Q}(t)}}{c(\bar{N}_i^{Q}(0))} \right\} \\ + \left\{ k^{Q}\left\{ d^{Q} + \gamma_{\tau}^{Q} \right\} \right\} \sum_{\mu=1}^{2} \frac{\tilde{\psi}_{\mu}^{K}\kappa_{\mu}^{K} e^{\varphi_{\mu}^{K}\kappa_{\mu}^{K}k^{Q}\left\{\sum_{t=1}^{\tau-1} Y^{Q}(t)\left\{d^{Q}+\gamma_t^{Q}\right\}\right\}}}{c(\bar{N}_{\mu}^{K}(0))} \end{array} \right\},$$

$$V^{L}_{\substack{WWT \\ CW+ME_{dep}}} = \left\{ \sum_{\nu=1}^{2} \frac{l_{\nu}\kappa_{\nu}^{CW}}{c(\bar{N}_{\nu}^{CW}(0))} + l^{CW} \right\} k_{CW}^{WWT} d^{CW} + \left\{ \sum_{o=1}^{2} \frac{l_{o}\kappa_{o}^{ME}}{c(\bar{N}_{o}^{ME}(0))} + l^{ME} \right\} k_{ME}^{WWT} d^{ME},$$

$$V^{L}_{\substack{WWT \\ CW+ME_{dev}}}(\tau) = \left\{ \left\{ \sum_{\nu=1}^{2} \frac{l_{\nu}\kappa_{\nu}^{CW}}{c(\bar{N}_{\nu}^{CW}(0))} + l^{CW} \right\} k_{CW}^{WWT} + \left\{ \sum_{o=1}^{2} \frac{l_{o}\kappa_{o}^{ME}}{c(\bar{N}_{o}^{ME}(0))} + l^{ME} \right\} k_{ME}^{WWT} \right\} \gamma_{\tau}^{WWT}.$$

We obtain the following marginal cost components for wastewater treatment.

D.1.1 Actual Marginal Human Labour Costs for Wastewater Treatment Caused by the Reduction of Wastewater Leakage During Wastewater Collection

$$\begin{aligned} MCL^{Q}_{\substack{WWT \\ WW_{inn}}}(\bar{\tau}) &= p_L(\bar{\tau}) \left| \left\{ -\varphi_r^{WWC} r_{WW}^{WWC}(\bar{\tau}) \right\} \right| Y^{Q}(\tau) | * \\ ...* U_{Q\;N^{Q}_{i=1+2}+N^{K}_{\mu=1,2}+^{WW}_{RP}}^{WWT}(\bar{\tau}) &\left[l^{WWT} + V^{L}_{\substack{WWT \\ CW+ME_{dep}}} + V^{L}_{\substack{WWT \\ CW+ME_{dev}}}(\bar{\tau}) \right]. \end{aligned} \qquad (10.38)$$

The wastewater leakage innovation effect is interpreted as reducing the social costs. This means that increased benefits lead to an increase of these marginal costs for wastewater treatment. As a result, the absolute value of this innovation effect should be taken into account when interpreting this type of actual marginal costs.

D.1.2 Actual Marginal Human Labour Costs for Wastewater Treatment as a Function of the Activity Level or Consumption Good Amounts

From the following expression, we learn that in the case, where the consumption level increases, the wastewater leakage is decreasing, which leads to an increase in the wastewater amounts to be treated. If the wastewater leakage rate is close to zero, the collected and treated wastewater amounts have nearly the same values.

$$
\begin{aligned}
& MCL^{Q}_{\underset{Y^Q}{WWT}}(\bar{\tau}) \\
& = p_L(\bar{\tau}) \left[1 - r^{WWC}_{WW}(0)\, e^{\varphi_r^{WWC} Y^Q(\tau)} \right] * \\
& \ldots * U^{WWT}_{\underset{N^Q_{i=1+2}+N^K_{\mu=1,2}+^{WW}_{RP}}{Q}}(\bar{\tau}) \left[l^{WWT} + V^L_{\underset{CW+ME_{dep}}{WWT}} + V^L_{\underset{CW+ME_{dev}}{WWT}}(\bar{\tau}) \right].
\end{aligned}
\tag{10.39}
$$

D.1.3 Inter-Temporal Marginal Human Labour Costs for Wastewater Treatment Caused by the Structural Change of the Wastewater Amounts

$$
\begin{aligned}
& ITMLC^{Q}_{\underset{N^Q_{i=1+2}}{WWT}\, \underset{N^K_{\mu=1+2}}{+^{WWT}}}(\bar{\tau} \to T) \\
& = \sum_{\tau^f=\bar{\tau}+1}^{T} \left\{ \begin{array}{l} p_L(\tau^f) \left[1 - r^{WWC}_{WW}(0)\, e^{\varphi_r^{WWC} Y^Q(\tau)} \right] Y^Q(\tau^f)^* \\ \ldots^* \left\{ \begin{array}{l} \sum\limits_{i=1}^{2} \dfrac{\tilde{\psi}_i^Q \kappa_i^Q \varphi_i^Q \kappa_i^Q}{c(\bar{N}_i^Q(\tau^f)} \\ +k^Q \left\{ d^Q + \gamma^Q_{\tau^f} \right\} k^Q \left\{ d^Q + \gamma^Q_{\tau^f} \right\} \sum\limits_{\mu=1}^{2} \dfrac{\tilde{\psi}_\mu^K \kappa_\mu^K \varphi_\mu^K \kappa_\mu^K}{c(\bar{N}_\mu^K(\tau^f)} \end{array} \right\} \end{array} \right\} * \\
& \ldots * \left[l^{WWT} + V^L_{\underset{CW+ME_{dep}}{WWT}} + V^L_{\underset{CW+ME_{dev}}{WWT}}(\bar{\tau}) \right]. \\
& ITMLC^{Q}_{\underset{N^Q_{i=1+2}}{WWT}}(\bar{\tau} \to T) \\
& = \sum_{\tau^f=\bar{\tau}+1}^{T} \left\{ \begin{array}{l} p_L(\tau^f) \left[1 - r^{WWC}_{WW}(\tau^f) \right] Y^Q(\tau^f)^* \\ \ldots^* \left\{ \sum\limits_{i=1}^{2} \dfrac{\tilde{\psi}_i^Q \kappa_i^Q \varphi_i^Q \kappa_i^Q}{c(\bar{N}_i^Q(\tau^f))} \right\} \left[l^{WWT} + V^L_{\underset{CW+ME_{dep}}{WWT}} + V^L_{\underset{CW+ME_{dev}}{WWT}}(\bar{\tau}) \right] \end{array} \right\}.
\end{aligned}
\tag{10.40}
$$

$$
\begin{aligned}
& ITMLC^{Q}_{\underset{N^K_{\mu=1+2}}{WWT}}(\bar{\tau} \to T) = \\
& \sum_{\tau^f=\bar{\tau}+1}^{T} \left\{ \begin{array}{l} p_L(\tau^f) \left[1 - r^{WWC}_{WW}(\tau^f) \right] Y^Q(\tau^f)^* \\ \ldots^* \left\{ k^Q \left\{ d^Q + \gamma^Q_{\tau^f} \right\} k^Q \left\{ d^Q + \gamma^Q_{\tau^f} \right\} \sum\limits_{\mu=1}^{2} \dfrac{\tilde{\psi}_\mu^K \kappa_\mu^K \varphi_\mu^K \kappa_\mu^K}{c(\bar{N}_\mu^K(\tau^f))} \right\} * \\ \ldots^* \left[l^{WWT} + V^L_{\underset{CW+ME_{dep}}{WWT}} + V^L_{\underset{CW+ME_{dev}}{WWT}}(\bar{\tau}) \right] \end{array} \right\}.
\end{aligned}
\tag{10.41}
$$

D.1.4 The Dependence of the Inter-Temporal Marginal Human Labour Costs for Wastewater on the Development Path of the Consumption Good Amounts

$$ITMLC^{Q}_{\gamma^{WWT}_{\bar{\tau}}}(\bar{\tau} \rightarrow T) = \sum_{\tau^f=\bar{\tau}+1}^{T} \left\{ \begin{array}{l} p_L(\tau^f)\left[1 - r^{WWC}_{WW}(\tau^f)\right] \; Y^Q(\tau^f)^* \\ \ldots^* U^{WWT}_{Q \atop N^Q_{i=1+2}+N^K_{\mu=1,2}+{}^{WW}_{RP}}(\tau^f))\, l_{k^{WWT}_{CW+ME}} \dfrac{\vartheta\, \gamma^{WWT}_{\tau^f}}{\vartheta\, Y^Q(\bar{\tau})} \end{array} \right\}. \tag{10.42}$$

This type of inter-temporal marginal human labour cost for wastewater treatment is influenced by the wastewater amounts of the reproduction activities.

D.2 *Marginal Energy Costs for Wastewater Treatment*

The total marginal energy costs for wastewater treatment show that the marginal energy costs for wastewater treatment include two innovation effects. The first effect is caused by reducing the leakage from the wastewater collection system. The second is the energy saving innovation effect for wastewater treatment.

$$MCE^{Q}_{WWT}(\bar{\tau} \mapsto T) = \frac{\vartheta}{\vartheta Y^Q(\bar{\tau})} \sum_{\tau=1}^{T} \left\{ \begin{array}{l} p_E(\tau)\left[1 - r^{WWC}_{WW}(0)\, e^{\varphi^{WWC}_r Y^Q(\tau)}\right] Y^Q(\tau)^* \\ \ldots^* U^{WWT}_{Q \atop N^Q_{i=1+2}+N^K_{\mu=1,2}+{}^{WW}_{RP}}(\tau)\left[V^E_{WWT \atop CW+ME_{dep}} + V^E_{WWT \atop CW+ME_{dev}}(\tau) + V^E_{WWT_{inn}}(\tau)\right] \end{array} \right\}.$$

$$MCE^{Q}_{WWT}(\bar{\tau} \mapsto T)$$

$$= \sum_{\tau=1}^{T} \left\{ \begin{array}{l} p_E(\tau)\left\{ \dfrac{\vartheta\left[1 - r^{WWC}_{WW}(0)\, e^{\varphi^{WWC}_r Y^Q(\tau)}\right]}{\vartheta Y^Q(\bar{\tau})} \right\} Y^Q(\tau)^* \\ \ldots^* U^{WWT}_{Q \atop N^Q_{i=1+2}+N^K_{\mu=1,2}+{}^{WW}_{RP}}(\tau)\left[V^E_{WWT \atop CW+ME_{dep}} + V^E_{WWT \atop CW+ME_{dev}}(\tau) + V^E_{WWT_{inn}}(\tau)\right] \end{array} \right\}$$

$$+ \sum_{\tau=1}^{T} \left\{ \begin{array}{l} p_E(\tau)\left[1 - r^{WWC}_{WW}(0)\, e^{\varphi^{WWC}_r Y^Q(\tau)}\right] \left\{ \dfrac{\vartheta Y^Q(\tau)}{\vartheta Y^Q(\bar{\tau})} \right\}^* \\ \ldots^* U^{WWT}_{Q \atop N^Q_{i=1+2}+N^K_{\mu=1,2}+{}^{WW}_{RP}}(\tau)\left[V^E_{WWT \atop CW+ME_{dep}} + V^E_{WWT \atop CW+ME_{dev}}(\tau) + V^E_{WWT_{inn}}(\tau)\right] \end{array} \right\}$$

$$+ \sum_{\tau=1}^{T} \left\{ \begin{array}{l} p_E(\tau)\left[1 - r^{WWC}_{WW}(0)\, e^{\varphi^{WWC}_r Y^Q(\tau)}\right] Y^Q(\tau)^* \left\{ \dfrac{\vartheta U^{WWT}_{Q \atop N^Q_{i=1+2}+N^K_{\mu=1,2}+{}^{WW}_{RP}}(\tau)}{\vartheta Y^Q(\bar{\tau})} \right\}^* \\ \ldots^* \left[V^E_{WWT \atop CW+ME_{dep}} + V^E_{WWT \atop CW+ME_{dev}}(\tau) + V^E_{WWT_{inn}}(\tau)\right] \end{array} \right\}$$

$$+\sum_{\tau=1}^{T}\left\{\begin{array}{l} p_E(\tau)\left[1-r_{WW}^{WWC}(0)\,e^{\varphi_r^{WWC}Y^Q(\tau)}\right]Y^Q(\tau)\;U^{WWT}_{\substack{Q\\ N^Q_{i=1+2}+N^K_{\mu=1,2}+^{WW}_{RP}}}(\tau)^{*}\\ \ldots^{*}\left\{\begin{array}{c}\dfrac{\vartheta\left[V^E_{\substack{WWT\\ CW+ME_{dep}}}+V^E_{\substack{WWT\\ CW+ME_{dev}}}(\tau)+V^E_{WWT_{inn}}(\tau)\right]}{\vartheta Y^Q(\bar{\tau})}\\ \end{array}\right\}\end{array}\right\}$$

with

$$U^{WWT}_{\substack{Q\\ N^Q_{i=1+2}+N^K_{\mu=1,2}}}(\tau):=\left\{\begin{array}{l}\left\{\rho^{WW}(\tau)\psi_Q^{RP}+\sum_{i=1}^{2}\dfrac{\tilde{\psi}_i^Q\kappa_i^Q e^{\varphi_i^Q\kappa_i^Q\sum_{t=1}^{\tau-1}Y^Q(t)}}{c(\bar{N}_i^Q(0))}\right\}\\ +\left\{k^Q\left\{d^Q+\gamma_\tau^Q\right\}\right\}\sum_{\mu=1}^{2}\dfrac{\tilde{\psi}_\mu^K\kappa_\mu^K e^{\varphi_\mu^K\kappa_\mu^K k^Q\left\{\sum_{t=1}^{\tau-1}Y^Q(t)\left\{d^Q+\gamma_t^Q\right\}\right\}}}{c(\bar{N}_\mu^K(0))}\end{array}\right\},$$

$$V^E_{\substack{WWT\\ CW+ME_{dep}}}=\left\{\sum_{\nu=1}^{2}\frac{e_\nu\,\kappa_\nu^{CW}}{c(\bar{N}_\nu^{CW}(0))}+\frac{ex^{CW}}{\eta^{CW}}\right\}k_{CW}^{WWT}\,d^{CW}+$$

$$\ldots+\left\{\sum_{o=1}^{2}\frac{e_o\,\kappa_o^{ME}}{c(\bar{N}_o^{ME}(0))}+\frac{ex^{ME}}{\eta^{ME}}\right\}k_{ME}^{WWT}\,d^{ME},$$

$$V^E_{\substack{WWT\\ CW+ME_{dev}}}(\tau)=\left\{\begin{array}{l}\left\{\sum_{\nu=1}^{2}\dfrac{e_\nu\,\kappa_\nu^{CW}}{c(\bar{N}_\nu^{CW}(0))}+\dfrac{ex^{CW}}{\eta^{CW}}\right\}k_{CW}^{WWT}+\\ \ldots\left\{\sum_{o=1}^{2}\dfrac{e_o\,\kappa_o^{ME}}{c(\bar{N}_o^{ME}(0))}+\dfrac{ex^{ME}}{\eta^{ME}}\right\}k_{ME}^{WWC}\end{array}\right\}\gamma_\tau^{WWT},$$

$$V^E_{WWT_{inn}}(\tau):=\tilde{\varepsilon}^{WWT}\left\{\ln\frac{c_Q^*(0)}{c_Q(0)}+\ln\frac{c_{\hat{j}}^*(0)}{c_{\hat{j}}(0)}\right\}\left[1+r_{an}^{WWT}(0)\,e^{\varphi_{an}^{WWT}Y^Q(\tau)}\right].$$

In detail, we obtain the following marginal energy costs components for wastewater treatment.

D.2.1 Actual Marginal Energy Costs for Wastewater Treatment Caused by Reducing Wastewater Leakage During Wastewater Collection

Reducing wastewater leakage leads to an increase in the wastewater treatment amounts and the corresponding actual marginal energy cost component.

$$\begin{array}{l} MCE^{Q}_{\substack{WWT\\ WW_{inn}}}(\bar{\tau})=\\ p_E(\bar{\tau})|\left\{-\varphi_r^{WWC}\right\}|r_{WW}^{WWC}(\bar{\tau})Y^Q(\bar{\tau})^{*}\\ \ldots^{*}\,U^{WWT}_{\substack{Q\\ N^Q_{i=1+2}+N^K_{\mu=1,2}+^{WW}_{RP}}}(\bar{\tau})\left[V^E_{\substack{WWT\\ CW+ME_{dep}}}+V^E_{\substack{WWT\\ CW+ME_{dev}}}(\bar{\tau})+V^E_{WWT_{innv}}(\bar{\tau})\right].\end{array}\quad(10.43)$$

D.2.2 The Dependence of the Actual Marginal Energy Costs for Wastewater Treatment on the Activity Level or Consumption Good Amounts

$$\begin{aligned} MCE^{Q}_{\substack{WWT\\ Y^Q}}(\bar{\tau}) &= p_E(\bar{\tau}) \left[1 - r^{WWC}_{WW}(\bar{\tau})\right] * \\ \ldots * U^{WWT}_{\substack{Q\\ N^Q_{i=1+2}+N^K_{\mu=1,2}+^{WW}_{RP}}}(\bar{\tau}) &\left[V^E_{\substack{WWT\\ CW+ME_{dep}}} + V^E_{\substack{WWT\\ CW+ME_{dev}}}(\bar{\tau}) + V^E_{WWT_{innv}}(\bar{\tau}) \right]. \end{aligned} \tag{10.44}$$

This type of actual marginal energy cost for wastewater treatment is lower when wastewater leakage rates are higher because the wastewater amounts are transported to the wastewater sector. These leaks, however, can also cause health risks and damage the natural environment.

D.2.3 Inter-Temporal Marginal Energy Costs for Wastewater Treatment Caused by the Structural Change of the Wastewater Amounts

Although the water coefficient and the wastewater generation rate for reproduction activities are both assumed time invariant, the inter-temporal marginal energy costs for wastewater treatment are only influenced by the structural change of the water input amounts If the civil work and mechanical & electrical capital stocks also rely on using non-renewable raw materials, which might also need water for their production, additional terms for this kind of inter-temporal marginal costs must be considered.

$$ITMEC^{Q}_{\substack{WWT\\ N^Q_{i=1+2}}\, +^{WWT}_{N^K_{\mu=1+2}}}(\bar{\tau} \to T) = \sum_{\tau^f=\bar{\tau}+1}^{T} \left\{ \begin{array}{l} p_E(\tau^f) \left[1 - r^{WWC}_{WW}(\tau^f)\right] Y^Q(\tau^f) * \\ \ldots * \left\{ \begin{array}{l} \left\{ \sum_{i=1}^{2} \frac{\tilde{\psi}^Q_i \kappa^Q_i \varphi^Q_i \kappa^Q_i}{c(\bar{N}^Q_i(\tau^f))} \right\} \\ +k^Q \left\{ d^Q + \gamma^Q_{\tau^f} \right\} k^Q \left\{ d^Q + \gamma^Q_{\tau^f} \right\} \sum_{\mu=1}^{2} \frac{\tilde{\psi}^K_\mu \kappa^K_\mu \varphi^K_\mu \kappa^K_\mu}{c(\bar{N}^K_\mu(\tau^f))} \end{array} \right\} * \\ \ldots * \left[V^E_{\substack{WWT\\ CW+ME_{dep}}} + V^E_{\substack{WWT\\ CW+ME_{dev}}}(\tau) + V^E_{WWT_{innv}}(\tau) \right] \end{array} \right\}.$$

$$ITMEC^{Q}_{\substack{WWT\\ N^Q_{i=1+2}}}(\bar{\tau} \to T) = \sum_{\tau^f=\bar{\tau}+1}^{T} \left\{ \begin{array}{l} p_E(\tau^f) \left[1 - r^{WWC}_{WW}(\tau^f)\right] Y^Q(\tau^f) \left\{ \sum_{i=1}^{2} \frac{\tilde{\psi}^Q_i \kappa^Q_i \varphi^Q_i \kappa^Q_i}{c(\bar{N}^Q_i(\tau^f))} \right\} * \\ \ldots * \left[V^E_{\substack{WWT\\ CW+ME_{dep}}} + V^E_{\substack{WWT\\ CW+ME_{dev}}}(\tau^f) + V^E_{WWT_{innv}}(\tau^f) \right] \end{array} \right\}. \tag{10.45}$$

$$ITMEC^{Q}_{\substack{WWT\\N^K_{\mu=1+2}}}(\bar{\tau} \rightarrow T) = \sum_{\tau^f=\bar{\tau}+1}^{T} \left\{ \begin{array}{l} p_E(\tau^f)\left[1 - r^{WWC}_{WW}(\tau^f)\right] Y^Q(\tau^f) \left\{ \begin{array}{l} k^Q\left\{d^Q + \gamma^Q_{\tau^f}\right\} k^Q\left\{d^Q + \gamma^Q_{\tau^f}\right\} * \\ \ldots * \sum\limits_{\mu=1}^{2} \frac{\tilde{\psi}^K_\mu \kappa^K_\mu \varphi^K_\mu \kappa^K_\mu}{c(\bar{N}^K_\mu(\tau^f))} \end{array} \right\} \\ \ldots * \left[V^E_{\substack{WWT\\CW+ME_{dep}}} + V^E_{\substack{WWT\\CW+ME_{dev}}}(\tau^f) + V^E_{WWT_{innv}}(\tau^f) \right] \end{array} \right\}. \quad (10.46)$$

D.2.4 The Dependence of the Actual and Inter-Temporal Marginal Energy Costs for Wastewater Treatment on the Development Path of the Consumption Good Amounts

From the fourth partial differentiation factor

$$\left\{ \frac{\vartheta}{\vartheta Y^Q(\bar{\tau})} \left[V^E_{\substack{WWT\\CW+ME_{dep}}} + V^E_{\substack{WWT\\CW+ME_{dev}}}(\tau) + V^E_{WWT_{innv}}(\tau) \right] \right\}$$
$$= \frac{\vartheta V^E_{\substack{WWT\\CW+ME_{dev}}}(\tau)}{\vartheta Y^Q(\bar{\tau})} + \frac{\vartheta V^E_{WWT_{innv}}(\tau)}{\vartheta Y^Q(\bar{\tau})},$$

we observe that dynamic changes in the specific energy input amounts are determined by two effects. The first effect is caused by the dynamics of the development path of the wastewater treatment amounts required for the production activities

$$\frac{\vartheta}{\vartheta Y^Q(\bar{\tau})} \left[V^E_{\substack{WWT\\CW+ME_{dev}}}(\tau) \right]$$
$$= \frac{\vartheta}{\vartheta Y^Q(\bar{\tau})} \left\{ \begin{array}{l} \left\{ \sum\limits_{\nu=1}^{2} \frac{e_\nu \kappa^{CW}_\nu}{c(\bar{N}^{CW}_\nu(0))} + \frac{ex^{CW}}{\eta^{CW}} \right\} k^{WWT}_{CW} \\ + \left\{ \sum\limits_{o=1}^{2} \frac{e_o \kappa^{ME}_o}{c(\bar{N}^{ME}_o(0))} + \frac{ex^{ME}}{\eta^{ME}} \right\} k^{WWT}_{ME} \end{array} \right\} \gamma^{WWT}_{\bar{\tau}} = e_{k^{WWT}_{CW+ME}} \frac{\vartheta \gamma^{WWT}_{\tau}}{\vartheta Y^Q(\bar{\tau})}$$

with

$$e_{k^{WWT}_{CW+ME}} := \left\{ \begin{array}{l} \left\{ \sum\limits_{\nu=1}^{2} \frac{e_\nu \kappa^{CW}_\nu}{c(\bar{N}^{CW}_\nu(0))} + \frac{ex^{CW}}{\eta^{CW}} \right\} k^{WWT}_{CW} \\ + \left\{ \sum\limits_{o=1}^{2} \frac{e_o \kappa^{ME}_o}{c(\bar{N}^{ME}_o(0))} + \frac{ex^{ME}}{\eta^{ME}} \right\} k^{WWT}_{ME} \end{array} \right\}.$$

Second is the energy saving innovation effect generated within the selected time interval, which can be called the actual marginal energy innovation effect for water treatment.

$$\frac{\vartheta\, V^{E}_{\substack{WWT\\E_{inn}}}(\tau)}{\vartheta Y^{Q}(\bar{\tau})}$$
$$= \frac{\vartheta}{\vartheta Y^{Q}(\bar{\tau})}\tilde{\varepsilon}^{WWT}\left\{\ln\frac{c^{*}_{Q}(0)}{c_{Q}(0)} + \ln\frac{c^{*}_{\hat{j}}(0)}{c_{\hat{j}}(0)}\right\}\left[1 + r^{WWT}_{an}(0)\, e^{\varphi^{WWT}_{an}\, Y^{Q}(\tau)}\right]$$
$$MEC^{Q}_{\substack{WWT\\E_{inn}}}(\bar{\tau}) = \tilde{\varepsilon}^{WWT}\left\{\ln\frac{c^{*}_{Q}(0)}{c_{Q}(0)} + \ln\frac{c^{*}_{\hat{j}}(0)}{c_{\hat{j}}(0)}\right\}\varphi^{WWT}_{an}\, r^{WD}_{an}(0)\, e^{\varphi^{WT}_{an}\, Y^{Q}(\bar{\tau})}$$
$$= \varphi^{WWT}_{an}\, e^{WWT}\, r^{WWT}_{an}(\bar{\tau})$$

with

$$e^{WWT} := \tilde{\varepsilon}^{WWT}\left\{\ln\frac{c^{*}_{Q}(0)}{c_{Q}(0)} + \ln\frac{c^{*}_{\hat{j}}(0)}{c_{\hat{j}}(0)}\right\}. \tag{10.47}$$

The inter-temporal marginal energy costs for wastewater treatment as a function of the development path of the wastewater or consumption good amounts are equal to

$$ITMEC^{Q}_{\gamma_{\tau}+\substack{WWT\\E_{inn}}}(\bar{\tau} \rightarrow T)$$
$$= \sum_{\tau^{f}=\bar{\tau}+1}^{T}\left\{\begin{array}{l} p_{E}(\tau^{f})\left[1 - r^{WWC}_{WW}(\tau^{f}\right]\, Y^{Q}(\tau^{f})U^{WWT}_{\substack{Q\\ N^{Q}_{i=1+2}+N^{K}_{\mu=1,2}+^{WWT}_{RP}}}(\tau^{f})^{*} \\ \ldots^{*}\left[e_{k^{WWT}_{CW+ME}}\dfrac{\vartheta\, \gamma^{WWT}_{\tau^{f}}}{\vartheta Y^{Q}(\bar{\tau})} + \varphi^{WWT}_{an}\, e^{WWT}\, r^{WWT}_{an}(\tau^{f})\right]\end{array}\right\}. \tag{10.48}$$

This means that the inter-temporal marginal costs for wastewater treatment are determined by the innovation effect for wastewater treatment. They are also determined by the development path of the consumption good amounts and the structural change of the wastewater treatment amounts generated by the production and reproduction activities.

When comparing the water supply with the wastewater sector, we can conclude that water losses lead to additional marginal energy and human labour costs for water treatment and distribution. The reduction of wastewater leakage leads to additional marginal energy and human labour costs for wastewater treatment. These can be interpreted as monetary benefits of prevented health risks and environmental damages otherwise caused by the wastewater leakage.

References

Atkinson T, Stiglitz J (1980) Lectures on public economics. McGraw-Hill, New York

Baumgärtner S (2000) Ambivalent joint production and natural environment: an economic and thermodynamic analysis. Springer, Heidelberg

Baumgärtner S, Dyckhoff H, Faber M, Proops J, Schiller J (2001) The concept of joint production and ecological economics. Ecol Econ 36:365–372

Baumgärtner S, Faber M, Proops J (2002) How environmental concern influences the investment decision? An application of capital theory. Ecol Econ 40:1–12

Baumgärtner S, Faber M, Schiller J (2006) Joint production and responsibility in ecological economics: on the foundations of environmental policy. Edward Elgar, Cheltenham

Beard TR, Lozada GA (2000) Economics, entropy and the environment: the extraordinary economics of Nicholas Georgescu Roegan. Edward Elgar, Cheltenham

Bekenstein JB (1974) Generalized second law of thermodynamics in black hole physics. Phys Rev D9:3292–3300

Belloc B (1980) Croissance économique et adaption du capital productif. Economica, Paris

Benndorf J (2003) Einführung in die Wasserforschung: DFG Wasserforschung im Spannungsfeld zwischen Gegenwartsbewältigung und Zukunftssicherung. In: Denkschrift, Deutsche Forschungsgemeinschaft (DFG), Wiley, Weinheim, pp 5–13

Binswanger M (1992) Information und Entropie. Ökologische Perspektiven des Übergangs zu einer Informationswirtschaft. Campus Frankfurtam Main

Breyer F (2004) Mikroökonomie: Eine Einführung (Microeconomics: an introduction), 2nd edn. Springer, Heidelberg

Cerbe G, Hoffmann HJ (1996) Einführung in die Thermodynamik (Introduction into thermodynamics), 11th edn. Carl Hanser, München

Claudius R (1850) Über die bewegende Kraft der Wärme und die Gesetze, welche sich daraus ableiten lassen (Towards the moving power of heat and the laws derivable). Planck M (ed) (1921), Akademische Verlagsgesellschaft, Leipzig

Faber M (1979) Introduction into modern Austrian capital theory. Springer, Heidelberg

Faber M (2007) How to be an ecological economist? Invited lecture to the 7th International Conference of the European Society for Ecological Economics, Integrating Natural and Social Sciences for Sustainability, UFZ, Leipzig, 5–8 June 2007

Faber M, Niemes H, Stephan G (1983) Entropie, Umweltschutz und Rohstoffverbrauch: Eine naturwissenschaftlich ökonomische Untersuchung (Entropy, environment and resources: a physico-economic analysis). Springer, Heidelberg

Faber M, Niemes H, Stephan G (1987) Entropy, environment and resources: an essay in physico-economics, 1st edn. Springer, Heidelberg

Faber M, Niemes H, Stephan G (1995) Entropy, environment and resources: an essay in physico-economics, 2nd edn. Springer, Heidelberg

Faber M, Winkler R (2006) Heterogeneity and time: from Neo-Austrian capital theory to ecological economics. Am J Econ Sociol 65:803–825

H. Niemes, M. Schirmer, *Entropy, Water and Resources*,
DOI 10.1007/978-3-7908-2416-2, © Springer-Verlag Berlin Heidelberg 2010

Feynman RP (1999) Es ist so einfach: Vom Vergnügen, Dinge zu entdecken (It is so easy: from the pleasure to discover things). Piper, München
Fujita M (1996) Urban land use theory in location theory. Fund Pure Appl Econ 5:73–149
Geistlinger H, Eisermann D, Beckmann A, Martienssen M, Schirmer M (2005) Mass-transfer-models: from bench-scale to field scale. Model Care Conference, Den Haag, The Hague 6–9 June 2005
GKW Consult (1993) Adana wastewater treatment plant feasibility study. Prepared for ASKI (Water and Sewage Company of Adana) and financed by the German Financing Agency Kreditanstalt fur Wiederaufbau (KfW)
GKW Consult (1996) Adana wastewater amendment study. Prepared for ASKI (Water and Sewage Company of Adana) and financed by the European Investment Bank (EIB)
GKW Consult (1997) Water and wastewater tariff study. Prepared for ASKI (Water and Sewage Company of Adana) and financed by the European Investment Bank (EIB)
GKW Consult (2004) Consultancy services for the construction, rehabilitation and extension of the Adana water supply system (1999–2004). Prepared for ASKI (Water and Sewage Company of Adana) and financed by the German Financing Agency Kreditanstalt für Wiederaufbau (KfW)
Glaeser E (1998) Are cities dying? J Econ Perspect 12(2):139–160
Görnitz T, Görnitz B (2006) Der Kreative Kosmos: Geist und Materie aus Information (The creative cosmos: spirit and mass from information). Spektrum Akademischer, Heidelberg
Graw KU, Maggio E (2001) Wasserver- und Abwasserentsorgung von Megastädten: Eine Aufgabe des neuen Jahrtausends (Water supply and wastewater disposal: a new challenge for the new millennium). Bauingenieur 1/2:26–29
Green C (2003) Handbook of water economics: principles and practice. Wiley, West Sussex
Hadley G (1964) Nonlinear and dynamic programming. Addison Wesley, Reading, Palo Alto, London
Harpke B, Müller H, Endrisczewitz M, Winter-Pellicioni M, Hasselwander E (2006) MTBE-Sanierung am Standort Leuna. Sicherung and Sanierung des Grundwasserabstroms der alten Raffinerie mit Dichtwand und Reinigungsanlage (MTBE sanitation at the site of Leuna. Protection and sanitation of the groundwater flow down streams of the old refinery by the protection wall and groundwater treatment plant). TerraTech 5:13–16
Hawkins SW (1975) Particle creation by black holes. Commun Math Phys 43:199–220
Hoekstra AY, Chapagain AK (2007) Water footprints of nations: water use by people as a function of their consumption pattern. Water Resour Manage 21(1):35–48
Jacobs J (1985) Cities and the wealth of nations. Vintage Books, Vancouver
Jaksch HJ (1975) Die Mehrergiebigkeit Längerer Produktionswege in einem Linearen Vielsektorenmodell. Zeitschrift Gesamte Staatswissenschaft 131:95–105
Jasper K, Calanca P, Fuhrer J (2006) Changes in summertime soil water patterns in complex terrain due to climate change. J Hydrol 327:550–563
Jasper K, Calanca P, Gyalistras D, Fuhrer J (2004) Differential impacts on the hydrology of two alpine river basins. Clim Res 26:113–129
Javanmardian M, Glasser HA (1997) In-situ biodegradation of MTBE using biosparging. J Am Chem Soc 37:424
Jörgensen SE, Svirezhev YM (2004) Towards a thermodynamic theory for ecological systems. Elsevier, Amsterdam
Krebs P (2003) Wasser im urbanen Raum. In: Wasserforschung im Spannungsfeld zwischen Gegenwartsbewältigung und Zukunftssicherung. In: Denkschrift, Deutsche Forschungsgemeinschaft (DFG), Wiley, Weinheim, pp 37–51
Lahmeyer International (LI) (1996) Feasibility study for construction, rehabilitation and extension of the Adana water supply system. Prepared for ASKI (Water and Sewage Company of Adana) and financed by German Financing Agency Kreditanstalt für Wiederaufbau (KfW)
Leeson A, Johnson PC, Johnson RL, Hinchee RE, McWorther DB (1999) Air sparging design paradigm. Batelle Memorial Institute, Columbus, OH
Lozada GA (2004) Entropy, free energy, work and other thermodynamic variables in economics. Ecol Econ 56:71–78

Martienssen M, Fabritius H, Kukla S, Balcke GU, Hasselwander E, Schirmer M (2006) Determination of naturally occurring MTBE biodegradation by analysing metabolites and biodegradation by-products. J Contam Hydrol 87:37–53

Mills E, Hamilton B (1994) Urban economics, 5th edn. Harper Collins, New York

Musgrave RA (1956) The theory of public finance. McGraw-Hill, New York

Nauges C (2003) Long-run study of residential water consumption: with an application to sample of Franc he communities. Environ Resour Econ 26:25–43

Newberry D (1990) Pricing and congestions: economics principles relevant to pricing roads. Oxf Rev Econ Pol 64:22–38

Niemes H (1981) Umwelt als Schadstoffempfänger: Die Wassergütewirtschaft als Beispiel (Natural environment as recipient of pollution: water quality economics as example). J.C.B. Mohr (Paul Siebeck), Tübingen

Niemes H (1995) Water quality simulation applied for the determination of the type and site for the wastewater treatment plant of Adana in Turkey. Technical reports about engineering hydrology and hydraulics, vol 53. Institute of Water Constructions at the Technical University of Darmstadt, pp. 187–216

Niemes H, Jöst F, Schirmer M (2007) Bi-directional relations between urbanization and water infrastructure: the urban centre of Adana in Turkey as case study. Paper presented at the 7th International Conference of the European Society for Ecological Economics, Integrating Natural and Social Sciences for Sustainability, UFZ, Leipzig, 5–8 June 2007

Salanitro JP, Johnson PC, Spinnler GE, Maner PM, Wisniewski HL, Bruce C (2000) Field scale demonstration of enhanced MTBE bio remediation through aquifer bio-augmentation and oxygenation. Int J Environ Sci Technol 34:4152–4162

Samuelson PA (1954) The theory of public expenditures. Rev Econ Stat 36:387–389

Schirmer M (1999) Das Verhalten des Benzininhaltstoffes Methyltertiärbutylether (MTBE) in Grundwasser. Grundwasser 4(3):95–102

Schirmer M, Martienssen M (2007) Enhanced natural attenuation of MTBE. In: Barcelo D (ed) The handbook of environmental chemistry, vol 5. Springer, Heidelberg, pp. 139–158

Schirmer M, Strauch G, Schirmer K, Reinstorf F (2007) Urbane Hydrogeologie: Herausforderungen für Forschung und Praxis (Urban hydrogeology: challenges for research and practice). Grundwasser 12(3):177–188

Smith AE, Hristova K, Wood I, Mackay DM, Lory E, Lorenzana D, Scow KM (2005) Comparison of bio stimulation versus bio augmentation with bacterial strain PM1 for treatment of groundwater contaminated with methyl tertiary butyl ether (MTBE). Environ Health Perspect 113:317–322.

Stephan G (1995) Introduction into capital theory: a Neo-Austrian perspective. Springer, Heidelberg

Tiebout C (1956) A pure theory of local expenditures. J Pol Econ 64(5):416–424

Ulmschneider P (2006) Intelligent life in the universe: principles and requirements behind its emergence. Springer, Heidelberg

United Nations (2005) World urbanization prospects. UN, New York

von Baeyer C (2005) Das informative Universum: Das neue Weltbild der Physik (The informative universe: the new world view in physics). C. H. Beck, München

Wilson RD, Mackay DM, Scow KM (2002) In situ MTBE bio degradation supported by diffusive oxygen release. Int J Environ Sci Technol 36:90–199

Winkler R (2005) Structural change with joint production of consumption and environmental pollution: a Neo-Austrian approach. Struct Change Econ Dyn 16:111–135

Index

A
Aggregation
  processes to sectors, 72–74
Aggregation of processes
  production sector, 162–163
  water and wastewater sector, 162, 164
Air stripping, 110

B
Beneficiary group, 111
BTEX (Benzene, Toluene, Ethylbenzene and Xylenes) aromatics, 107

C
Capital stock, 44–45
Capital stock dynamic, 46
Chemical systems, 33
Concentration
  extracted raw material, 26
Concentration of by-product, 41
Constraint
  capital stock, 71, 157
  consumption good, 151, 157
    amount, 45
  free energy, 64, 154, 157
  human labour, 155, 157
    input, 68–71
  raw materials, 152, 154
    amount, 64
  water and wastewater, 152, 157
    amount, 57
Coordinative work of human labour, 15

D
Desired good, 27
Deterioration of capital good, 36
Deterioration and consumption rate, 27
Dilution, 112
Discount rate, 16
Dissipative work, 12
Drinking water work, 111
Durable good, 21
Dynamic prime costs, 115, 128

E
Economy of scale, 108
Embedded information
  capital stock, 14
Emission standard, 36
Energy constraint, 157, 159
Energy efficiency
  water infrastructure, 141–142
Energy efficiency coefficient, 50, 52
Energy, water, global warming, 177–178
Enhanced natural attenuation (ENA), 110
Entropy free work, 12
Entropy and information
  Bekenstein, Hawkins, 9
Entropy and irreversible processes, 13
Entropy notion
  Boltzmann, 8
  Clausius, 8
  Shannon, 9
Exergy change, 28, 30
Exergy of human labour, 38
Exergy and information
  Kullback measure, 12
Exergy roundaboutness, 21
Exergy superiority, 20
Extraction of raw material, 39

F
Feasible consumption bundles, 17
Free energy production, 25

G
Global roundaboutness, 20
Global superiority, 19

Groundwater flow, 112
Groundwater flow velocity, 107
Groundwater pollution, 106

H
High exergy, 13
High exergy natural resource, 27
Hydraulic conductivity, 107

I
Ignorance of social costs, 111
Industrial evolution process, 15
Information
  social education, 13
Inter-temporal externalities
  extracting raw material, 62
Inter-temporal welfare function, 17, 86–87

J
Joint by-product, 25

K
Kuhn-Tucker, 76

L
LAGRANGE-function, 76, 172–173
LAGRANGE-multiplier, 76, 83, 172
Leakage Rate
  wastewater collection, 143–144
Leibniz rule, 78
Local public good, 107, 128

M
Marginal cost
  actual and inter-temporal, 75, 78
Mega-city, 119
Methyl Tertiary-Butyl Ether (MTBE), 106

N
Neo-Austrian capital theory, 1
Net free energy, 51–52
Non-profit condition
  production sector, 74, 166, 174
  water and wastewater sector, 83, 101
  water sector, 166, 167
No storage assumption, 62
Number of goods, 40

O
Optimality condition, 78–79
  demand side, 77
  production side, 78, 148, 172, 177
Optimization concept, 75–77
Oxygen injection, 110

P
Physical work of human labour, 15
Planning and implementation
  wastewater, 121, 125
Planning and implementation concept, 121
  water supply, 125
Polluter-pay-principle, 111, 124
Process with capital input, 47–48
Process coefficient
  CW&ME production, 140–141
  extraction of raw materials, 57–58
  producing free energy, 42–43
  production sector, 46
  reproduction sector, 53
  wastewater treatment, 55
  water infrastructure, 136
Production of free energy, 39
Public good, 118

Q
Quality of free energy, 43

R
Reduction of variables
  by capital stock substitution, 147–149
Required raw materials, 40
Roundaboutness, 19

S
Self-purification capability, 108
Shadow price, 76, 172–173
Social welfare function, 16
Stripping technology, 109
Structure
  water infrastructure model, 132–136
Superiority, 19

T
Target concentration, 29
Target group, 111
Technical progress
  human labour, 43
Thermodynamic equilibrium, 11
Thermodynamic laws, 2
Time duration
  transformation process, 24
Time structure, 39
Transformation curve, 17
Transformation process, 13
Treatment technology, 108

U
Urbanization, 117
Usage of water, 41

W
Wastewater generation rate, 41, 54, 120, 142, 151, 199, 200, 209
Wastewater treatment activity, 35
Water concentration of by-product, 33
Water Infrastructure, 41, 101, 105, 117
Water loss rate
  water distribution, 152–153
Water production, 41
Water treatment shifting factor, 142
Water usage, 26
Water and wastewater constraint
  after substitution, 158, 159
Work and exergy, 9

Kolibri 360 GmbH
Ferdinand-Jühlke-Straße 7
99095 Erfurt, Deutschland
produktsicherheit@kolibri360.de